Communications and Control Engineering

Springer
London
Berlin
Heidelberg
New York
Hong Kong
Milan
Paris
Tokyo

Roberto Tempo, Giuseppe Calafiore
and Fabrizio Dabbene

Randomized Algorithms for Analysis and Control of Uncertain Systems

With 54 Figures

 Springer

Roberto Tempo, PhD
Fabrizio Dabbene, PhD
IEIIT-CNR, Politecnico di Torino, Corso Duca degli Abruzzi 24, 10129 Torino, Italy

Giuseppe Calafiore, PhD
Dipartimento di Automatica e Informatica, Politecnico di Torino,
Corso Duca degli Abruzzi 24, 10129 Torino, Italy

Series Editors
E.D. Sontag • M. Thoma • A. Isidori • J.H. van Schuppen

British Library Cataloguing in Publication Data
Tempo, R.
    Randomized algorithms for analysis and control of uncertain
systems
    1. Robust control 2. Algorithms 3. Uncertainty (Information
theory)
    I. Title II. Giuseppe, Calafiore III. Dabbene, Fabrizio
629.8'312

Library of Congress Cataloging-in-Publication Data
Tempo, R.
    Randomized algorithms for analysis and control of uncertain systems / Roberto Tempo,
Giuseppe Calafiore, and Fabrizio Dabbene.
        p. cm. — (Communications and control engineering, ISSN 0178-5354)
    Includes bibliographical references and index.

    1. Control theory. 2. System analysis. 3. Stochastic processes. 4. Algorithms.
    I. Calafiore, Giuseppe, 1969– II. Dabbene, Fabrizio, 1968– III. Title. IV. Series.
QA402.T43 2003
003—dc21   2002070728

Communications and Control Engineering Series ISSN 0178-5354
ISBN 978-1-84996-882-9   e-ISBN 978-1-84628-052-8
Springer is a part of Springer Science+Business Media
springeronline.com

69/3830-543210   Printed on acid-free paper

to Chicchi and Giulia for their remarkable endurance

R.T.

to Anne, to my family, and to the memory of my grandfather

G.C.

to Paola and my family for their love and support

F.D.

It follows that the Scientist, like the Pilgrim, must wend a straight and narrow path between the Pitfalls of Over-simplification and the Morass of Overcomplication.

Richard Bellman, 1957

Foreword

The subject of control system synthesis, and in particular robust control, has had a long and rich history. Since the 1980s, the topic of robust control has been on a sound mathematical foundation. The principal aim of robust control is to ensure that the performance of a control system is satisfactory, or nearly optimal, even when the system to be controlled is itself not known precisely. To put it another way, the objective of robust control is to assure satisfactory performance even when there is "uncertainty" about the system to be controlled.

During the two past two decades, a great deal of thought has gone into modeling the "plant uncertainty." Originally the uncertainty was purely "deterministic," and was captured by the assumption that the "true" system belonged to some sphere centered around a nominal plant model. This nominal plant model was then used as the basis for designing a robust controller. Over time, it became clear that such an approach would often lead to rather conservative designs. The reason is that in this model of uncertainty, every plant in the sphere of uncertainty is deemed to be equally likely to occur, and the controller is therefore obliged to guarantee satisfactory performance for every plant within this sphere of uncertainty. As a result, the controller design will trade off optimal performance at the nominal plant condition to assure satisfactory performance at off-nominal plant conditions.

To avoid this type of overly conservative design, a recent approach has been to assign some notion of probability to the plant uncertainty. Thus, instead of assuring satisfactory performance at every single possible plant, the aim of controller design becomes one of maximizing the expected value of the performance of the controller. With this reformulation, there is reason to believe that the resulting designs will often be much less conservative than those based on deterministic uncertainty models.

A parallel theme has its beginnings in the early 1990s, and is the notion of the complexity of controller design. The tremendous advances in robust control synthesis theory in the 1980s led to very neat-looking problem formulations, based on very advanced concepts from functional analysis, in particular, the theory of Hardy spaces. As the research community began to apply these methods to large-sized practical problems, some researchers began to study the rate at which the computational complexity of robust control synthesis methods grew as a function of the problem size. Somewhat to everyone's surprise, it was soon established that several problems of practical interest were in fact NP-hard. Thus, if one makes the reasonable assumption that P $\neq$ NP, then there do not exist polynomial-time algorithms for solving many reasonable-looking problems in robust control.

In the mainstream computer science literature, for the past several years researchers have been using the notion of randomization as a means of tackling difficult computational problems. Thus far there has not been any instance of a problem that is intractable using deterministic algorithms, but which becomes tractable when a randomized algorithm is used. However, there are several problems (for example, sorting) whose computational complexity reduces significantly when a randomized algorithm is used instead of a deterministic algorithm. When the idea of randomization is applied to control-theoretic problems, however, there appear to be some NP-hard problems that do indeed become tractable, provided one is willing to accept a somewhat diluted notion of what constitutes a "solution" to the problem at hand.

With all these streams of thought floating around the research community, it is an appropriate time for a book such as this. The central theme of the present work is the application of randomized algorithms to various problems in control system analysis and synthesis. The authors review practically all the important developments in robustness analysis and robust controller synthesis, and show how randomized algorithms can be used effectively in these problems. The treatment is completely self-contained, in that the relevant notions from elementary probability theory are introduced from first principles, and in addition, many advanced results from probability theory and from statistical learning theory are also presented. A unique feature of the book is that it provides a comprehensive treatment of the issue of sample generation. Many papers in this area simply assume that independent identically distributed (iid) samples generated according to a specific distribution are available, and do not bother themselves about the difficulty of generating these samples. The trade-off between the nonstandardness of the distribution and the difficulty of generating iid samples is clearly brought out here. If one wishes to apply randomization to practical problems, the issue of sample generation becomes very significant. At the same time, many of the results presented here on sample generation are not readily accessible to the control theory community. Thus the authors render a signal service to the research community by discussing the topic at the length they do. In addition to traditional problems in robust controller synthesis, the book also contains applications of the theory to network traffic analysis, and the stability of a flexible structure.

All in all, the present book is a very timely contribution to the literature. I have no hesitation in asserting that it will remain a widely cited reference work for many years.

M. Vidyasagar

Hyderabad, India
June 2004

Acknowledgments

This book has been written with substantial help from many friends and colleagues. In particular, we are grateful to B. Ross Barmish, Yasumasa Fujisaki, Constantino Lagoa, Harald Niederreiter, Yasuaki Oishi, Carsten Scherer and Valery Ugrinovskii for suggesting many improvements on preliminary versions, as well as for pointing out various inaccuracies and errors. We are also grateful to Tansu Alpcan and Hideaki Ishii for their careful reading of Sections 19.1 and 19.3.

During the spring semester of the academic year 2002, part of this book was taught as a special-topic graduate course at CSL, University of Illinois at Urbana-Champaign, and during the fall semester of the same year at Politecnico di Milano, Italy. We warmly thank Tamer Başar and Patrizio Colaneri for the invitations to teach at their respective institutions and for the insightful discussions. Seminars on parts of this book were presented at the EECS Department, University of California at Berkeley, during the spring term 2003. We thank Laurent El Ghaoui for his invitation, as well as Elijah Polak and Pravin Varaya for stimulating discussions. Some parts of this book have been utilized for a NATO lecture series delivered during spring 2003 in various countries, and in particular at Università di Bologna, Forlì, Italy, Escola Superior de Tecnologia de Setúbal, Portugal, and University of Southern California, Los Angeles. We thank Constantine Houpis for the direction and supervision of these events.

We are pleased to thank the National Research Council (CNR) of Italy for generously supporting for various years the research reported here, and to acknowledge funding from the Italian Ministry of Instruction, University and Research (MIUR) through an FIRB research grant.

Roberto Tempo
Giuseppe Calafiore
Fabrizio Dabbene

Torino, Italy
June 2004

Contents

1. Overview

Don't assume the worst-case scenario.
It's emotionally draining and probably
won't happen anyway.

Anonymous

The main objective of this book is to introduce the reader to the fundamentals of the emerging research area of probabilistic and randomized methods for analysis and design of uncertain systems. This area is fairly recent, even though its roots lie in the robustness techniques for handling complex control systems developed in the 1980s.

1.1 Uncertainty and robustness

The presence of uncertainty in the system description has always been a critical issue in control theory and applications. The earliest attempt to deal with uncertainty in a rigorous fashion was based on a *stochastic* approach, which led to classical optimal control results related to linear quadratic Gaussian (LQG) control and Kalman filtering. In this theory, uncertainty is considered only in the form of exogenous disturbances having a stochastic characterization, while the plant dynamics are assumed to be exactly known. In this framework, LQG theory combines in an optimal way the deterministic information on the plant dynamics with probabilistic information on the stochastic disturbances.

Since the early 1980s, there has been a successful attempt to overcome this paradigm, introducing uncertainty directly in the plant description. The design objective hence becomes to determine solutions that are guaranteed against all possible uncertainty realizations, i.e. *worst-case* (or *robust*) solutions. In this approach, a plant family is characterized by a set-theoretic description, often called an unknown but bounded model [241]. Hence, a controller is designed with the aim of guaranteeing a specified performance for all plants that are compatible with the uncertainty description. A major stepping stone in the robustness direction was the formulation in 1981 by Zames of the $\mathcal{H}_\infty$ problem [293]. This formulation surmounted some drawbacks of classical optimal control, such as the lack of guaranteed margins of LQG. In the subsequent fifteen years, the research in robust control evolved in various directions, each based on diverse problem formulations and mathematical tools. Even though several subareas played a major role within robustness, we feel that $\mathcal{H}_\infty$ deserves the credit for its centrality and also for its connections with classical optimal control. However, other successful methods to handle uncertainty have been developed. In particular, we recall the methods based on the structured singular value, also known as μ theory [211], the approach dealing with systems affected by parametric uncertainty, or Kharitonov theory [24, 40], the optimization-based methods based on linear matrix inequalities [49], the ℓ_1 optimal control theory [75] and the so-called quantitative feedback theory (QFT) [138, 139].

In the late 1980s, robust control became a well-known discipline so that the technical results and the algorithms developed were successfully used in various industrial applications, including aerospace, chemical, electrical and mechanical engineering. A few years later, probably in the early 1990s, researchers in robust control realized more fully some of its theoretical limitations, which can be roughly summarized as the issues of conservatism and computational complexity. In fact, when compared with classical stochastic methods, the worst-case paradigm may lead to problems whose exact solution cannot be determined in polynomial time, see e.g. [42]. Therefore, relaxation techniques are typically introduced so that the resulting problem can be solved with numerically efficient algorithms. Clearly, this entails a compromise between tightness of the solution and numerical complexity.

1.2 Probability and robustness of uncertain systems

The so-called *probabilistic approach* to system robustness has emerged as a complementary method for handling uncertain systems with the objective of addressing the issues described above. In contrast to previous deterministic robustness techniques, its main ingredient is the use of probabilistic concepts. One of the goals of this research endeavor is to provide a reapprochement between the classical stochastic and robust paradigms, combining worst-case bounds with probabilistic information, thus potentially reducing the conservatism inherent in the worst-case design. In this way, the control engineer gains additional insight that may help bridge the gap between theory and applications.

The algorithms derived in the probabilistic context are based on uncertainty randomization and are usually called *randomized algorithms*. For control systems analysis, these algorithms have low complexity and are associated with robustness bounds that are generally less conservative than the classical ones, obviously at the expense of a probabilistic risk of failure.

Randomized algorithms have been used successfully in various areas, including computer science, computational geometry and optimization, see e.g. [192, 195]. In these areas, several problems have been efficiently solved using randomization, such as data structuring and search trees, graph algorithms and related optimization issues, as well as problems involving linear programs. However, the mathematical tools developed therein are different in spirit from those studied in this book for control systems.

1.3 Some historical notes

From the historical point of view, probabilistic methods for robustness made some early appearances in the 1980s, but they did not receive adequate attention in the systems and control literature at that time. In particular, the notion of "probability of instability," which is crucial for probabilistic robustness, was first introduced in the context of flight control in 1980 by Stengel [252]. Similar ideas were subsequently revisited in Stengel's book on stochastic optimal control [253] in 1986. In 1989, the paper titled "Probabilistic robust controller design" [87] was published in the Proceedings of the IEEE Conference on Decision and Control. This is presumably the first paper with a title containing both words "probabilistic" and "robust." Stengel and co-workers further pursued the line of research on stochastic robustness, publishing several interesting results. In particular, those papers explored various techniques, mainly based on Monte Carlo, for the computation of the probability of instability, and related performance concepts, with specific attention to flight dynamics applications within aerospace engineering. However, the absence of newly developed mathematical tools basically limited these attempts to merge probability and robustness to *analysis* problems.

A few years later, in 1996 the papers [154, 263] (developed independently by Khargonekar and Tikku and by Tempo, Bai and Dabbene) proposed an approach based on explicit sample size bounds, thus refuelling enthusiasm on randomized techniques. Subsequently, the study of statistical learning theory and its application to robust control conducted by Vidyasagar [282, 283] provided additional impetus and also exposed researchers of robust control to a different viewpoint based on solid mathematical foundations. This formulation led to the development of randomized algorithms for control system *design*. More recently, at the beginning of the new millennium, randomized algorithms and probabilistic methods are becoming increasingly popular within the control community. We hope that this book will further contribute to their development in the years to come.

1.4 Structure of the book

We now briefly describe the structure of the book. Chapter 2 deals with basic elements of probability theory and introduces the notions of random variables and matrices used in the rest of the book. Classical univariate and multivariate densities are also listed. The book can be roughly divided into five parts:

- Introduction to uncertain systems and limits of worst-case paradigm

  Chapter 3: Uncertain Linear Systems and Robustness
  Chapter 4: Linear Robust Control Design
  Chapter 5: Some Limits of the Robustness Paradigm

The first part of the book contains an introduction to robust control and its limitations. This part may be skipped by the reader familiar with these topics. Chapters 3 and 4 present a rather general and "dry" summary of the key results regarding robustness analysis and design. In Chapter 3, after introducing norms, balls and signals, the standard $M-\Delta$ model for uncertainty description of linear time-invariant systems is studied. The small gain theorem (in various forms), μ theory and its connections with real parametric uncertainty, and the computation of robustness margins constitute the backbone of the chapter.

Chapter 4 deals with $\mathcal{H}_\infty$ and $\mathcal{H}_2$ design methods following a classical approach based on linear matrix inequalities. Special attention is also devoted to linear quadratic Gaussian, linear quadratic regulator and guaranteed-cost control of uncertain systems.

In Chapter 5, some limitations of classical robust control are outlined. First, a summary of concepts and results on computational complexity is presented and a number of NP-hard problems within systems and control are listed. Second, the issue of conservatism in the robustness margin computation is discussed. Third, a classical example regarding discontinuity of the robustness margin is revisited. This chapter provides a launching point for the probabilistic methods discussed next.

- Probabilistic techniques for robustness, Monte Carlo and quasi-Monte Carlo methods, and randomized algorithms

  Chapter 6: Probabilistic Methods for Robustness
  Chapter 7: Monte Carlo Methods
  Chapter 8: Randomized Algorithms in Systems and Control

In Chapter 6, the key ideas of probabilistic methods for systems and control are discussed. Basic concepts such as the so-called "good set" and "bad set" are introduced and three different problems, which are the probabilistic counterparts of classical robustness analysis problems, are presented. This chapter also includes many specific examples showing that these problems can sometimes be solved in closed form without resorting to randomization.

The first part of Chapter 7 deals with Monte Carlo methods and provides a general overview of classical methods for both integration and optimization. The laws of large numbers for empirical mean, empirical probability and empirical maximum are reported. The second part of the chapter concentrates on quasi-Monte Carlo, which is a deterministic version of Monte Carlo methods. In this case, deterministic sequences for integration and optimization, together with specific error bounds, are discussed.

Chapter 8 formally defines *randomized algorithms* and provides a connection between general Monte Carlo methods and specific tools for systems and control problems. A clear distinction between analysis and synthesis is made. For analysis, two probabilistic robustness problems previously introduced in Chapter 6 are revisited and meta-algorithms for their solution are presented. For control synthesis, two different philosophies are studied: the first approach is aimed at designing a controller which performs well "on average," and the second one is focused on design for probabilistic robustness. Specific meta-algorithms are provided in both cases. The chapter ends with a formal definition of efficient randomized algorithms.

- Probability inequalities, sample size estimates and statistical learning theory for control systems
  Chapter 9: Probability Inequalities
  Chapter 10: Statistical Learning Theory and Control Design

These two chapters address the crucial issue of finite-time convergence of randomized algorithms. In the first part of Chapter 9, classical probability inequalities such as Markov and Chebychev are studied. Extensions to deviation inequalities are subsequently considered, deriving the Hoeffding inequality. These inequalities are then specified for the probability estimation problem, obtaining Chernoff and related bounds.

Chapter 10 is divided into two parts. In the first part, the main definitions and results regarding statistical learning theory are given. These results include the well-known Vapnik–Chervonenkis and Pollard bounds for uniform convergence of empirical means. In the second part of the chapter, a specific application of this theory to average performance synthesis problems is shown, and the connections with the meta-algorithms presented in Chapter 8 are given. Examples regarding static output feedback and average sensitivity minimization conclude the chapter.

- Randomized algorithms for probabilistic robust design, sequential methods and scenario approach

  Chapter 11: Sequential Algorithms for Probabilistic Robust Design
  Chapter 12: Sequential Algorithms for LPV Systems
  Chapter 13: Scenario Approach for Probabilistic Robust Design

In this part of the book, we move on to robust controller design of uncertain systems with probabilistic techniques. The main point of Chapters 11 and 12 is the development of iterative stochastic algorithms, which represent specific instances of the meta-algorithms given in Chapter 8. The distinction with classical methods studied in the stochastic approximation literature are discussed. In particular, in Chapter 11, using the standard setting of linear quadratic regulators, we analyze sequential algorithms based on gradient iterations for designing a controller which quadratically stabilizes an uncertain system with probability one in a finite number of steps. A version of this algorithm which finds a probabilistic solution and explicitly provides the number of required steps is also given. Finally, randomized algorithms for solving uncertain linear matrix inequalities are presented. In Chapter 12, similar methods for linear parameter varying (LPV) systems are analyzed. However, since this chapter deals with parameters, rather than uncertainty, we keep it distinct from Chapter 11.

Chapter 13 studies a different non sequential methodology for dealing with robust design in a probabilistic setting. In the scenario approach, the design problem is reformulated as a "chance-constrained" optimization problem, which is then solved by means of a one-shot convex optimization involving a finite number of sampled uncertainty instances (the scenarios). The results obtained include an explicit formula for the number of scenarios required by the randomized algorithm.

- Random number and variate generations, statistical theory of random vectors and matrices, and related algorithms

  Chapter 14: Random Number and Variate Generation
  Chapter 15: Statistical Theory of Radial Random Vectors
  Chapter 16: Vector Randomization Methods
  Chapter 17: Statistical Theory of Radial Random Matrices
  Chapter 18: Matrix Randomization Methods

The main objective of this part of the book is the development of sampling schemes for different classes of uncertainty structures, analyzed in Chapters 3 and 4, affecting the control system. This requires the study of specific techniques for generation of independent and identically distributed vector and matrix samples within various sets of interest in robust control. Chapters 15 and 17 address statistical properties of random vectors and matrices respectively. They are very technical, especially the latter, which is focused on random matrices. The reader interested in specific randomized algorithms for sampling within various norm-bounded sets may skip these chapters and concentrate instead on Chapters 16 and 18.

Chapter 14 deals with the topic of random number and variate generation. This chapter begins with an overview of classical linear and nonlinear congruential methods and includes results regarding random variate transformations. Extensions to multivariate problems, as well as rejection methods and techniques based on the conditional density method, are also analyzed. Finally, a brief account of asymptotic techniques, including the so-called Markov chain Monte Carlo method, is given.

Chapter 15 deals with statistical properties of radial random vectors. In particular, some general results for radially symmetric density functions are presented. Chapter 16 studies specific algorithms which make use of the theoretical results of the previous chapter for random sample generation within ℓ_p norm balls. In particular, efficient algorithms (which do not require rejection or asymptotic techniques) based on the so-called generalized Gamma density are developed.

Chapter 17 is focused on the statistical properties of random matrices. Various norms are considered, but specific attention is devoted to the spectral norm, owing to its interest in robust control. In this chapter, methods based on the singular value decomposition (SVD) of real and complex random matrices are studied. The key point is to compute the distributions of the SVD factors of a random matrix. This provides meaningful extensions of the results currently available in the area denoted as theory of random matrices.

In Chapter 18, specific randomized algorithms for real and complex matrices are constructed by means of the conditional density method. One of the main points of this chapter is to develop algebraic tools for the closed-form computation of the marginal density, which is required in the application of this method.

Chapter 19 presents three specific applications of randomized algorithms for systems and control problems: congestion control of high-speed communication networks, probabilistic robustness of a flexible structure, and stability of quantized sampled-data systems. Finally, the Appendix includes results regarding transformations between random matrices, Jacobians of transformations and the Selberg and Dyson–Mehta integrals.

2. Elements of Probability Theory

In this chapter, we formally review some basic concepts of probability theory. Most of this material is standard and available in classical references, such as [69, 214]; more advanced material on multivariate statistical analysis can be found in [12]. The definitions introduced here are instrumental to the study of randomized algorithms presented in subsequent chapters.

2.1 Probability, random variables and random matrices

2.1.1 Probability space

Given a sample space Ω and a σ-algebra $\mathcal{S}$ of subsets S of Ω (the events), a probability $\mathrm{PR}\{S\}$ is a real-valued function on $\mathcal{S}$ satisfying:

1. $\mathrm{PR}\{S\} \in [0,1]$;
2. $\mathrm{PR}\{\Omega\} = 1$;
3. If the events S_i are mutually exclusive ($S_i \cap S_k = \emptyset$ for $i \neq k$), then

$$\mathrm{PR}\left\{\bigcup_{i \in \mathcal{I}} S_i\right\} = \sum_{i \in \mathcal{I}} \mathrm{PR}\{S_i\}$$

where $\mathcal{I}$ is a countable[1] set of positive integers.

The triple $(\Omega, \mathcal{S}, \mathrm{PR}\{S\})$ is called a *probability space*.

A *discrete probability space* is a probability space where Ω is countable. In this case, $\mathcal{S}$ is given by subsets of Ω and the probability $\mathrm{PR} : \Omega \to [0,1]$ is such that

$$\sum_{\omega \in \Omega} \mathrm{PR}\{w\} = 1.$$

---

[1] By countable we mean finite (possibly empty) or countably infinite.

2.1.2 Real and complex random variables

We denote with $\mathbb{R}$ and $\mathbb{C}$ the real and complex field respectively. The symbol $\mathbb{F}$ is also used to indicate either $\mathbb{R}$ or $\mathbb{C}$. A function $f : \Omega \to \mathbb{R}$ is said to be measurable with respect to a σ-algebra $\mathcal{S}$ of subsets of Ω if $f^{-1}(A) \in \mathcal{S}$ for every Borel set $A \subseteq \mathbb{R}$.

A *real random variable* $\mathbf{x}$ defined on a probability space $(\Omega, \mathcal{S}, \text{PR}\{S\})$ is a measurable function mapping Ω into $\mathcal{Y} \subseteq \mathbb{R}$, and this is indicated with shorthand notation $\mathbf{x} \in \mathcal{Y}$. The set $\mathcal{Y}$ is called the range or *support* of the random variable $\mathbf{x}$. A *complex random variable* $\mathbf{x} \in \mathbb{C}$ is a sum $\mathbf{x} = \mathbf{x}_{\mathbb{R}} + j\mathbf{x}_{\mathbb{I}}$, where $\mathbf{x}_{\mathbb{R}} \in \mathbb{R}$ and $\mathbf{x}_{\mathbb{I}} \in \mathbb{R}$ are real random variables, and $j \doteq \sqrt{-1}$. If the random variable $\mathbf{x}$ maps the sample space Ω into a subset $[a, b] \subset \mathbb{R}$, we write $\mathbf{x} \in [a, b]$. If Ω is a discrete probability space, then $\mathbf{x}$ is a *discrete* random variable mapping Ω into a countable set.

Distribution and density functions. The (cumulative) *distribution function* (cdf) of a random variable $\mathbf{x}$ is defined as

$$F_{\mathbf{x}}(x) \doteq \text{PR}\{\mathbf{x} \leq x\}.$$

The function $F_{\mathbf{x}}(x)$ is nondecreasing, right continuous, and $F_{\mathbf{x}}(x) \to 0$ for $x \to -\infty$, $F_{\mathbf{x}}(x) \to 1$ for $x \to \infty$. Associated with the concept of distribution function, we define the α *percentile* of a random variable

$$x_{\alpha} = \inf\{x : F_{\mathbf{x}}(x) \geq \alpha\}.$$

For random variables of continuous type, if there exists a Lebesgue measurable function $f_{\mathbf{x}}(x) \geq 0$ such that

$$F_{\mathbf{x}}(x) = \int_{-\infty}^{x} f_{\mathbf{x}}(x)\mathrm{d}x$$

then the cdf $F_{\mathbf{x}}(x)$ is said to be absolutely continuous, and

$$f_{\mathbf{x}}(x) = \frac{\mathrm{d}F_{\mathbf{x}}(x)}{\mathrm{d}x}$$

holds except possibly for a set of measure zero. The function $f_{\mathbf{x}}(x)$ is called the probability *density function* (pdf) of the random variable $\mathbf{x}$.

For discrete random variables, the cdf is a staircase function, i.e. $F_{\mathbf{x}}(x)$ is constant except at a countable number of points $x_1, x_2, \ldots$ having no finite limit point. The total probability is hence distributed among the "mass" points $x_1, x_2, \ldots$ at which the "jumps" of size

$$f_{\mathbf{x}}(x_i) \doteq \lim_{\epsilon \to 0} F_{\mathbf{x}}(x_i + \epsilon) - F_{\mathbf{x}}(x_i - \epsilon) = \text{PR}\{\mathbf{x} = x_i\}$$

occur. The function $f_{\mathbf{x}}(x_i)$ is called the *mass density* of the discrete random variable $\mathbf{x}$. The definition of random variables is extended to real and complex random matrices in the next section.

2.1.3 Real and complex random matrices

Given n random variables $\mathbf{x}_1, \ldots, \mathbf{x}_n$, their *joint distribution* is defined as

$$F_{\mathbf{x}_1,\ldots,\mathbf{x}_n}(x_1,\ldots,x_n) \doteq \mathrm{PR}\left\{\mathbf{x}_1 \le x_1, \ldots, \mathbf{x}_n \le x_n\right\}.$$

When the above distribution is absolutely continuous, we can define the joint density function $f_{\mathbf{x}_1,\ldots,\mathbf{x}_n}(x_1,\ldots,x_n)$

$$f_{\mathbf{x}_1,\ldots,\mathbf{x}_n}(x_1,\ldots,x_n) \doteq \frac{\partial^n F_{\mathbf{x}_1,\ldots,\mathbf{x}_n}(x_1,\ldots,x_n)}{\partial x_1 \cdots \partial x_n}.$$

The random variables $\mathbf{x}_1, \ldots, \mathbf{x}_n$ are said to be *independent* if

$$F_{\mathbf{x}_1,\ldots,\mathbf{x}_n}(x_1,\ldots,x_n) = \prod_{i=1}^{n} F_{\mathbf{x}_i}(x_i)$$

where $F_{\mathbf{x}_i}(x_i) = \mathrm{PR}\left\{\mathbf{x}_i \le x_i\right\}$.

A real random matrix $\mathbf{X} \in \mathbb{R}^{n,m}$ is a measurable function $\mathbf{X} : \Omega \to \mathcal{Y} \subseteq \mathbb{R}^{n,m}$. That is, the entries of $\mathbf{X}$ are real random variables $[\mathbf{X}]_{i,k}$ for $i = 1, \ldots, n$ and $k = 1, \ldots, m$. A complex random matrix $\mathbf{X} \in \mathbb{C}^{n,m}$ is defined as the sum $\mathbf{X} = \mathbf{X}_{\mathbb{R}} + j\mathbf{X}_{\mathbb{I}}$, where $\mathbf{X}_{\mathbb{R}}$ and $\mathbf{X}_{\mathbb{I}}$ are real random matrices. A random matrix is *discrete* if its entries are discrete random variables.

The distribution function $F_{\mathbf{X}}(X)$ of a real random matrix $\mathbf{X}$ is the joint cdf of the entries of $\mathbf{X}$. If $\mathbf{X}$ is a complex random matrix, then its cdf is the joint cdf of $\mathbf{X}_{\mathbb{R}}$ and $\mathbf{X}_{\mathbb{I}}$. The pdf $f_{\mathbf{X}}(X)$ of a real or complex random matrix is analogously defined as the joint pdf of the real and imaginary parts of its entries. The notation $\mathbf{X} \sim f_{\mathbf{X}}(X)$ means that $\mathbf{X}$ is a random matrix with probability density function $f_{\mathbf{X}}(X)$.

Let $\mathbf{X} \in \mathbb{F}^{n,m}$ be a real or complex random matrix (of continuous type) with pdf $f_{\mathbf{X}}(X)$ and support $\mathcal{Y} \subseteq \mathbb{F}^{n,m}$. Then, if $Y \subseteq \mathcal{Y}$, we have

$$\mathrm{PR}\left\{\mathbf{X} \in Y\right\} = \int_{Y} f_{\mathbf{X}}(X)\mathrm{d}X.$$

Clearly, $\mathrm{PR}\left\{\mathbf{X} \in \mathcal{Y}\right\} = \int_{\mathcal{Y}} f_{\mathbf{X}}(X)\mathrm{d}X = 1$. When needed, to further emphasize that the probability is relative to the random matrix $\mathbf{X}$, we explicitly write $\mathrm{PR}_{\mathbf{X}}\left\{\mathbf{X} \in Y\right\}$.

2.1.4 Expected value and covariance

Let $\mathbf{X} \in \mathcal{Y} \subseteq \mathbb{F}^{n,m}$ be a random matrix and let $J : \mathbb{F}^{n,m} \to \mathbb{R}^{p,q}$ be a Lebesgue measurable function. The *expected value* of the random matrix $J(\mathbf{X})$ is defined as

$$\mathrm{E}_{\mathbf{X}}(J(\mathbf{X})) \doteq \int_{\mathcal{Y}} J(X) f_{\mathbf{X}}(X)\mathrm{d}X$$

where $\mathcal{Y}$ is the support of $\mathbf{X}$. We make use of the symbol $\mathrm{E}_{\mathbf{X}}(J(\mathbf{X}))$ to emphasize the fact that the expected value is taken with respect to $\mathbf{X}$. The suffix is omitted when clear from the context.

If $\mathbf{X} \in \mathbb{F}^{n,m}$ is a discrete random matrix with countable support $\mathcal{Y} = \{X_1, X_2, \ldots\}$, $X_i \in \mathbb{F}^{n,m}$ and $Y \subseteq \mathcal{Y}$, then

$$\mathrm{PR}\{\mathbf{X} \in Y\} = \sum_{X_i \in Y} f_{\mathbf{X}}(X_i) = \sum_{X_i \in Y} \mathrm{PR}\{\mathbf{X} = X_i\}.$$

The expected value of $J(\mathbf{X})$ is defined as

$$\mathrm{E}(J(\mathbf{X})) \doteq \sum_{X_i \in \mathcal{Y}} J(X_i) f_{\mathbf{X}}(X_i).$$

The expected value of $\mathbf{X} \in \mathbb{R}^{n,m}$ is usually called the *mean*. The *covariance matrix* of $\mathbf{x} \in \mathbb{R}^n$ is defined as

$$\mathrm{Cov}(\mathbf{x}) \doteq \mathrm{E}_{\mathbf{x}}\big((\mathbf{x} - \mathrm{E}_{\mathbf{x}}(\mathbf{x}))^T (\mathbf{x} - \mathrm{E}_{\mathbf{x}}(\mathbf{x}))\big)$$

where X^T denotes the transpose of X. The covariance of $\mathbf{x} \in \mathbb{R}$ is called the *variance* and is given by

$$\mathrm{Var}(\mathbf{x}) \doteq \mathrm{E}_{\mathbf{x}}\big((\mathbf{x} - \bar{x})^2\big).$$

The square root of the variance $(\mathrm{Var}(\mathbf{x}))^{1/2}$ is called the *standard deviation*.

2.2 Marginal and conditional densities

Consider a random vector $\mathbf{x} = [\mathbf{x}_1 \cdots \mathbf{x}_n]^T \in \mathbb{R}^n$ with joint density function

$$f_{\mathbf{x}}(x) = f_{\mathbf{x}_1,\ldots,\mathbf{x}_n}(x_1,\ldots,x_n).$$

The *marginal density* of the first i components of the random vector $\mathbf{x} = [\mathbf{x}_1 \cdots \mathbf{x}_n]^T$ is defined as

$$f_{\mathbf{x}_1,\ldots,\mathbf{x}_i}(x_1,\ldots,x_i) \doteq \int \cdots \int f_{\mathbf{x}}(x_1,\ldots,x_n)\mathrm{d}x_{i+1}\cdots\mathrm{d}x_n. \qquad (2.1)$$

The *conditional density* $f_{\mathbf{x}_i|x_1,\ldots,x_{i-1}}(x_i|x_1,\ldots,x_{i-1})$ of the random variable $\mathbf{x}_i$ conditioned to the event $\mathbf{x}_1 = x_1,\ldots,\mathbf{x}_{i-1} = x_{i-1}$ is given by the ratio of marginal densities

$$f_{\mathbf{x}_i|x_1,\ldots,x_{i-1}}(x_i|x_1,\ldots,x_{i-1}) \doteq \frac{f_{\mathbf{x}_1,\ldots,\mathbf{x}_i}(x_1,\ldots,x_i)}{f_{\mathbf{x}_1,\ldots,\mathbf{x}_{i-1}}(x_1,\ldots,x_{i-1})}. \qquad (2.2)$$

2.3 Univariate and multivariate density functions

We present a list of classical univariate and multivariate density functions.

Binomial density. The binomial density is defined as

$$B_n(x) \doteq \binom{n}{x} p^x (1-p)^{n-x}, \quad x \in \{0, 1, \ldots, n\}$$

where $\binom{n}{x}$ indicates the binomial coefficient $\binom{n}{x} = \frac{n!}{x!(n-x)!}$.

Normal density. The normal (Gaussian) density with mean $\bar{x} \in \mathbb{R}$ and variance $\sigma^2 \in \mathbb{R}$ is defined as

$$\mathcal{N}_{\bar{x}, \sigma^2}(x) \doteq \frac{1}{\sigma \sqrt{2\pi}} e^{-\frac{1}{2}(x-\bar{x})^2/\sigma^2}, \quad x \in \mathbb{R}. \tag{2.3}$$

Multivariate normal density. The multivariate normal density with mean $\bar{x} \in \mathbb{R}^n$ and symmetric positive definite covariance matrix $W \in \mathbb{S}^n$, $W \succ 0$, is defined as

$$\mathcal{N}_{\bar{x}, W}(x) \doteq (2\pi)^{-n/2} |W|^{-1/2} e^{-\frac{1}{2}(x-\bar{x})^T W^{-1}(x-\bar{x})}, \quad x \in \mathbb{R}^n. \tag{2.4}$$

Uniform density. The uniform density on the interval $[a, b]$ is defined as

$$\mathcal{U}_{[a,b]}(x) \doteq \begin{cases} \dfrac{1}{b-a} & \text{if } x \in [a, b]; \\ 0 & \text{otherwise.} \end{cases} \tag{2.5}$$

Uniform density over a set. Let S be a Lebesgue measurable set of nonzero volume (see Section 3.1.3 for a precise definition of volume). The uniform density over S is defined as

$$\mathcal{U}_S(X) \doteq \begin{cases} \dfrac{1}{\text{Vol}(S)} & \text{if } X \in S; \\ 0 & \text{otherwise.} \end{cases} \tag{2.6}$$

If instead S is a finite discrete set, i.e. it consists of a finite number of elements $S = \{X_1, X_2, \ldots, X_N\}$, then the uniform density over S is defined as

$$\mathcal{U}_S(X) \doteq \begin{cases} \dfrac{1}{\text{Card}(S)} & \text{if } X \in S; \\ 0 & \text{otherwise} \end{cases}$$

where $\text{Card}(S)$ is the cardinality of S.

Chi-square density. The unilateral chi-square density with $n > 0$ degrees of freedom is defined as

$$\chi_n^2(x) \doteq \frac{1}{\Gamma(n/2) 2^{n/2}} x^{n/2-1} e^{-x/2}, \quad x \in \mathbb{R}_+ \tag{2.7}$$

where $\Gamma(\cdot)$ is the Gamma function

$$\Gamma(x) \doteq \int_0^\infty \xi^{x-1} e^{-\xi} d\xi, \quad x > 0.$$

Weibull density. The Weibull density with parameter $a > 0$ is defined as

$$W_a(x) \doteq ax^{a-1}e^{-x^a}, \quad x \in \mathbb{R}. \tag{2.8}$$

Laplace density. The unilateral Laplace (or exponential) density with parameter $\lambda > 0$ is defined as

$$L_\lambda(x) \doteq \lambda e^{-\lambda x}, \quad x \in \mathbb{R}_+. \tag{2.9}$$

Gamma density. The unilateral Gamma density with parameters $a > 0$, $b > 0$ is defined as

$$G_{a,b}(x) \doteq \frac{1}{\Gamma(a)b^a} x^{a-1}e^{-x/b}, \quad x \in \mathbb{R}_+. \tag{2.10}$$

Generalized Gamma density. The unilateral generalized Gamma density with parameters $a > 0$, $c > 0$ is defined as

$$\overline{G}_{a,c}(x) \doteq \frac{c}{\Gamma(a)} x^{ca-1}e^{-x^c}, \quad x \in \mathbb{R}_+. \tag{2.11}$$

2.4 Convergence of random variables

We now briefly recall the formal definitions of convergence almost everywhere (or almost sure convergence) and convergence in probability. Other convergence concepts not discussed here include vague convergence, convergence of moments and convergence in distribution, see e.g. [69].

Definition 2.1 (Convergence almost everywhere). *A sequence of random variables* $\mathbf{x}^{(1)}, \mathbf{x}^{(2)}, \ldots$ *converges* almost everywhere *(a.e.) (or with probability one) to the random variable* $\mathbf{x}$ *if*

$$\mathrm{PR}\left\{ \lim_{N \to \infty} \mathbf{x}^{(N)} = \mathbf{x} \right\} = 1.$$

Definition 2.2 (Convergence in probability). *A sequence of random variables* $\mathbf{x}^{(1)}, \mathbf{x}^{(2)}, \ldots$ *converges* in probability *to the random variable* $\mathbf{x}$ *if, for any* $\epsilon > 0$, *we have*

$$\lim_{N \to \infty} \mathrm{PR}\left\{ |\mathbf{x} - \mathbf{x}^{(N)}| > \epsilon \right\} = 0.$$

It can be shown that convergence a.e. implies convergence in probability, but the converse is not necessarily true.

3. Uncertain Linear Systems and Robustness

This chapter presents a summary of some classical results regarding robust-ness analysis. Synthesis problems are subsequently presented in Chapter 4. In these two chapters, we concentrate on linear, continuous and time-invariant systems and assume that the reader is familiar with the basics of linear al-gebra and systems and control theory, see e.g. [65, 228]. We do not attempt to provide a comprehensive treatment of robust control, which is discussed in depth for instance in [98, 123, 245, 295]. Advanced material may be also found in the special issues [164, 231], and specific references are listed in [91].

3.1 Norms, balls and volumes

3.1.1 Vector norms and balls

Let $x \in \mathbb{F}^n$, where $\mathbb{F}$ is either the real or the complex field, then the ℓ_p norm of vector x is defined as

$$\|x\|_p \doteq \left(\sum_{i=1}^{n} |x_i|^p \right)^{1/p}, \quad p \in [1, \infty) \tag{3.1}$$

and the ℓ_∞ norm of x is

$$\|x\|_\infty \doteq \max_i |x_i|.$$

The ℓ_2 norm is usually called the Euclidean norm. We define the ball of radius ρ in the ℓ_p norm as

$$\mathcal{B}_{\|\cdot\|_p}(\rho, \mathbb{F}^n) \doteq \{x \in \mathbb{F}^n : \|x\|_p \leq \rho\} \tag{3.2}$$

and its boundary as

$$\partial\mathcal{B}_{\|\cdot\|_p}(\rho, \mathbb{F}^n) \doteq \{x \in \mathbb{F}^n : \|x\|_p = \rho\}. \tag{3.3}$$

When clear from the context, we simply write $\mathcal{B}_{\|\cdot\|_p}(\rho)$ and $\partial\mathcal{B}_{\|\cdot\|_p}(\rho)$ to denote $\mathcal{B}_{\|\cdot\|_p}(\rho, \mathbb{F}^n)$ and $\partial\mathcal{B}_{\|\cdot\|_p}(\rho, \mathbb{F}^n)$, respectively. Moreover, for balls of unit radius, we write $\mathcal{B}_{\|\cdot\|_p}(\mathbb{F}^n)$ and $\partial\mathcal{B}_{\|\cdot\|_p}(\mathbb{F}^n)$, or in brief as $\mathcal{B}_{\|\cdot\|_p}$ and $\partial\mathcal{B}_{\|\cdot\|_p}$.

We introduce further the weighted ℓ_2 norm of a real vector $x \in \mathbb{R}^n$. For a symmetric, positive definite matrix $W \succ 0$, the weighted ℓ_2 norm, denoted by ℓ_2^W, is defined as

$$\|x\|_2^W \doteq \left(x^T W^{-1} x\right)^{1/2}. \tag{3.4}$$

Clearly, if we compute the Cholesky decomposition $W^{-1} = R^T R$, then we have $\|x\|_2^W = \|Rx\|_2$. The ball of radius ρ in the ℓ_2^W norm is

$$\mathcal{B}_{\|\cdot\|_2^W}(\rho, \mathbb{R}^n) \doteq \{x \in \mathbb{R}^n : \|x\|_2^W \leq \rho\}. \tag{3.5}$$

This ball is an ellipsoid in the standard ℓ_2 metric. In fact, if we denote the ellipsoid of center $\bar{x}$ and shape matrix $W \succ 0$ as

$$\mathcal{E}(\bar{x}, W) \doteq \{x \in \mathbb{R}^n : (x - \bar{x})^T W^{-1}(x - \bar{x}) \leq 1\} \tag{3.6}$$

then $\mathcal{B}_{\|\cdot\|_2^W}(\rho, \mathbb{R}^n) = \mathcal{E}\left(0, \rho^2 W\right)$.

3.1.2 Matrix norms and balls

Two different classes of norms can be introduced when dealing with matrix variables: the so-called Hilbert–Schmidt norms, based on the isomorphism between the matrix space $\mathbb{F}^{n,m}$ and the vector space $\mathbb{F}^{nm}$, and the induced norms, where the matrix is viewed as an operator between vector spaces.

Hilbert–Schmidt matrix norms. The (generalized) Hilbert–Schmidt ℓ_p norm of a matrix $X \in \mathbb{F}^{n,m}$ is defined as (see e.g. [137])

$$\|X\|_p \doteq \left(\sum_{i=1}^n \sum_{k=1}^m |[X]_{i,k}|^p\right)^{1/p}, \quad p \in [0, \infty); \tag{3.7}$$

$$\|X\|_\infty \doteq \max_{i,k} |[X]_{i,k}|$$

where $[X]_{i,k}$ is the (i, k) entry of matrix X. We remark that for $p = 2$ the Hilbert–Schmidt ℓ_p norm corresponds to the well-known Frobenius matrix norm

$$\|X\|_2 = \sqrt{\operatorname{Tr} X X^*}$$

where Tr denotes the trace and X^* is the conjugate transpose of X. Given a matrix $X \in \mathbb{F}^{n,m}$, we introduce the column vectorization operator

$$\operatorname{vec}(X) \doteq \begin{bmatrix} \xi_1 \\ \vdots \\ \xi_m \end{bmatrix} \tag{3.8}$$

where $\xi_1, \ldots, \xi_m$ are the columns of X. Then, using (3.7) the Hilbert–Schmidt ℓ_p norm of X can be written as

$$\|X\|_p = \|\mathrm{vec}(X)\|_p.$$

In analogy to vectors, we denote the ℓ_p Hilbert–Schmidt norm ball in $\mathbb{F}^{n,m}$ of radius ρ as

$$\mathcal{B}_{\|\cdot\|_p}(\rho, \mathbb{F}^{n,m}) \doteq \{X \in \mathbb{F}^{n,m} : \|X\|_p \leq \rho\}.$$

When clear from the context, we write $\mathcal{B}_{\|\cdot\|_p}(\rho)$ to denote $\mathcal{B}_{\|\cdot\|_p}(\rho, \mathbb{F}^{n,m})$ and $\mathcal{B}_{\|\cdot\|_p}(\mathbb{F}^{n,m})$ or $\mathcal{B}_{\|\cdot\|_p}$ for unit radius balls.

Induced matrix norms. The ℓ_p induced norm of a matrix $X \in \mathbb{F}^{n,m}$ is defined as

$$\|X\|_p \doteq \max_{\|\xi\|_p=1} \|X\xi\|_p, \quad \xi \in \mathbb{F}^m. \tag{3.9}$$

The ℓ_1 induced norm of a matrix $X \in \mathbb{F}^{n,m}$ turns out to be the maximum of the ℓ_1 norms of its columns, that is

$$\|X\|_1 = \max_{i=1,\dots,m} \|\xi_i\|_1 \tag{3.10}$$

where $\xi_1, \cdots, \xi_m$ are the columns of X. Similarly, the ℓ_∞ induced norm is equal to the maximum of the ℓ_1 norms of the rows of X, i.e.

$$\|X\|_\infty = \max_{i=1,\dots,n} \|\eta_i\|_1$$

where $\eta_1^T, \dots, \eta_n^T$ are the rows of X.

The ℓ_2 induced norm of a matrix is called the *spectral norm* and is related to the *singular value decomposition* (SVD), see for instance [137]. The SVD of a matrix $X \in \mathbb{F}^{n,m}$, $m \geq n$, is given by

$$X = U\Sigma V^*$$

where $\Sigma = \mathrm{diag}\,([\sigma_1 \cdots \sigma_n])$, with $\sigma_1 \geq \cdots \geq \sigma_n \geq 0$, $U \in \mathbb{F}^{n,n}$ is unitary, and $V \in \mathbb{F}^{m,n}$ has orthonormal columns.

The elements of Σ are called the *singular values* of X, and Σ is called the singular values matrix. The maximum singular value σ_1 of X is denoted by $\bar{\sigma}(X)$. The ℓ_2 induced norm of a matrix X is equal to

$$\|X\|_2 = \bar{\sigma}(X). \tag{3.11}$$

The ℓ_p induced norm ball of radius ρ in $\mathbb{F}^{n,m}$ is denoted by

$$\mathcal{B}_{\|\cdot\|_p}(\rho, \mathbb{F}^{n,m}) \doteq \{X \in \mathbb{F}^{n,m} : \|X\|_p \leq \rho\}. \tag{3.12}$$

For simplicity, we denote a ball of radius ρ in the spectral norm as

$$\mathcal{B}_\sigma(\rho, \mathbb{F}^{n,m}) \doteq \mathcal{B}_{\|\cdot\|_2}(\rho, \mathbb{F}^{n,m}). \tag{3.13}$$

When clear from the context, we write $\mathcal{B}_{\|\cdot\|_p}(\rho)$ and $\mathcal{B}_\sigma(\rho)$ to denote the balls $\mathcal{B}_{\|\cdot\|_p}(\rho, \mathbb{F}^{n,m})$ and $\mathcal{B}_\sigma(\rho, \mathbb{F}^{n,m})$ respectively. Similarly, $\mathcal{B}_{\|\cdot\|_p}(\mathbb{F}^{n,m})$ or $\mathcal{B}_{\|\cdot\|_p}$, and $\mathcal{B}_\sigma(\mathbb{F}^{n,m})$ or $\mathcal{B}_\sigma$ denote unit radius balls.

3.1.3 Volumes

Consider the field $\mathbb{F}^{n,m}$. The dimension d of $\mathbb{F}^{n,m}$ is $d = nm$ if $\mathbb{F} \equiv \mathbb{R}$, and $d = 2nm$ if $\mathbb{F} \equiv \mathbb{C}$. Let $S \subset \mathbb{F}^{n,m}$ be a Lebesgue measurable set and let $\mu_d(\cdot)$ denote the d-dimensional Lebesgue measure, then the *volume* of S is defined as

$$\text{Vol}(S) \doteq \int_S \mathrm{d}\mu_d(X). \tag{3.14}$$

Similarly, we indicate by $\text{Surf}(S)$ the *surface* of S, that is the $(d - 1)$-dimensional Lebesgue measure of S. In particular, for norm balls, volume and surface measures depend on the ball radius according to the relations (see for instance [22])

$$\text{Vol}(\mathcal{B}_{\|\cdot\|_p}(\rho)) = \text{Vol}(\mathcal{B}_{\|\cdot\|_p})\rho^d;$$
$$\text{Surf}(\mathcal{B}_{\|\cdot\|_p}(\rho)) = \text{Surf}(\mathcal{B}_{\|\cdot\|_p})\rho^{d-1}.$$

3.2 Signals

In this section, we briefly introduce some concepts related to signals, see e.g. [166] for a comprehensive treatment of this topic.

3.2.1 Deterministic signals

A *deterministic signal* $v(t) : \mathbb{R} \to \mathbb{R}^n$ is a Lebesgue measurable function of the time variable $t \in \mathbb{R}$. The set

$$\mathcal{V}^+ = \{v(t) \in \mathbb{R}^n : v \text{ is Lebesgue measurable}, \ v(t) = 0 \text{ for all } t < 0\}$$

is the linear space of *causal* signals. For $p \in [1, \infty)$, the infinite-horizon $\mathcal{L}_p^+$ space is defined as the space of signals $v \in \mathcal{V}^+$ such that the integral

$$\left(\int_0^\infty \|v(t)\|_p^p \mathrm{d}t \right)^{1/p} \tag{3.15}$$

exists and is bounded. In this case, (3.15) defines a signal norm, which is denoted by $\|v\|_p$. For $p = \infty$, we have $\|v\|_\infty \doteq \text{ess sup}_t \, v(t)$.

For the important special case $p = 2$, $\mathcal{L}_2^+$ is a Hilbert space, equipped with the standard inner product

$$\langle x, y \rangle = \int_0^\infty y^T(t)x(t)\mathrm{d}t$$

where $x, y \in \mathcal{L}_2^+$. Signals in $\mathcal{L}_2^+$ are therefore causal signals with finite total energy. These are typically *transient* signals which decay to zero as $t \to \infty$.

We now discuss some fundamental results related to the Laplace transform of signals in $\mathcal{L}_2^+$. The $\mathcal{H}_2^n$ space (see Definition 3.3) is the space of functions of complex variable $g(s) : \mathbb{C} \to \mathbb{C}^n$ which are analytic[1] in $\text{Re}(s) > 0$ and for which the integral

$$\left(\frac{1}{2\pi} \int_{-\infty}^{\infty} g^*(j\omega) g(j\omega) d\omega \right)^{1/2} \tag{3.16}$$

exists and is bounded. In this case, (3.16) defines a norm denoted by $\|g\|_2$. Define further the unilateral Laplace transform of the signal $v \in \mathcal{V}^+$ as

$$\zeta(s) = \mathcal{L}(v) \doteq \int_0^{\infty} v(t) e^{-st} dt$$

and the inverse Laplace transform

$$v(t) = \mathcal{L}^{-1}(\zeta) \doteq \lim_{\omega \to \infty} \frac{1}{2\pi j} \int_{c-j\omega}^{c+j\omega} \zeta(s) e^{st} ds.$$

Then, if $v \in \mathcal{L}_2^+$, its Laplace transform is in $\mathcal{H}_2^n$. Conversely, by the Paley-Wiener theorem, see e.g. [98], for any $\zeta \in \mathcal{H}_2^n$ there exists a causal signal $v \in \mathcal{L}_2^+$ such that $\zeta = \mathcal{L}(v)$. Notice also that $\mathcal{H}_2^n$ is a Hilbert space, equipped with the inner product

$$\langle g, h \rangle = \frac{1}{2\pi} \int_{-\infty}^{\infty} g^*(j\omega) h(j\omega) d\omega$$

for $g, h \in \mathcal{H}_2^n$. Finally, we recall the Parseval identity, see e.g. [123], which relates the inner product of the signals $v, w \in \mathcal{L}_2^+$ to the inner product of their Laplace transforms

$$\langle v, w \rangle = \langle \mathcal{L}(v), \mathcal{L}(w) \rangle.$$

3.2.2 Stochastic signals

The performance specifications of control systems are sometimes expressed in terms of *stochastic*, rather than deterministic, signals. In this section, we summarize some basic definitions related to stochastic signals. For formal definitions of random variables and matrices and their statistics, the reader can refer to Chapter 2 and to [88, 214] for further details on stochastic processes.

---

[1] Let $S \subset \mathbb{C}$ be an open set. A function $f : S \to \mathbb{C}$ is said to be *analytic* at a point $s_0 \in S$ if it is differentiable for all points in some neighborhood of s_0. The function is analytic in S if it is analytic for all $s \in S$. A matrix-valued function is analytic if every element of the matrix is analytic.

Denote with $\mathbf{v}(t)$ a zero-mean, stationary stochastic process. The *autocorrelation* of $\mathbf{v}(t)$ is defined as

$$R_{\mathbf{v},\mathbf{v}}(\tau) \doteq E_{\mathbf{v}}\big(\mathbf{v}(t)\mathbf{v}^T(t+\tau)\big)$$

where $E_{\mathbf{v}}(\cdot)$ denotes the expectation with respect to the stochastic process. The power spectral density (psd) $\Phi_{\mathbf{v},\mathbf{v}}(\omega)$ of $\mathbf{v}$ is defined as the Fourier transform of $R_{\mathbf{v},\mathbf{v}}(\tau)$. A frequently used measure of a stationary stochastic signal is its root-mean-square (rms) value

$$\|\mathbf{v}\|_{\mathrm{rms}}^2 = E_{\mathbf{v}}\big(\mathbf{v}^T(t)\mathbf{v}(t)\big) = \mathrm{Tr}\,R_{\mathbf{v},\mathbf{v}}(0).$$

The rms value measures the *average power* of the stochastic signal, and it is a steady-state measure of the behavior of the signal, i.e. it is not affected by transients. By the Parseval identity, the average power can alternatively be computed as an integral over frequency of the power spectral density

$$\|\mathbf{v}\|_{\mathrm{rms}}^2 = \frac{1}{2\pi}\int_{-\infty}^{\infty} \mathrm{Tr}\,\Phi_{\mathbf{v},\mathbf{v}}(\omega)\mathrm{d}\omega.$$

If the process $\mathbf{v}(t)$ is *ergodic*, then its moments can be equivalently computed as time-domain averages of a single realization $v(t)$ of the process. With probability one, the rms norm is given by

$$\|\mathbf{v}\|_{\mathrm{rms}}^2 = \lim_{T\to\infty} \frac{1}{T}\int_0^T v^T(t)v(t)\mathrm{d}t.$$

3.3 Linear time-invariant systems

Consider a linear time-invariant (LTI), proper system described in standard state space form

$$\dot{x} = Ax + Bw; \tag{3.17}$$
$$z = Cx + Dw$$

where $A \in \mathbb{R}^{n_s,n_s}$, $B \in \mathbb{R}^{n_s,q}$, $C \in \mathbb{R}^{p,n_s}$, $D \in \mathbb{R}^{p,q}$. This state space system is *stable*, or *Hurwitz*, if $\mathrm{Re}(\lambda_i(A)) < 0$, $i = 1,\ldots,n_s$, where $\lambda_i(A)$ denote the eigenvalues of A.

Assuming $x(0) = 0$, system (3.17) defines a proper linear operator $\mathcal{G}$ mapping the input signal space into the output signal space. In the space of Laplace transforms, the operator $\mathcal{G}$ is represented by the *transfer-function matrix*, or simply *transfer matrix*

$$G(s) = C(sI - A)^{-1}B + D.$$

The system (3.17) is indicated compactly by means of the matrix quadruple

$$\Omega_G \doteq \left[\begin{array}{c|c} A & B \\ \hline C & D \end{array}\right].$$

The operator $\mathcal{G}$ related to system (3.17) is *stable* if and only if it maps $\mathcal{L}_2^+$ into $\mathcal{L}_2^+$. A necessary and sufficient stability condition for $\mathcal{G}$ is that its transfer matrix $G(s)$ has all its poles in the open left-half plane.

Definition 3.1 ($\mathcal{RH}_\infty$ space). *The space $\mathcal{RH}_\infty^{p,q}$ is defined as the space of proper, rational functions with real coefficients $G : \mathbb{C} \to \mathbb{C}^{p,q}$ that are analytic in the open right-half plane.*

From this definition, it follows that the operator $\mathcal{G}$ is stable if and only if its transfer matrix $G(s)$ belongs to $\mathcal{RH}_\infty$.

Assuming $\mathcal{G}$ stable, since $\mathcal{G}$ maps $\mathcal{L}_2^+$ into $\mathcal{L}_2^+$, it is natural to define its $\mathcal{L}_2^+$-gain as

$$\|\mathcal{G}\|_{\mathcal{L}_2^+ \to \mathcal{L}_2^+} \doteq \sup_{0 \neq w \in \mathcal{L}_2^+} \frac{\|\mathcal{G}w\|_2}{\|w\|_2}.$$

If $\mathcal{G}$ is represented in the frequency domain by the transfer matrix $G(s)$, then it can be shown that its $\mathcal{L}_2^+$-gain coincides with the so-called $\mathcal{H}_\infty$ norm of $G(s)$, defined as

$$\|G(s)\|_\infty \doteq \operatorname*{ess\,sup}_{\omega \in \mathbb{R}} \bar{\sigma}(G(j\omega)) \tag{3.18}$$

where $\bar{\sigma}(G(j\omega))$ denotes the largest singular value of $G(j\omega)$, i.e.

$$\|G(s)\|_\infty = \|\mathcal{G}\|_{\mathcal{L}_2^+ \to \mathcal{L}_2^+}. \tag{3.19}$$

In a more general setting, one can define the space of functions $G : \mathbb{C} \to \mathbb{C}^{p,q}$ (not necessarily rational), for which the norm (3.18) is bounded.

Definition 3.2 ($\mathcal{H}_\infty$ space). *The space $\mathcal{H}_\infty^{p,q}$ is defined as the space of functions $G : \mathbb{C} \to \mathbb{C}^{p,q}$ that are analytic and bounded in the open right-half plane.*

From this definition it follows immediately that $\mathcal{RH}_\infty \subset \mathcal{H}_\infty$.

Remark 3.1 ($\mathcal{H}_\infty$ norm interpretations). The $\mathcal{H}_\infty$ norm of a stable system may be interpreted from (3.19) as the maximum energy gain of the system. In the case of stochastic signals, it has an alternative interpretation as the rms gain of the system, i.e. it denotes the maximum average power amplification from input to output. We also remark that the $\mathcal{H}_\infty$ norm is *submultiplicative*, i.e.

$$\|GH\|_\infty \leq \|G\|_\infty \|H\|_\infty.$$

For stable single-input single-output (SISO) systems, (3.18) indicates that the value of the $\mathcal{H}_\infty$ norm coincides with the peak of the magnitude of the Bode plot of the transfer function of the system.   ◁

Another frequently used measure of a system "gain" is the $\mathcal{H}_2$ norm. This norm and the corresponding linear space of transfer matrices are now defined.

Definition 3.3 ($\mathcal{H}_2$ and $\mathcal{RH}_2$ spaces). *The space $\mathcal{H}_2^{p,q}$ is defined as the space of functions $G : \mathbb{C} \to \mathbb{C}^{p,q}$ that are analytic in the open right-half plane and such that the integral*

$$\left(\frac{1}{2\pi} \int_{-\infty}^{\infty} \operatorname{Tr} G^*(j\omega) G(j\omega) d\omega \right)^{1/2} \tag{3.20}$$

exists and is bounded. In this case, (3.20) defines the $\mathcal{H}_2$ norm of G, which is denoted by $\|G\|_2$. The space $\mathcal{RH}_2^{p,q}$ is then defined as

$$\mathcal{RH}_2^{p,q} \doteq \{G \in \mathcal{H}_2^{p,q} : G \text{ is real rational}\}.$$

Notice that, according to the above definition, a rational transfer matrix $G(s)$ belongs to $\mathcal{RH}_2$ if and only if it is stable and strictly proper.

Remark 3.2 ($\mathcal{H}_2$ norm interpretations). The $\mathcal{H}_2$ norm of a stable system has two interpretations. First, we notice that $\|G(s)\|_2^2$ can be computed in the time domain using the Parseval identity

$$\|G\|_2^2 = \int_0^{\infty} \operatorname{Tr} g^T(t) g(t) dt$$

where $g(t) = \mathcal{L}^{-1}(G(s))$ is the impulse response matrix. The $\mathcal{H}_2$ norm can hence be interpreted as the energy of the impulse response of the system.

Secondly, the $\mathcal{H}_2$ norm can be viewed as a measure of the average power of the steady-state output, when the system is driven by white noise input, see for instance [48]. In fact, when a stochastic signal $\mathbf{w}$ with power spectral density $\Phi_{\mathbf{w},\mathbf{w}}(\omega)$ enters a stable and strictly proper system with transfer matrix G, then the output $\mathbf{z}$ has spectral density given by

$$\Phi_{\mathbf{z},\mathbf{z}}(\omega) = G(j\omega) \Phi_{\mathbf{w},\mathbf{w}}(\omega) G^*(j\omega)$$

and the average output power is $\|\mathbf{z}\|_{\mathrm{rms}}$. When $\mathbf{w}$ is white noise, then $\Phi_{\mathbf{w},\mathbf{w}}(\omega) = I$, and $\|\mathbf{z}\|_{\mathrm{rms}} = \|G\|_2$.   ◁

3.4 Linear matrix inequalities

Many of the analysis and design specifications for control systems may be expressed in the form of satisfaction of a positive (or negative) definiteness condition for a matrix function which depends affinely on the decision variables of the problem. Such matrix "inequalities" are commonly known under the name of linear matrix inequalities (LMIs), and are now briefly defined.

Let $x \in \mathbb{R}^m$ be a vector of decision variables. An LMI condition on x is a matrix inequality of the form

$$F(x) \succ 0 \tag{3.21}$$

with

$$F(x) = F_0 + \sum_{i=1}^{m} x_i F_i \tag{3.22}$$

and where $F_i \in \mathbb{S}^n$, $i = 0, 1, \ldots, m$ are given symmetric matrices. Inequality (3.21) is called a *strict* matrix inequality, because strict positive definiteness is required by the condition. Non strict LMIs are defined analogously, by requiring only positive semidefiniteness of matrix $F(x)$, and are indicated with the notation $F(x) \succeq 0$. The *feasible set* of the LMI (3.21) is defined as the set of x that satisfy the matrix inequality

$$\mathcal{X} = \{x \in \mathbb{R}^m : F(x) \succ 0\}.$$

The most notable feature of LMIs is that the feasible set $\mathcal{X} \in \mathbb{R}^m$ is a *convex set*, meaning that for all $x_1, x_2 \in \mathcal{X}$ and all $\lambda \in [0, 1]$ it holds that

$$\lambda x_1 + (1 - \lambda) x_2 \in \mathcal{X}.$$

This fact can be easily understood by noticing that the condition $F(x) \succ 0$ is equivalent to the condition

$$\xi^T F(x) \xi > 0, \text{ for all non-zero } \xi \in \mathbb{R}^n.$$

Indeed, for any given non-zero $\xi \in \mathbb{R}^n$, the set $\{x : \xi^T F(x) \xi > 0\}$ is an open half-space, hence a convex set, and $\mathcal{X}$ is the (infinite) intersection of such half-spaces. LMI conditions are often used as constraints in optimization problems. In particular, mathematical programs having linear objective and an LMI constraint

$$\min_{x \in \mathbb{R}^m} c^T x \text{ subject to } F(x) \succ 0$$

are known as semidefinite programs (SDPs), see e.g. [266, 278]. Clearly, SDPs are convex optimization problems, and encompass linear, as well as convex quadratic and conic programs.

The representation of control analysis and design problems by means of SDPs has had enormous success in recent years, owing to the availability of efficient numerical algorithms (interior point algorithms in particular, see [199]) for the solution of SDPs. We refer the reader to [49] for an introduction to LMIs and SDPs in systems and control. The LMI representation for control problems is extensively used in subsequent chapters.

Finally, we remark that in applications we often encounter LMIs where the decision variables are in matrix rather than in vector form as in the standard representation of (3.21) and (3.22). The first and most notable example is the Lyapunov inequality

$$AX + XA^T \prec 0 \tag{3.23}$$

where $A \in \mathbb{R}^{n,n}$ is a given matrix, and $X \in \mathbb{S}^n$ is the decision matrix. Such LMIs in matrix variables can, however, be converted in the standard form (3.22) by introducing a vector x containing the free variables of X and exploiting the linearity of the representation. For example, the LMI (3.23) is rewritten in standard form by first introducing vector $x \in \mathbb{R}^m$, $m = n(n - 1)/2$, containing the free elements of the symmetric matrix X. Then, one writes $X = \sum_{i=1}^m x_i S_i$, where $S_i \in \mathbb{R}^{n,n}$ represents an element of the standard basis of symmetric matrices, and therefore (3.23) takes the standard form

$$F_0 + \sum_{i=1}^m x_i F_i \succ 0$$

with $F_0 = 0_{n,n}$, $F_i = -(AS_i + S_iA^T)$, $i = 1, \ldots, m$.

3.5 Computing $\mathcal{H}_2$ and $\mathcal{H}_\infty$ norms

Let $G(s) = C(sI - A)^{-1}B \in \mathcal{RH}_2^{p,q}$ be a strictly proper transfer matrix, and assume that A is stable. Then, we have

$$\|G\|_2^2 = \operatorname{Tr} CW_c C^T$$

where W_c is the controllability Gramian of the system. The controllability Gramian is positive semidefinite $W_c \succeq 0$ and is the unique solution of the Lyapunov equation

$$AW_c + W_cA^T + BB^T = 0.$$

Equivalently, in the dual formulation we obtain

$$\|G\|_2^2 = \operatorname{Tr} B^T W_o B$$

where the observability Gramian $W_o \succeq 0$ is the unique solution of the Lyapunov equation

$$A^T W_o + W_o A + C^T C = 0.$$

For the monotonicity property of the Lyapunov equation, we can also express the $\mathcal{H}_2$ norm in terms of a Lyapunov *inequality*. This characterization in terms of LMIs is stated in the next lemma, see for instance [236].

Lemma 3.1 ($\mathcal{H}_2$ norm characterization). *Let* $G(s) = C(sI - A)^{-1}B + D$ *and* $\gamma > 0$. *The following three statements are equivalent:*

1. *A is stable and* $\|G(s)\|_2^2 < \gamma$;
2. $D = 0$, *and there exist* $S \succ 0$ *such that*

$$AS + SA^T + BB^T \prec 0;$$
$$\operatorname{Tr} CSC^T < \gamma;$$

3. $D = 0$, *and there exist* $P \succ 0$ *and* $Q \succ 0$ *such that*

$$\begin{bmatrix} PA + A^T P & PB \\ B^T P & -I \end{bmatrix} \prec 0;$$

$$\begin{bmatrix} P & C^T \\ C & Q \end{bmatrix} \succ 0;$$

$$\operatorname{Tr} Q < \gamma.$$

The next well-known lemma, often denoted as *bounded real lemma,* gives a characterization of the $\mathcal{H}_\infty$ norm of a system.

Lemma 3.2 (Bounded real lemma). *Let* $G(s) = C(sI - A)^{-1}B + D$ *and* $\gamma > 0$. *The following two statements are equivalent:*

1. *A is stable and* $\|G(s)\|_\infty < \gamma$;
2. *There exist* $P \succ 0$ *such that*

$$\begin{bmatrix} PA + A^T P & PB & C^T \\ B^T P & -\gamma I & D^T \\ C & D & -\gamma I \end{bmatrix} \prec 0. \tag{3.24}$$

A detailed proof of the bounded real lemma in this form may be found in [234]. There is also a non strict characterization of the $\mathcal{H}_\infty$ norm, given in the next lemma, see [234].

Lemma 3.3 (Non strict bounded real lemma). *Let* $G(s) = C(sI - A)^{-1}B + D$, *with A stable and* (A, B) *controllable, and let* $\gamma \geq 0$. *The following two statements are equivalent:*

1. $\|G(s)\|_\infty \leq \gamma$;
2. *There exist* $P = P^T$ *such that*

$$\begin{bmatrix} PA + A^T P & PB & C^T \\ B^T P & -\gamma I & D^T \\ C & D & -\gamma I \end{bmatrix} \preceq 0.$$

From the computational point of view, checking whether the $\mathcal{H}_\infty$ norm is less than γ amounts to solving equation (3.24) with respect to P, which is a convex feasibility problem with LMI constraints.

3.6 Modeling uncertainty of linear systems

In this section, we present a general model that is adopted to represent various sources of uncertainty that may affect a dynamic system. In particular, we follow a standard approach based on the so-called M–Δ model, which is frequently used in modern control theory, see e.g. [295], for a systematic discussion on this topic.

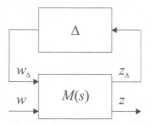

Fig. 3.1. M–Δ model. $M(s)$ is the known part of the system, consisting of the interconnection of plant and controller, Δ represents the uncertain part, w includes noise, disturbances and reference signals and z represents controlled signals and tracking errors.

In Figure 3.1, $M \in \mathcal{RH}_\infty^{c,r}$ represents the transfer matrix of the known part of the system, which consists of the extended plant and the controller. In this description, $\Delta \in \mathcal{RH}_\infty^{r_\Delta,c_\Delta}$ encompasses all time-invariant uncertainties acting on the system. This uncertainty is assumed to belong to a block-diagonal *structured set* $\widetilde{\mathbb{D}}$ of the form

$$\widetilde{\mathbb{D}} \doteq \{\Delta \in \mathcal{RH}_\infty^{r_\Delta,c_\Delta} : \Delta = \mathrm{bdiag}(q_1 I_{m_1}, \ldots, q_\ell I_{m_\ell}, \Delta_1, \ldots, \Delta_b)\} \quad (3.25)$$

where $q \doteq [q_1 \cdots q_\ell]^T$ represents (real or complex) uncertain parameters q_i, with multiplicity m_i, $i = 1, \ldots, \ell$, and Δ_i, $i = 1, \ldots, b$, denote general full-block stable and proper transfer matrices of size $r_i \times c_i$. Moreover, a bound ρ on the magnitude of the uncertainty is imposed. In particular, we assume that $\Delta \in \mathcal{B}_{\widetilde{\mathbb{D}}}(\rho)$, where

$$\mathcal{B}_{\widetilde{\mathbb{D}}}(\rho) \doteq \left\{\Delta \in \widetilde{\mathbb{D}} : \|q\|_p \leq \rho, \|\Delta_i\|_\infty \leq \rho, i = 1, \ldots, b\right\} \quad (3.26)$$

and $\|\cdot\|_p$ is an ℓ_p vector norm.

Following classical literature on the subject, one can associate to $\widetilde{\mathbb{D}}$ a corresponding *matrix* structure, where Δ_i are (real or complex) matrices

$$\mathbb{D} \doteq \{\Delta \in \mathbb{F}^{r_\Delta,c_\Delta} : \Delta = \mathrm{bdiag}(q_1 I_{m_1}, \ldots, q_\ell I_{m_\ell}, \Delta_1, \ldots, \Delta_b)\}. \quad (3.27)$$

We remark that if q_i in $\widetilde{\mathbb{D}}$ is real (complex), then the corresponding parameter in $\mathbb{D}$ is also real (complex). Similarly, if a full block Δ_i of size $r_i \times c_i$ in $\widetilde{\mathbb{D}}$ is a static real (complex) matrix gain, then the corresponding block in $\mathbb{D}$ is also a real (complex) matrix of size $r_i \times c_i$. If instead a full block $\Delta_i \in \mathcal{RH}_\infty^{r_i,c_i}$ in $\widetilde{\mathbb{D}}$ is a dynamic operator, then the corresponding block in $\mathbb{D}$ is a static complex block of size $r_i \times c_i$. The related norm-bounded set is defined as

$$\mathcal{B}_{\mathbb{D}}(\rho) \doteq \{\Delta \in \mathbb{D} : \|q\|_p \leq \rho, \bar{\sigma}(\Delta_i) \leq \rho, i = 1, \ldots, b\}. \tag{3.28}$$

We simply write $\mathcal{B}_{\widetilde{\mathbb{D}}}$ and $\mathcal{B}_{\mathbb{D}}$ to denote the unit balls $\mathcal{B}_{\widetilde{\mathbb{D}}}(1)$ and $\mathcal{B}_{\mathbb{D}}(1)$ respectively. We remark that if no dynamic block appears in (3.25), then the sets $\mathcal{B}_{\widetilde{\mathbb{D}}}(\rho)$ and $\mathcal{B}_{\mathbb{D}}(\rho)$ coincide. In the more general situation of dynamic blocks, $\mathcal{B}_{\mathbb{D}}(\rho)$ may be viewed as a "snapshot" of $\mathcal{B}_{\widetilde{\mathbb{D}}}(\rho)$ at a fixed frequency.

The signal w in Figure 3.1 usually represents disturbances of various nature entering the system, such as white (or colored) noise or a deterministic, norm-bounded signal, and z describes errors or other quantities that should be kept small in some sense. The transfer matrix $M(s)$ is partitioned as

$$\begin{bmatrix} z_\Delta(s) \\ z(s) \end{bmatrix} = \begin{bmatrix} M_{11}(s) \; M_{12}(s) \\ M_{21}(s) \; M_{22}(s) \end{bmatrix} \begin{bmatrix} w_\Delta(s) \\ w(s) \end{bmatrix}$$

so that the transfer matrix of the performance channel $w \to z$ can be expressed in terms of the *upper* linear fractional transformation (LFT)

$$\mathcal{F}_u(M, \Delta) \doteq M_{22} + M_{21}\Delta(I - M_{11}\Delta)^{-1}M_{12}. \tag{3.29}$$

This LFT is well defined whenever the matrix M_{11} satisfies a *well-posedness* condition, i.e. $(I - M_{11}(\infty)\Delta(\infty))$ is nonsingular for all $\Delta \in \mathcal{B}_{\widetilde{\mathbb{D}}}(\rho)$. Moreover, the condition $(I - M_{11}\Delta)^{-1} \in \mathcal{RH}_\infty^{c_\Delta,c_\Delta}$ guarantees that the interconnection between $M \in \mathcal{RH}_\infty^{c,r}$ and $\Delta \in \mathcal{RH}_\infty^{r_\Delta,c_\Delta}$ is *internally stable*, see [295].

Two key requirements are typically imposed on the interconnection in Figure 3.1: (i) the interconnection is well posed and internally stable, for all $\Delta \in \mathcal{B}_{\widetilde{\mathbb{D}}}(\rho)$; (ii) the influence of the disturbances w on the controlled outputs z is below a desired level for all uncertainties $\Delta \in \mathcal{B}_{\widetilde{\mathbb{D}}}(\rho)$. In particular, condition (i) is a *robust stability* condition (see details in Section 3.7), and condition (ii) typically expresses performance requirements imposed as bounds on the gain of the $w \to z$ channel.

Remark 3.3 (LFT representation lemma). Notice that the LFT representation in (3.29) is general enough to encompass uncertainty entering the transfer matrix in a generic polynomial or rational manner, provided that the transfer matrix has no singularities at zero. This result is known as the LFT representation lemma, see for instance [105]. In [295], constructive rules for building the LFT representation starting from basic algebraic operations on LFTs are given.   ◁

In the following, we present some examples, involving different uncertainty configurations, and show how we can express them in the M–Δ framework.

Example 3.1 (Real unstructured uncertainty).
Consider a linear time-invariant system expressed in state space form

$$\dot{x} = A(\Delta)x + Bw; \tag{3.30}$$
$$z = Cx + Dw$$

where the matrix $A(\Delta) \in \mathbb{R}^{n_s, n_s}$ depends on the uncertainty $\Delta \in \mathbb{R}^{r_\Delta, c_\Delta}$ in a simple additive form, i.e.

$$A(\Delta) = A + L\Delta R \tag{3.31}$$

with $L \in \mathbb{R}^{n_s, r_\Delta}$ and $R \in \mathbb{R}^{c_\Delta, n_s}$. In this case, since Δ is a full real block, $\mathcal{B}_{\widetilde{\mathbb{D}}}(\rho) = \mathcal{B}_{\mathbb{D}}(\rho) = \{\Delta \in \mathbb{R}^{r_\Delta, c_\Delta} : \bar{\sigma}(\Delta) \leq \rho\}$.

The uncertainty structure in (3.31) has been extensively studied in the literature, see Section 3.7. In particular, the smallest value of ρ such that there exists a value of Δ that makes the system (3.30) unstable is called the *real stability radius* of the system, see Definition 3.4 and Theorem 3.3 for its computation. It can be easily verified that the uncertain system (3.30) may be represented in M–Δ form with

$$M(s) = \begin{bmatrix} R \\ C \end{bmatrix} (sI - A)^{-1} \begin{bmatrix} L & B \end{bmatrix} + \begin{bmatrix} 0 & 0 \\ 0 & D \end{bmatrix}.$$

★

Example 3.2 (Real parametric uncertainty).
Consider a system described by the transfer function

$$G(s, q) = \frac{s + 3 + q_1}{s^2 + (2 + q_1)s + 5 + q_2} \tag{3.32}$$

where $q = [q_1 \ q_2]^T$ represents a two-dimensional vector of real parametric uncertainty. The vector q is assumed to be bounded in the set $\{q : \|q\|_\infty \leq \rho\}$. An M–Δ representation of the uncertain system in (3.32) may be derived by writing the system in controllable canonical form and *pulling out* the uncertainty, see Figure 3.2. We therefore obtain

$$M(s) = \begin{bmatrix} 0 & \dfrac{-1}{s^2 + 2s + 5} & \dfrac{-1}{s^2 + 2s + 5} & \dfrac{1}{s^2 + 2s + 5} \\[2mm] 0 & \dfrac{-s}{s^2 + 2s + 5} & \dfrac{-s}{s^2 + 2s + 5} & \dfrac{s}{s^2 + 2s + 5} \\[2mm] 0 & \dfrac{-1}{s^2 + 2s + 5} & \dfrac{-1}{s^2 + 2s + 5} & \dfrac{1}{s^2 + 2s + 5} \\[2mm] 1 & \dfrac{-(s+3)}{s^2 + 2s + 5} & \dfrac{-(s+3)}{s^2 + 2s + 5} & \dfrac{s+3}{s^2 + 2s + 5} \end{bmatrix}$$

and the matrix Δ belongs to the set $\widetilde{\mathbb{D}} \equiv \mathbb{D} = \{\Delta = \mathrm{bdiag}(q_1, q_2 I_2)\}$, with the bound $\|q\|_\infty \leq \rho$. The resulting M–Δ interconnection is shown in Figure 3.3.

★

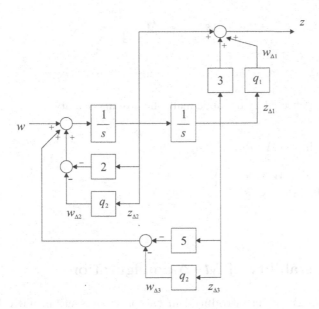

Fig. 3.2. $G(s, q)$ of Example 3.2 expressed in controllable canonical form.

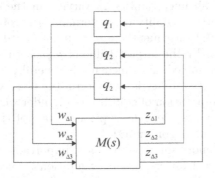

Fig. 3.3. M–Δ representation for Example 3.2.

Example 3.3 (Unmodeled dynamics).

Consider a transfer matrix $G(s, \Delta)$ affected by multiplicative uncertainty of the type

$$G(s, \Delta) = (I + W_1(s)\Delta(s)W_2(s))G(s) \tag{3.33}$$

where the transfer matrices $W_1(s), W_2(s)$ weight the uncertainty over frequency, as shown in Figure 3.4.

In this case, $G(s)$ is the nominal model and the uncertainty is constituted by unknown dynamics, expressed in terms of an uncertain stable transfer matrix $\Delta(s)$ that is assumed to belong to the set $\mathcal{B}_{\widetilde{\mathbb{D}}}(\rho) = \{\Delta(s) \in \mathcal{RH}_\infty^{r_\Delta, c_\Delta} : \|\Delta\|_\infty \le \rho\}$. The multiplicative model (3.33) is immedi-

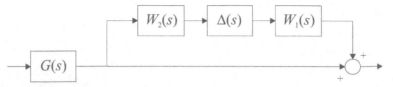

Fig. 3.4. System affected by dynamic multiplicative uncertainty.

ately rewritten in M–Δ form, letting

$$M(s) = \begin{bmatrix} 0 & W_2(s)G(s) \\ W_1(s) & G(s) \end{bmatrix}.$$

$\star$

3.7 Robust stability of M–Δ configuration

Consider the M–Δ model introduced in the previous section, with $M(s) \in \mathcal{RH}_\infty^{c,r}$ and $\Delta \in \mathcal{B}_{\widehat{\mathbb{D}}}(\rho)$. For fixed $\rho > 0$, we say that the system is *robustly stable* if it is stable for all uncertainties Δ varying in the set $\mathcal{B}_{\widehat{\mathbb{D}}}(\rho)$. More generally, a given property (stability or performance) is robustly satisfied if it holds for all $\Delta \in \mathcal{B}_{\widehat{\mathbb{D}}}(\rho)$. Consequently, a measure of the "degree" of robustness of a system is given by the largest value of ρ such that the considered property is robustly guaranteed. This measure is generally called the *robustness margin*. When robust stability is of concern, the term *stability radius* is frequently used. The computation of stability radii under various uncertainty structures and for different system descriptions is one of the main areas of research in the robust control community.

In this section, we report some of the most important results regarding robust stability for different uncertainty configurations. Since the main focus of the section is on stability, to simplify our discussion we refer to systems described by the configuration of Figure 3.5, where the performance channel $w \to z$ is neglected. In this case, the dimensions of Δ are compatible with M, i.e. $\Delta \in \mathcal{RH}_\infty^{r,c}$.

For historical reasons, we begin our discussion with the case when Δ is an unknown stable transfer matrix bounded in the $\mathcal{H}_\infty$ norm, and show how this problem is equivalent to a "static" problem with a *complex* matrix Δ.

3.7.1 Dynamic uncertainty and stability radii

Consider the configuration of Figure 3.5, in which $M(s) \in \mathcal{RH}_\infty^{c,r}$ and Δ is an unknown transfer matrix $\Delta(s) \in \mathcal{RH}_\infty^{r,c}$. We define the stability radius $r_{\text{LTI}}(M(s))$ of $M(s)$ under LTI perturbations as the smallest value of $\|\Delta(s)\|_\infty$ such that well-posedness or internal stability are violated.

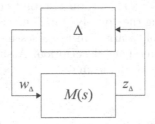

Fig. 3.5. M–Δ configuration for robust stability.

The radius of stability may be computed by invoking a fundamental result of robustness analysis known as the *small gain theorem*, see e.g. [295].

Theorem 3.1 (Small gain). *Consider the interconnected system in Figure 3.5, with $M \in \mathcal{RH}_\infty^{c,r}$, and $\rho > 0$. Then, the interconnection is well posed and internally stable for all $\Delta \in \mathcal{RH}_\infty^{r,c}$ with $\|\Delta\|_\infty \leq \rho$ if and only if $\|M\|_\infty < 1/\rho$.*

From this result, it follows immediately that the stability radius under LTI perturbations is given by

$$r_{\mathrm{LTI}}(M(s)) = \frac{1}{\|M(s)\|_\infty} = \frac{1}{\sup_\omega \bar{\sigma}(M(j\omega))}.$$

Remark 3.4 (Extensions of the small gain theorem). It should be observed that the small gain theorem holds for larger classes of uncertainty, see [98]. In fact, this result follows as a special case of the *contraction mapping* theorem in Banach spaces. In particular, it can be reformulated for nonlinear operators from $\mathcal{L}_p^+ \to \mathcal{L}_p^+$, provided that the (Lipschitz) incremental gains of the system and of the uncertainty are used in place of the $\mathcal{H}_\infty$ norm, see also [123]. This fact is useful in robustness analysis, since it allows one to model certain classes of nonlinearities as gain-bounded uncertainties. ◁

Corollary 3.1. *Consider the interconnected system in Figure 3.5, with $M \in \mathcal{RH}_\infty^{c,r}$, and $\rho > 0$. The following statements are equivalent:*

1. *The interconnection is well posed and internally stable for all $\Delta \in \mathcal{H}_\infty^{r,c}$ with $\|\Delta\|_\infty \leq \rho$;*
2. *The interconnection is well posed and internally stable for all $\Delta \in \mathcal{RH}_\infty^{r,c}$ with $\|\Delta\|_\infty \leq \rho$;*
3. *The interconnection is well posed and internally stable for all matrices $\Delta \in \mathbb{C}^{r,c}$ with $\|\Delta\| \leq \rho$;*
4. *$M \in \mathcal{RH}_\infty^{c,r}$ and $\|M\|_\infty < 1/\rho$.*

An interesting conclusion drawn from this corollary is that checking robust stability of the M–Δ interconnection under purely static uncertainties is necessary and sufficient for robust stability under rather general dynamic perturbations. As a consequence of this fact, when the system is affected by dynamic uncertainty with $\widetilde{\mathbb{D}} = \mathcal{RH}_\infty^{r,c}$, robust stability of the interconnection can be detected by considering the "purely static" matrix structure $\mathbb{D} = \mathbb{C}^{r,c}$.

A related robustness problem arises in the assessment of robust stability for systems described in state space description. Assume for instance that the uncertain linear system is of the form

$$\dot{x} = A(\Delta)x + Bw; \tag{3.34}$$
$$z = Cx + Dw$$

where $A(\Delta) = A + L\Delta R$, $A \in \mathbb{R}^{n_s,n_s}$ is stable and $\Delta \in \mathbb{F}^{r,c}$ is a full block either real (as in Example 3.1) or complex.

Definition 3.4 (Real and complex stability radii). *Consider the uncertain system (3.34), with*

$$A(\Delta) = A + L\Delta R \tag{3.35}$$

where $A \in \mathbb{R}^{n_s,n_s}$ is stable, $L \in \mathbb{R}^{n_s,r}$, $R \in \mathbb{R}^{c,n_s}$ and $\Delta \in \mathbb{F}^{r,c}$. Then, the stability radius of the triple A, L, R is defined as

$$r_{\mathbb{F}}(A, L, R) \doteq \inf\{\bar{\sigma}(\Delta) : \Delta \in \mathbb{F}^{r,c} \text{ and } A(\Delta) \text{ is unstable}\}. \tag{3.36}$$

In particular, for $\Delta \in \mathbb{R}^{r,c}$, $r_{\mathbb{R}}(A, L, R)$ is called the *real stability radius*. Similarly, for $\Delta \in \mathbb{C}^{r,c}$, $r_{\mathbb{C}}(A, L, R)$ is called the *complex stability radius*.[2]

The next theorem, which is a direct consequence of the small gain theorem, relates the complex stability radius to the computation of the $\mathcal{H}_\infty$ norm of a certain operator.

Theorem 3.2 (Complex stability radius). *Let $A(\Delta) = A + L\Delta R$, where $A \in \mathbb{R}^{n_s,n_s}$ is stable, $L \in \mathbb{R}^{n_s,r}$, $R \in \mathbb{R}^{c,n_s}$ and $\Delta \in \mathbb{C}^{r,c}$. Then, the* complex stability radius *is given by*

$$r_{\mathbb{C}}(A, L, R) = \frac{1}{\|M\|_\infty}$$

where $M(s) = R(sI - A)^{-1}L$.

Proof. Since A is stable, then $M \in \mathcal{RH}_\infty^{c,r}$. Therefore, by the small gain theorem, the closed-loop interconnection of M with $\Delta \in \mathbb{C}^{r,c}$ is stable for all $\|\Delta\| \leq \rho$ if and only if $\|M\|_\infty < 1/\rho$. The statement then follows, noticing that the dynamic matrix of this interconnection is indeed $A + L\Delta R$.   □

When the uncertain matrix $A(\Delta)$ is expressed in the form (3.35), but the uncertainty Δ is real, the above result is clearly conservative. In this case, a formula for computing exactly the real stability radius is derived in [224].

---

[2] Real and complex stability radii can also be defined in more general form for complex matrices A, L, R, see for instance [133].

Theorem 3.3 (Real stability radius). *Let $A(\Delta) = A + L\Delta R$, where $A \in \mathbb{R}^{n_s,n_s}$ is stable, $L \in \mathbb{R}^{n_s,r}$, $R \in \mathbb{R}^{c,n_s}$ and $\Delta \in \mathbb{R}^{r,c}$. Then, the real stability radius is given by*

$$r_{\mathbb{R}}(A, L, R) = \left\{ \sup_{\omega} \inf_{\gamma \in (0,1]} \sigma_2 \left(\begin{bmatrix} \mathrm{Re}(M(j\omega)) & -\gamma\,\omega\,\mathrm{Im}(M(j\omega)) \\ \gamma^{-1}\,\omega\,\mathrm{Im}(M(j\omega)) & \mathrm{Re}(M(j\omega)) \end{bmatrix} \right) \right\}^{-1}$$
(3.37)

where $M(s) = R(sI - A)^{-1}L$, and $\sigma_2(\cdot)$ denotes the second largest singular value.

We remark that the minimization over γ in (3.37) can be easily performed since the function to be optimized is unimodal with respect to $\gamma \in (0,1]$, see further details in [224].

3.7.2 Structured singular value and μ analysis

In this section, we consider the general case when the matrix Δ in Figure 3.5 belongs to the structured set defined in (3.25)

$$\widetilde{\mathbb{D}} = \{\Delta \in \mathcal{RH}_\infty^{r,c} : \Delta = \mathrm{bdiag}(q_1 I_{m_1}, \ldots, q_\ell I_{m_\ell}, \Delta_1, \ldots, \Delta_b)\}$$

where $q_i \in \mathbb{F}$, $i = 1, \ldots, \ell$ and $\Delta_i \in \mathcal{RH}_\infty^{r_i,c_i}$, $i = 1, \ldots, b$. To the operator structure $\widetilde{\mathbb{D}}$, we associate the matrix structure (3.27)

$$\mathbb{D} = \{\Delta \in \mathbb{F}^{r,c} : \Delta = \mathrm{bdiag}(q_1 I_{m_1}, \ldots, q_\ell I_{m_\ell}, \Delta_1, \ldots, \Delta_b)\}.$$

Letting $M \in \mathcal{RH}_\infty^{c,r}$, we consider the complex matrix $M(j\omega) \in \mathbb{C}^{c,r}$ obtained by evaluating the transfer matrix $M(s)$ for $s = j\omega$, with $\omega \in \mathbb{R}_+$. In this setting, the *multivariable stability margin* of $M(j\omega)$ for a system with diagonal perturbations $\mathbb{D}$ is discussed in [230], and its inverse, the *structured singular value* of $M(j\omega)$ with respect to $\mathbb{D}$, is studied in [92]. We now formally define the structured singular value $\mu_{\mathbb{D}}(M(j\omega))$.

Definition 3.5 (Structured singular value). *Let $M(j\omega) \in \mathbb{C}^{c,r}$, the structured singular value of $M(j\omega)$ with respect to $\mathbb{D}$ is defined as*

$$\mu_{\mathbb{D}}(M(j\omega)) \doteq \frac{1}{\min\{\bar{\sigma}(\Delta) : \det(I - M(j\omega)\Delta) = 0, \Delta \in \mathbb{D}\}}$$
(3.38)

unless no $\Delta \in \mathbb{D}$ makes $I - M(j\omega)\Delta$ singular, in which case $\mu_{\mathbb{D}}(M(j\omega)) \doteq 0$.

An alternative expression for $\mu_{\mathbb{D}}(M(j\omega))$ is

$$\mu_{\mathbb{D}}(M(j\omega)) = \max_{\Delta \in \mathcal{B}_{\mathbb{D}}(1)} \rho_\lambda(M(j\omega)\Delta)$$

where $\rho_\lambda(\cdot)$ denotes the spectral radius, i.e. the maximum modulus of the eigenvalues.

The theorem stated next is a fundamental result for robust stability under structured perturbations and constitutes a generalization of the small gain theorem.

Theorem 3.4 (Small μ). *Consider the interconnected system in Figure 3.5, with $M \in \mathcal{RH}_\infty^{c,r}$, and $\rho > 0$. The interconnection is well posed and internally stable for all $\Delta \in \widetilde{\mathbb{D}}$ with $\|\Delta\|_\infty < \rho$ if and only if*

$$\sup_{\omega \in \mathbb{R}} \mu_{\mathbb{D}}(M(j\omega)) \leq \frac{1}{\rho}. \tag{3.39}$$

Remark 3.5 (Equivalence between dynamic and static perturbations). From this result, we see that checking robust stability under dynamic perturbations is equivalent to checking robust stability against *purely static* perturbations, since only the static perturbations set $\mathbb{D}$ enters in condition (3.39). ◁

Lemma 3.4. *Let $M(s) = C(sI - A)^{-1}B + D \in \mathcal{RH}_\infty^{c,r}$. Then, the interconnection in Figure 3.5 is well posed and internally stable for all $\Delta \in \widetilde{\mathbb{D}}$ with $\|\Delta\|_\infty < \rho$ if and only if the two conditions:*

1. *$(I - D\Delta)$ is nonsingular;*
2. *$A + B\Delta(I - D\Delta)^{-1}C$ is stable*

hold for all $\Delta \in \mathbb{D}$ with $\bar{\sigma}(\Delta) < \rho$.

Proof. From Theorem 3.4, it follows that the interconnection is well posed and robustly stable for all $\Delta \in \widetilde{\mathbb{D}}$ with $\|\Delta\|_\infty < \rho$ if and only if it is well posed and robustly stable for all $\Delta \in \mathbb{D}$ with $\bar{\sigma}(\Delta) < \rho$. Then, closing the loop on M with $w_\Delta = \Delta z_\Delta$ with $\Delta \in \mathbb{D}$, we have

$$\dot{x} = Ax + B\Delta z_\Delta;$$
$$(I - D\Delta)z_\Delta = Cx.$$

If $(I - D\Delta)$ is nonsingular, then the dynamic matrix of the closed loop is $A + B\Delta(I - D\Delta)^{-1}C$ and the result follows immediately. □

The previous lemma permits us to extend the notion of stability radius to the case of structured static perturbations.

Definition 3.6 (Stability radius under structured perturbations). *Let*

$$A(\Delta) = A + B\Delta(I - D\Delta)^{-1}C$$

where $A \in \mathbb{R}^{n_s,n_s}$ is stable, $B \in \mathbb{R}^{n_s,r}$, $C \in \mathbb{R}^{c,n_s}$, $D \in \mathbb{R}^{c,r}$ and $\Delta \in \mathbb{D}$. Then, the stability radius under structured perturbations *is defined as*

$$r_{\mathbb{D}}(A,B,C,D) \doteq \inf\{\bar{\sigma}(\Delta), \Delta \in \mathbb{D} : A(\Delta) \text{ is unstable or } I - D\Delta \text{ is singular}\}. \tag{3.40}$$

As a direct consequence of Theorem 3.4 and Lemma 3.4, we have the result presented next.

Theorem 3.5 (Stability radius under structured perturbations). *Let* $A(\Delta) = A + B\Delta(I - D\Delta)^{-1}C$, *where* $A \in \mathbb{R}^{n_s,n_s}$ *is stable,* $B \in \mathbb{R}^{n_s,r}$, $C \in \mathbb{R}^{c,n_s}$, $D \in \mathbb{R}^{c,r}$ *and* $\Delta \in \mathbb{D}$. *Moreover, let* $M(s) = C(sI - A)^{-1}B + D$, $M \in \mathcal{RH}_\infty^{c,r}$. *Then, the* stability radius *under structured perturbations is given by*

$$r_\mathbb{D}(M) = r_\mathbb{D}(A, B, C, D) = \frac{1}{\sup_{\omega \in \mathbb{R}} \mu_\mathbb{D}(M(j\omega))}.$$

3.7.3 Computation of bounds on $\mu_\mathbb{D}$

The computation of $\mu_\mathbb{D}$ under general uncertainty structures $\mathbb{D}$ is a difficult nonconvex problem and many results have appeared in the literature in this sense. The interested reader is addressed to Section 5.1 and to the survey paper [42] focused on computational complexity in systems and control. For these reasons, research on computation of the structured singular value mainly concentrated on establishing upper and lower bounds on $\mu_\mathbb{D}$. In particular, we now consider the purely complex uncertainty structure

$$\mathbb{D} = \{\Delta \in \mathbb{C}^{r,c} : \Delta = \mathrm{bdiag}(q_1 I_{m_1}, \dots, q_\ell I_{m_\ell}, \Delta_1, \dots, \Delta_b)\}$$

consisting of complex repeated scalars $q_i \in \mathbb{C}$ and square full complex blocks $\Delta_i \in \mathbb{C}^{r_i,r_i}$. In addition, let $M \in \mathbb{C}^{r,r}$, where $r = \sum_{i=1}^{\ell} m_i + \sum_{i=1}^{b} r_i$ and introduce the two scalings sets

$$\mathcal{V} \doteq \{V \in \mathbb{D} : VV^* = I_r\}$$

and

$$\mathcal{D} \doteq \{D : D = \mathrm{bdiag}(D_1, \dots, D_\ell, d_1 I_{r_1}, \dots, d_{b-1} I_{r_{b-1}}, I_{r_b}),$$
$$D_i \in \mathbb{C}^{m_i,m_i}, D_i \succ 0, d_i \in \mathbb{R}, d_i > 0\}.$$

Then, the following bounds hold, see e.g. [295]

$$\max_{V \in \mathcal{V}} \rho_\lambda(VM) \leq \mu_\mathbb{D}(M) \leq \inf_{D \in \mathcal{D}} \bar{\sigma}(DMD^{-1}).$$

The lower bound in this equation is actually an equality, but unfortunately no efficient algorithm with guaranteed global convergence is currently available for its computation. In contrast, the upper bound can be computed efficiently by solving a convex optimization problem (therefore achieving the global optimum), but the bound coincides with $\mu_\mathbb{D}(M)$ only for special uncertainty structures for which $2\ell + b \leq 3$, see [211] for details and also see [92] for proof of the case $\ell = 0, b = 3$. For $2\ell + b > 3$ the gap between $\mu_\mathbb{D}$ and its upper bound can be arbitrarily large, but computational practice shows that this gap often remains quite small.

Similar upper bounds can also be constructed for the more general uncertainty structure (3.27). These upper bounds are still convex with respect to suitably selected scaling matrices and, therefore, can be efficiently computed. However, in general there is no guarantee that they are close to the actual $\mu_\mathbb{D}$.

3.7.4 Rank-one μ problem and Kharitonov theory

There are special cases in which $\mu_{\mathbb{D}}$ can also be efficiently computed for general uncertainty structures $\mathbb{D}$. One such case is the so-called *rank-one μ problem*, where $M \in \mathbb{C}^{r,r}$ is a rank-one matrix

$$M = uv^*, \quad u, v \in \mathbb{C}^r.$$

Then, it has been shown in [292] that under rather general uncertainty structures, which may include real or complex repeated scalars, and full complex blocks, $\mu_{\mathbb{D}}$ and its upper bound actually coincide.

The interest in the rank-one μ problem also resides in the fact that it provides a connection to Kharitonov-type results for uncertain polynomials discussed in the next section. To explain this connection more precisely, we assume that the transfer matrix $M(s)$ is of the form

$$M(s) = u(s)v^T(s)$$

where $u(s), v(s) \in \mathcal{RH}_\infty^{r,1}$ and that the structured set $\mathbb{D}$ is given by

$$\mathbb{D} = \{\Delta \in \mathbb{R}^{r,r} : \Delta = \mathrm{diag}\left([q_1 \cdots q_r]\right), q_i \in \mathbb{R}, i = 1, \ldots, r\}. \tag{3.41}$$

Subsequently, we define

$$D(s, q) \doteq \det(I + M(s)\Delta) = 1 + \sum_{i=1}^{r} u_i(s)v_i(s)q_i$$

where $u_i(s)$ and $v_i(s)$, $i = 1, \ldots, r$, are rational functions of the form

$$u_i(s) = n_{ui}(s)/d_{ui}(s), \quad v_i(s) = n_{vi}(s)/d_{vi}(s)$$

and $n_{ui}(s), d_{ui}(s), n_{vi}(s), d_{vi}(s)$ are coprime polynomials in s. The assumption $u(s), v(s) \in \mathcal{RH}_\infty^{r,1}$ implies that $d_{ui}(s), d_{vi}(s)$ are Hurwitz polynomials, i.e. all their roots lie in the open left half plane.

Then, assuming a bound $\|q\|_\infty \leq 1$ on the uncertainty q, checking robust stability amounts to verifying if $D(j\omega, q) \neq 0$ for all $\omega \in \mathbb{R}$ and all q within this bound. We notice that

$$D(s, q) = \frac{1}{p_0(s)} \left(p_0(s) + \sum_{i=1}^{r} p_i(s)q_i \right)$$

where

$$p_0(s) = \prod_{i=1}^{r} d_{ui}(s)d_{vi}(s);$$

$$p_i(s) = n_{ui}(s)n_{vi}(s) \prod_{\substack{k=1 \\ k \neq i}}^{r} d_{uk}(s)d_{vk}(s), \quad i = 1, \ldots, r.$$

Since $p_0(s)$ is Hurwitz, $p_0(j\omega) \neq 0$ for all $\omega \in \mathbb{R}$. Then, $D(j\omega, q) \neq 0$ if and only if $p(s, q) \neq 0$, where the affine polynomial $p(s, q)$ is given by

$$p(s, q) = p_0(s) + \sum_{i=1}^{r} p_i(s)q_i.$$

Therefore, robust stability is guaranteed if and only if

$$p(j\omega, q) \neq 0$$

for all $\omega \in \mathbb{R}$ and $\|q\|_\infty \leq 1$. Now, since $p(j\omega, 0) \neq 0$, by simple continuity arguments, this condition is satisfied if and only if no root of $p(s, q)$ crosses the imaginary axis for $\|q\|_\infty \leq 1$, i.e. if and only if $p(s, q)$ is Hurwitz for all q. Conversely, one can show that for any polynomial $p(s, q)$ affine in q, there exists a rational matrix of the form $M(s) = u(s)v^T(s)$ with $u(s)$ and $v(s)$ rational, such that $p(s, q) = \det(I + M(s)\Delta)$, with Δ in (3.41).

3.8 Robustness analysis with parametric uncertainty

In this section, we consider systems affected by real parametric uncertainty of the type

$$q \doteq [q_1 \; \cdots \; q_\ell]^T$$

where each q_i, $i = 1, \ldots, \ell$, is bounded in the interval $[q_i^-, q_i^+]$. That is, the uncertainty vector q is assumed to belong to the set

$$\mathcal{B}_q \doteq \{q \in \mathbb{R}^\ell : q_i \in [q_i^-, q_i^+], \; i = 1, \ldots, \ell\}. \tag{3.42}$$

The set $\mathcal{B}_q$ is a hyperrectangle whose vertices $q^1, \ldots, q^{2^\ell}$ are obtained considering the 2^ℓ combinations of either $q_i = q_i^-$ or $q_i = q_i^+$, for $i = 1, \ldots, \ell$. As described in Example 3.1, systems affected by bounded parametric uncertainty can be represented in the general M–Δ form. In this case, it can be easily shown that the uncertainty Δ in the M–Δ representation depends on the vector

$$\bar{q} \doteq [\bar{q}_1 \; \cdots \; \bar{q}_\ell]^T$$

where

$$\bar{q}_i = \frac{2}{q_i^+ - q_i^-}q_i - \frac{q_i^+ + q_i^-}{q_i^+ - q_i^-}, \quad i = 1, \ldots, \ell.$$

Consequently, the uncertainty Δ is bounded in the set

$$\mathcal{B}_\mathbb{D} = \{\Delta \in \mathbb{D} : \|\Delta\|_\infty \leq 1\}$$

where the structure $\mathbb{D}$ has the simple form

$$\mathbb{D} = \left\{ \Delta \in \mathbb{R}^{r,c} : \Delta = \mathrm{bdiag}(\bar{q}_1 I_{m_1}, \ldots, \bar{q}_\ell I_{m_\ell}), \bar{q} \in \mathbb{R}^\ell \right\}.$$

We notice that once the uncertain system has been rewritten in the M–Δ form, the robustness results presented in Section 3.7 can be applied directly.

In parallel with the μ analysis approach, in the 1980s and early 1990s a framework for studying stability of systems affected by real parametric uncertainty was developed independently. This framework does not necessarily require rewriting the system in M–Δ form and is based upon a direct representation of the SISO uncertain plant in the transfer function form

$$G(s, q) \doteq \frac{N_G(s, q)}{D_G(s, q)}$$

where $N_G(s, q)$ and $D_G(s, q)$ are the numerator and denominator plant polynomials whose coefficients depend on the uncertainty q. Subsequently, for a given controller

$$K(s) \doteq \frac{N_K(s)}{D_K(s)}$$

with numerator and denominator controller polynomials $N_K(s)$ and $D_K(s)$, we construct the closed-loop polynomial

$$\begin{aligned} p(s, q) &= N_K(s) N_G(s, q) + D_K(s) D_G(s, q) \\ &= a_0(q) + a_1(q)s + \cdots + a_n(q)s^n \end{aligned}$$

where the coefficients $a_i(q)$ of $p(s, q)$ are functions of q. An illustration of the parametric approach, borrowed from [265], is given in the next example.

Example 3.4 (DC electric motor with uncertain parameters).
    Consider the system in Figure 3.6, representing an armature-controlled DC electric motor with independent excitation. The voltage to angle transfer function $G(s) = \Theta(s)/V(s)$ is given by

$$G(s) = \frac{K_m}{LJs^3 + (RJ + BL)s^2 + (K_m^2 + RB)s}$$

where L is the armature inductance, R is the armature resistance, K_m is the motor electromotive force–speed constant, J is the moment of inertia and B is the mechanical friction. Then, taking a unitary feedback controller $K(s) = 1$, we write the closed-loop polynomial, obtaining

$$p(s) = K_m + (K_m^2 + RB)s + (RJ + BL)s^2 + LJs^3.$$

Clearly, the values of some of the motor parameters may be uncertain. For example, the moment of inertia and the mechanical friction are functions of the load. Therefore, if the load is not fixed, then the values of J and B are not precisely known. Similarly, the armature resistance R is a parameter that can be measured very accurately but which is subject to temperature variations, and the motor constant K_m is a function of the field magnetic flow, which may vary. To summarize, it is reasonable to say that the motor parameters, or a subset of them, may be unknown but bounded within given intervals. More precisely, we can identify uncertain parameters

$$q_1 = L, \quad q_2 = R, \quad q_3 = K_m, \quad q_4 = J, \quad q_5 = B$$

and specify a given range of variation $[q_i^-, q_i^+]$ for each $q_i, i = 1, \ldots, 5$. Then, instead of $G(s)$, we write

$$G(s, q) = \frac{q_3}{q_1 q_4 s^3 + (q_2 q_4 + q_1 q_5)s^2 + (q_3^2 + q_2 q_5)s}$$

and the closed-loop uncertain polynomial becomes

$$p(s, q) = q_3 + (q_3^2 + q_2 q_5)s + (q_2 q_4 + q_1 q_5)s^2 + q_1 q_4 s^3. \tag{3.43}$$

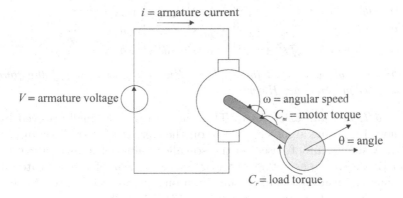

$i =$ armature current

$V =$ armature voltage

$\omega =$ angular speed

$C_m =$ motor torque

$\theta =$ angle

$C_r =$ load torque

Fig. 3.6. DC-electric motor of Example 3.4.

However, not necessarily all motor parameters are uncertain. For example, we may assume that the armature inductance L, the armature resistance R and the constant K_m are fixed and that the moment of inertia J and the mechanical friction B are unknown. Then, we take $q_1 = J$ and $q_2 = B$ as uncertain parameters. In this case, the closed-loop polynomial has an *affine uncertainty* structure

$$p(s, q) = K_m + (K_m^2 + Rq_2)s + (Rq_1 + Lq_2)s^2 + Lq_1 s^3.$$

On the other hand, if L, K_m and B are fixed, and R and J are uncertain, then we identify q_1 and q_2 with R and J respectively. In this case, the closed-loop polynomial coefficients are no longer affine functions of the uncertainties, instead they are *multiaffine*[3] functions of q

$$p(s,q) = K_m + (K_m^2 + Bq_1)s + (q_1q_2 + BL)s^2 + Lq_2s^3.$$

In the general case when all motor parameters are uncertain, the polynomial $p(s,q)$ in (3.43) has a *polynomial uncertainty* structure. $\star$

The study of stability of polynomials with various uncertainty structures is one of the main goals of *parametric stability*. The first result we present is the celebrated Kharitonov theorem [156] on the stability of interval polynomials, i.e. polynomials $p(s,q)$ whose coefficients $a_i(q) = q_i$ are independent and bounded in given intervals $[q_i^-, q_i^+]$.

Theorem 3.6 (Kharitonov). *Consider the interval polynomial family $\mathcal{P}$*

$$\mathcal{P} = \{p(s,q) = q_0 + q_1s + \cdots + q_ns^n : q_i \in [q_i^-, q_i^+], i = 0, 1, \ldots, n\}$$

and the four fixed Kharitonov polynomials

$$p_1(s) \doteq q_0^- + q_1^- s + q_2^+ s^2 + q_3^+ s^3 + q_4^- s^4 + q_5^- s^5 + q_6^+ s^6 + \cdots ;$$
$$p_2(s) \doteq q_0^+ + q_1^+ s + q_2^- s^2 + q_3^- s^3 + q_4^+ s^4 + q_5^+ s^5 + q_6^- s^6 + \cdots ;$$
$$p_3(s) \doteq q_0^+ + q_1^- s + q_2^- s^2 + q_3^+ s^3 + q_4^+ s^4 + q_5^- s^5 + q_6^- s^6 + \cdots ;$$
$$p_4(s) \doteq q_0^- + q_1^+ s + q_2^+ s^2 + q_3^- s^3 + q_4^- s^4 + q_5^+ s^5 + q_6^+ s^6 + \cdots .$$

Then, the interval polynomial family $\mathcal{P}$ is Hurwitz if and only if the four Kharitonov polynomials are Hurwitz.

Remark 3.6 (Proof and extensions). This theorem was originally proved by Kharitonov [156] using arguments based on the Hermite-Bieler theorem, see e.g. [117]. A simpler proof based on the so-called value set approach can be found in [191]. We observe that the original statement of the Kharitonov theorem has an invariant degree assumption on the interval family; this hypothesis has been subsequently removed in [131, 147, 288].   ◁

The Kharitonov theorem is computationally attractive since it requires checking stability of only four "extreme" polynomials, regardless of the degree of the polynomial. However, it is based on the hypothesis that the polynomial coefficients vary within independent intervals. This hypothesis is generally not satisfied by generic uncertain systems, and therefore the Kharitonov theorem is usually applied by first overbounding the uncertainty with independent intervals at the expense of conservatism, see Chapter 5.

---

[3] A function $f : \mathbb{R}^\ell \to \mathbb{R}$ is said to be multiaffine if the following condition holds: if all components $q_1, \ldots, q_\ell$ except one are fixed, then f is affine. For example, $f(q) = 3q_1q_2q_3 - 6q_1q_3 + 4q_2q_3 + 2q_1 - 2q_2 + q_3 - 1$ is multiaffine.

A more general result, which takes into account the dependence of the polynomial coefficients on the uncertain parameters is known as the *edge theorem* [29], is discussed next. In the edge theorem, it is assumed that the coefficients of the polynomial are *affine* functions of the uncertainty vector $q = [q_1 \cdots q_\ell]^T$ bounded in the hyperrectangle $\mathcal{B}_q$ defined in (3.42). The polynomial family is therefore expressed as

$$\mathcal{P} = \{p(s, q) = a_0(q) + a_1(q)s + \cdots + a_{n-1}(q)s^{n-1} + s^n :$$

$$a_i(q) = a_{i0} + \sum_{k=1}^{\ell} a_{ik}q_k, q \in \mathcal{B}_q, i = 0, 1, \ldots, n-1\}. \quad (3.44)$$

The family $\mathcal{P}$ is often called a *polytope of polynomials*, whose vertices are the polynomials $p^i(s)$ corresponding to the 2^ℓ vertices of the hyperrectangle $\mathcal{B}_q$. We define

$$[p^i, p^k] \doteq \{p(s) = \lambda p^i(s) + (1 - \lambda)p^k(s), \; \lambda \in [0, 1]\}$$

as the convex combination of the two vertex polynomials $p^i(s)$ and $p^k(s)$. Then, $[p^i, p^k]$ is said to be an *edge* of the polytope $\mathcal{P}$ if $p^i(s), p^k(s)$ are such that, for any polynomials $p_a(s), p_b(s) \in \mathcal{P}$ with $p_a(s), p_b(s) \notin [p^i, p^k]$, it follows that $[p_a, p_b] \cap [p^i, p^k] = \emptyset$.

The edge theorem is due to [29]. A simplified version of this result in the special case of Hurwitz stability is reported below.

Theorem 3.7 (Edge theorem). *Consider the polytope of polynomials $\mathcal{P}$ defined in (3.44). The family $\mathcal{P}$ is Hurwitz if and only if all edges of $\mathcal{P}$ are Hurwitz.*

Remark 3.7 (Edge theorem for D-stability). The edge theorem can be stated in more general form for various root confinement regions of the complex plane, which include the unit disc (Schur stability) and other simply connected sets $\mathcal{D}$. In this case, we deal with the so-called $\mathcal{D}$-stability. ◁

Remark 3.8 (Edge redundancy). Notice that every polynomial belonging to an edge of $\mathcal{P}$ is obtained from one of the $\ell 2^{\ell-1}$ edges of the hyperrectangle $\mathcal{B}_q$, but not necessarily vice versa, [25]. When applying the edge theorem, it is often easier to work with the edges of $\mathcal{B}_q$ instead of the edges of $\mathcal{P}$. In doing this, we accept the possibility of redundancy in the robustness test. ◁

In the context of real parametric uncertainty, for an uncertain polynomial $p(s, q)$, the robustness margin discussed at the beginning of Section 3.7 is formally defined as

$$r_q \doteq \inf\{\rho : p(s, q) \text{ not Hurwitz for some } q \in \mathcal{B}_q(\rho)\}$$

where

$$\mathcal{B}_q(\rho) \doteq \{q \in \mathbb{R}^\ell : q_i \in [\rho q_i^-, \rho q_i^+], \; i = 1, \ldots, \ell\}. \quad (3.45)$$

Remark 3.9 (Interval matrices). An obvious question that was studied in depth in the area of parametric robustness is whether the stability results for interval and polytopic polynomials can be extended to uncertain matrices. In particular, a natural generalization of the interval polynomial framework leads to the notion of an *interval matrix*, i.e. a matrix whose entries are bounded in given intervals. Formally, a family $\mathcal{A}$ of interval matrices is defined as

$$\mathcal{A} \doteq \left\{ A \in \mathbb{R}^{n,n} : [A]_{i,k} \in [a_{ik}^-, a_{ik}^+], i, k = 1, \ldots, n \right\}. \tag{3.46}$$

Unfortunately, it has been shown that extensions of Kharitonov-like results to stability of interval matrices fail, see for instance [24]. Moreover, this problem has been shown to be computationally intractable, see Section 5.1.   ◁

Finally, we notice that one of the objectives of parametric stability is to establish "extreme point" and "edge-like" results for special classes of uncertain polynomials and uncertain feedback systems. An important tool for analyzing stability of uncertain systems in the frequency domain is the so-called *value set* (or "template," see [138] for specific discussions on this concept) and efficient algorithms have been developed for its construction. The literature on parametric stability is very broad and the interested reader is redirected, for instance, to the books [24, 40] and the surveys [25, 265].

4. Linear Robust Control Design

This chapter continues the study of robustness of uncertain systems initiated in Chapter 3. In particular, we now focus on robust synthesis, i.e. we discuss the problem of designing a controller K such that the interconnection in Figure 4.1 achieves robust stability with respect to uncertainty and (nominal or robust) performance in the $w \to z$ channel.

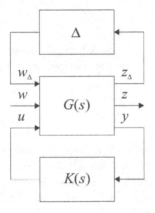

Fig. 4.1. Interconnection for control design guaranteeing robust stability and (nominal or robust) performance.

4.1 $\mathcal{H}_\infty$ design

The first and fundamental problem that we study is the design of a controller such that the interconnection shown in Figure 4.2 is robustly stable for $\widetilde{\mathbb{D}} = \mathcal{RH}_\infty^{r,c}$, $\Delta \in \mathcal{B}_{\widetilde{\mathbb{D}}}(\rho)$, where $\widetilde{\mathbb{D}}$ and $\mathcal{B}_{\widetilde{\mathbb{D}}}(\rho)$ are defined in (3.25) and (3.26). This yields the classical $\mathcal{H}_\infty$ design problem discussed in this section. In particular, we consider an extended plant G with state space representation

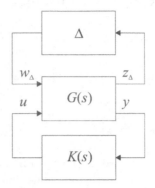

Fig. 4.2. Interconnection for control design guaranteeing robust stability.

$$\dot{x} = Ax + B_1 w_\Delta + B_2 u; \tag{4.1}$$
$$z_\Delta = C_1 x + D_{11} w_\Delta + D_{12} u;$$
$$y = C_2 x + D_{21} w_\Delta$$

where $A \in \mathbb{R}^{n_s, n_s}, B_1 \in \mathbb{R}^{n_s, r}, B_2 \in \mathbb{R}^{n_s, n_i}, C_1 \in \mathbb{R}^{c, n_s}, D_{11} \in \mathbb{R}^{c, r}, D_{12} \in \mathbb{R}^{c, n_i}, C_2 \in \mathbb{R}^{n_o, n_s}, D_{21} \in \mathbb{R}^{n_o, r}$. In this section, we make the following standing assumptions

(A, B_2) stabilizable; $\tag{4.2}$
(A, C_2) detectable.

The small gain theorem implies that the well posedness and internal stability of the interconnection in Figure 4.2 for all $\Delta \in \mathcal{B}_{\widetilde{\mathbb{D}}}(\rho)$ are equivalent to the existence of a controller K such that the closed loop between G and K is well posed and internally stable, and the transfer matrix T_{w_Δ, z_Δ} of the channel $w_\Delta \to z_\Delta$ satisfies the bound $\|T_{w_\Delta, z_\Delta}\|_\infty < 1/\rho$. Partitioning the transfer matrix G as

$$\begin{bmatrix} z_\Delta \\ y \end{bmatrix} = \begin{bmatrix} G_{11} & G_{12} \\ G_{21} & G_{22} \end{bmatrix} \begin{bmatrix} w_\Delta \\ u \end{bmatrix}$$

we write the transfer matrix T_{w_Δ, z_Δ} in terms of the *lower* linear fractional transformation

$$T_{w_\Delta, z_\Delta} = \mathcal{F}_l(G, K) \doteq G_{11} + G_{12} K (I - G_{22} K)^{-1} G_{21}. \tag{4.3}$$

Letting $\gamma = 1/\rho$, the (γ-suboptimal) $\mathcal{H}_\infty$ robust design problem is now reformulated.

Problem 4.1 (γ-suboptimal $\mathcal{H}_\infty$ design). *For fixed $\gamma > 0$, find a controller K such that the closed loop between G and K is well posed and internally stable, and $\|T_{w_\Delta, z_\Delta}\|_\infty < \gamma$.*

In the following, for simplicity, we consider full-order controllers, i.e. controllers of the same order n_s of the plant. Let the controller K, with state space representation

$$\dot{x}_K = A_K x_K + B_K y;$$
$$u_K = C_K x_K + D_K y$$

be described by the quadruple

$$\Omega_K \doteq \left[\begin{array}{c|c} A_K & B_K \\ \hline C_K & D_K \end{array} \right]. \tag{4.4}$$

Then, the closed-loop transfer matrix $T_{w_\Delta, z_\Delta} = \mathcal{F}_l(G, K)$ has state space representation given by the block matrix

$$\begin{aligned}
\Omega_T &\doteq \left[\begin{array}{c|c} A_{cl} & B_{cl} \\ \hline C_{cl} & D_{cl} \end{array} \right] \\
&= \left[\begin{array}{cc|c} A & 0 & B_1 \\ 0 & 0 & 0 \\ \hline C_1 & 0 & D_{11} \end{array} \right] + \left[\begin{array}{cc} 0 & B_2 \\ I & 0 \\ \hline 0 & D_{12} \end{array} \right] \Omega_K \left[\begin{array}{cc|c} 0 & I & 0 \\ C_2 & 0 & D_{21} \end{array} \right]. \tag{4.5}
\end{aligned}$$

From the bounded real lemma (Lemma 3.2), the closed loop is stable and $\|T_{w_\Delta, z_\Delta}\|_\infty < \gamma$ if and only if there exist a positive definite matrix $P_{cl} \succ 0$ such that the linear matrix inequality

$$\left[\begin{array}{ccc} 0 & 0 & 0 \\ 0 & -\gamma I & 0 \\ 0 & 0 & -\gamma I \end{array} \right] + \left[\begin{array}{cc} P_{cl} & 0 \\ 0 & 0 \\ 0 & I \end{array} \right] \Omega_T \left[\begin{array}{ccc} I & 0 & 0 \\ 0 & I & 0 \end{array} \right] + \left[\begin{array}{cc} I & 0 \\ 0 & I \\ 0 & 0 \end{array} \right] \Omega_T^T \left[\begin{array}{ccc} P_{cl} & 0 & 0 \\ 0 & 0 & I \end{array} \right] \prec 0$$

holds. This matrix inequality is rewritten more compactly as

$$Z(P_{cl}) + \Phi(P_{cl})\Omega_K \Psi + \Psi^T \Omega_K^T \Phi^T(P_{cl}) \prec 0 \tag{4.6}$$

where

$$Z(P_{cl}) = \left[\begin{array}{ccc} 0 & 0 & 0 \\ 0 & -\gamma I & 0 \\ 0 & 0 & -\gamma I \end{array} \right] + \left[\begin{array}{cc} P_{cl} & 0 \\ 0 & 0 \\ 0 & I \end{array} \right] \left[\begin{array}{cc|c} A & 0 & B_1 \\ 0 & 0 & 0 \\ C_1 & 0 & D_{11} \end{array} \right] \left[\begin{array}{ccc} I & 0 & 0 \\ 0 & I & 0 \end{array} \right]$$
$$+ \left(\left[\begin{array}{cc} P_{cl} & 0 \\ 0 & 0 \\ 0 & I \end{array} \right] \left[\begin{array}{cc|c} A & 0 & B_1 \\ 0 & 0 & 0 \\ C_1 & 0 & D_{11} \end{array} \right] \left[\begin{array}{ccc} I & 0 & 0 \\ 0 & I & 0 \end{array} \right] \right)^T ;$$

$$\Phi(P_{cl}) = \left[\begin{array}{cc} P_{cl} & 0 \\ 0 & 0 \\ 0 & I \end{array} \right] \left[\begin{array}{cc} 0 & B_2 \\ I & 0 \\ 0 & D_{12} \end{array} \right] ;$$

$$\Psi = \left[\begin{array}{cc|c} 0 & I & 0 \\ C_2 & 0 & D_{21} \end{array} \right] \left[\begin{array}{ccc} I & 0 & 0 \\ 0 & I & 0 \end{array} \right].$$

Notice that the matrix inequality (4.6) is *not* jointly linear in P_{cl} and Ω_K. However, (4.6) is an LMI in the controller matrix Ω_K for fixed P_{cl}. A way to overcome this difficulty has been proposed in [116, 143]. Here, we adopt the representation of [116]. First, we partition P_{cl} and P_{cl}^{-1} as

$$P_{cl} = \begin{bmatrix} S & N \\ N^T & \# \end{bmatrix}, \quad P_{cl}^{-1} = \begin{bmatrix} R & M \\ M^T & \# \end{bmatrix} \tag{4.7}$$

with $M, N \in \mathbb{R}^{n_s, n_s}$, $R, S \in \mathbb{S}^{n_s}$ and $\#$ means "it doesn't matter." Then, the variable Ω_K is eliminated from condition (4.6) using suitable projections. This leads to the existence theorem stated below, see [116].

Theorem 4.1 ($\mathcal{H}_\infty$ LMI solvability conditions). *Consider the extended plant (4.1) with assumptions (4.2). Let N_{12} and N_{21} be orthogonal bases of the null spaces of $[B_2^T \ \ D_{12}^T]$ and $[C_2 \ \ D_{21}]$ respectively. Then, there exists a controller matrix Ω_K such that the closed loop is internally stable and $\|T_{w_\Delta, z_\Delta}\|_\infty < \gamma$ if and only if there exist symmetric matrices $R, S \in \mathbb{S}^{n_s}$ such that the following system of LMIs is feasible*

$$\begin{bmatrix} N_{12} & 0 \\ \hline 0 & I \end{bmatrix}^T \begin{bmatrix} AR + RA^T & RC_1^T & B_1 \\ C_1 R & -\gamma I & D_{11} \\ \hline B_1^T & D_{11}^T & -\gamma I \end{bmatrix} \begin{bmatrix} N_{12} & 0 \\ \hline 0 & I \end{bmatrix} \prec 0; \tag{4.8}$$

$$\begin{bmatrix} N_{21} & 0 \\ \hline 0 & I \end{bmatrix}^T \begin{bmatrix} A^T S + SA & SB_1 & C_1^T \\ B_1^T S & -\gamma I & D_{11}^T \\ \hline C_1 & D_{11} & -\gamma I \end{bmatrix} \begin{bmatrix} N_{21} & 0 \\ \hline 0 & I \end{bmatrix} \prec 0; \tag{4.9}$$

$$\begin{bmatrix} R & I \\ I & S \end{bmatrix} \succ 0. \tag{4.10}$$

We now make a few comments regarding the use of this result for constructing an $\mathcal{H}_\infty$ stabilizing controller, see [116]. Once a feasible pair R, S for (4.8)–(4.10) is found, a full-order controller can be determined as follows:

1. Compute via SVD two invertible matrices $M, N \in \mathbb{R}^{n_s, n_s}$ such that $MN^T = I - RS$;
2. The matrix $P_{cl} \succ 0$ in the bounded real lemma is uniquely determined as the solution of the linear equations

$$P_{cl} \begin{bmatrix} R & I \\ M^T & 0 \end{bmatrix} = \begin{bmatrix} I & S \\ 0 & N^T \end{bmatrix};$$

3. For P_{cl} determined as above, any Ω_K that is feasible for the controller LMI (4.6) yields a stabilizing γ-suboptimal controller for the system (4.1).

Remark 4.1 (Reduced-order $\mathcal{H}_\infty$ controller). The LMI condition (4.10) in Theorem 4.1 can be relaxed to non strict inequality. The resulting possible rank drop can be exploited to determine a controller of reduced order

$k < n_s$. In particular, there exists a γ-suboptimal controller of order $k < n_s$ if and only if (4.8), (4.9) and the inequality

$$\begin{bmatrix} R & I \\ I & S \end{bmatrix} \succeq 0$$

hold for some R, S which further satisfy

$$\text{rank}(I - RS) \leq k.$$

We also remark that in the third step of the above procedure for determining the controller matrix Ω_K it is not actually necessary to solve numerically the controller LMI, since analytic formulas are given in [115, 143]. ◁

4.1.1 Regular $\mathcal{H}_\infty$ problem

In the so-called regular $\mathcal{H}_\infty$ problem discussed in classical references such as [93], the following standard simplifying assumptions are made

$$D_{11} = 0, \quad D_{12}^T \begin{bmatrix} C_1 & D_{12} \end{bmatrix} = \begin{bmatrix} 0 & I \end{bmatrix}, \quad D_{21} \begin{bmatrix} B_1^T & D_{21}^T \end{bmatrix} = \begin{bmatrix} 0 & I \end{bmatrix}. \quad (4.11)$$

With these assumptions, it may be easily verified[1] that, letting $X \doteq \gamma R^{-1}$ and $Y \doteq \gamma S^{-1}$, and using the Schur complement rule, conditions (4.8)-(4.10) are equivalent to

$$A^T X + XA + X(\gamma^{-2}B_1 B_1^T - B_2 B_2^T)X + C_1^T C_1 \prec 0; \quad (4.12)$$
$$AY + YA^T + Y(\gamma^{-2}C_1^T C_1 - C_2^T C_2)Y + B_1 B_1^T \prec 0; \quad (4.13)$$
$$X \succ 0, \quad Y \succ 0, \quad \rho_\lambda(XY) < \gamma^2. \quad (4.14)$$

The left-hand sides of inequalities (4.12) and (4.13) coincide with the expressions arising in the standard Riccati-based $\mathcal{H}_\infty$ formulae [93]. In particular, the connection with the algebraic Riccati equations (AREs) is detailed in the next lemma.

Lemma 4.1 (ARE based $\mathcal{H}_\infty$ solution). *Consider the extended plant (4.1) satisfying the following regularity conditions:*

1. $D_{11} = 0$, $D_{12}^T \begin{bmatrix} C_1 & D_{12} \end{bmatrix} = \begin{bmatrix} 0 & I \end{bmatrix}$, $D_{21} \begin{bmatrix} B_1^T & D_{21}^T \end{bmatrix} = \begin{bmatrix} 0 & I \end{bmatrix}$;
2. $\begin{bmatrix} A - j\omega I & B_2 \\ C_1 & D_{12} \end{bmatrix}$ *has full column rank for all ω;*

---

[1] To this end, take $N_{21} = \begin{bmatrix} I & 0 \\ -D_{21}^T C_2 & D_{21}^{T\perp} \end{bmatrix}$, where $D_{21}^{T\perp}$ is the orthogonal comple-
ment of D_{21}^T, i.e. $D_{21}D_{21}^{T\perp} = 0$ and $D_{21}^{T\perp T}D_{21}^{T\perp} = I$. Since $D_{21}B_1^T = 0$ we may write $B_1^T = D_{21}^{T\perp}Z$ for some matrix Z and, therefore, $B_1 D_{21}^{T\perp} D_{21}^{T\perp T} B_1^T = B_1 B_1^T$. Everything follows in a similar way for the first inequality, choosing $N_{12} = \begin{bmatrix} I & 0 \\ -B_2^T D_{12} & D_{12}^T \end{bmatrix}$.

3. $\begin{bmatrix} A - j\omega I & B_1 \\ C_2 & D_{21} \end{bmatrix}$ *has full row rank for all* ω.

Suppose that the two algebraic Riccati inequalities (ARIs)

$$A^T X + X A + X(\gamma^{-2} B_1 B_1^T - B_2 B_2^T)X + C_1^T C_1 \prec 0; \tag{4.15}$$
$$AY + Y A^T + Y(\gamma^{-2} C_1^T C_1 - C_2^T C_2)Y + B_1 B_1^T \prec 0 \tag{4.16}$$

admit positive definite solutions $X_0, Y_0 \succ 0$. *Then, the corresponding AREs*

$$A^T X + X A + X(\gamma^{-2} B_1 B_1^T - B_2 B_2^T)X + C_1^T C_1 = 0; \tag{4.17}$$
$$AY + Y A^T + Y(\gamma^{-2} C_1^T C_1 - C_2^T C_2)Y + B_1 B_1^T = 0 \tag{4.18}$$

have stabilizing solutions[2] X_∞, Y_∞ *satisfying*

$$0 \preceq X_\infty \prec X_0;$$
$$0 \preceq Y_\infty \prec Y_0.$$

Moreover, if $\rho_\lambda(X_0 Y_0) < \gamma^2$, *then* $\rho_\lambda(X_\infty Y_\infty) < \gamma^2$.

Conversely, if the AREs (4.17) and (4.18) admit stabilizing solutions $X_\infty \succeq 0$, $Y_\infty \succeq 0$ *satisfying* $\rho_\lambda(X_\infty Y_\infty) < \gamma^2$, *then there exist feasible solutions* $X_0, Y_0 \succ 0$ *of the ARIs (4.15) and (4.16) such that* $\rho_\lambda(X_0 Y_0) < \gamma^2$.

Remark 4.2 (Controller formulae). Under the simplifying assumptions (4.11), for any pair X, Y which is feasible for (4.12)–(4.14), a γ-suboptimal controller may be constructed as

$$A_K = A + (\gamma^{-2} B_1 B_1^T - B_2 B_2^T)X + (\gamma^{-2} Y X - I)^{-1} Y C_2^T C_2;$$
$$B_K = -(\gamma^{-2} Y X - I)^{-1} Y C_2^T;$$
$$C_K = -B_2^T X;$$
$$D_K = 0.$$

Moreover, if $X_\infty \succeq 0$, $Y_\infty \succeq 0$ satisfying $\rho_\lambda(X_\infty Y_\infty) < \gamma^2$ are stabilizing solutions of the AREs (4.17) and (4.18), then substituting formally $X = X_\infty$, $Y = Y_\infty$ in the above controller formulae, we obtain the expression of the so-called *central* controller given in [93]. Riccati-based solutions for the general $\mathcal{H}_\infty$ problem, without simplifying assumptions, are given in [237, 238] and in [232]. ◁

4.1.2 Alternative LMI solution for $\mathcal{H}_\infty$ design

An alternative approach to determine synthesis LMIs for the $\mathcal{H}_\infty$ problem is based on a systematic technique for transforming analysis LMI conditions into synthesis LMIs using a suitable linearizing change of controller variables and a congruence transformation, see [182, 236]. The next lemma is based on [236].

---

[2] $X \in \mathbb{S}^n$ is a stabilizing solution of the ARE $A^T X + X A + X R X + Q = 0$ if it satisfies the equation and $A + RX$ is stable.

Lemma 4.2 (Linearizing change of variables). *Let P_{cl}, P_{cl}^{-1} be partitioned as in (4.7), where $R, S \in \mathbb{S}^n$ and $M, N \in \mathbb{R}^{n_s, n_s}$. Let*

$$
\begin{aligned}
\widehat{A} &\doteq NA_K M^T + NB_K C_2 R + SB_2 C_K M^T + S(A + B_2 D_K C_2)R; \\
\widehat{B} &\doteq NB_K + SB_2 D_K; \\
\widehat{C} &\doteq C_K M^T + D_K C_2 R; \\
\widehat{D} &\doteq D_K
\end{aligned}
\tag{4.19}
$$

and

$$
\Pi_1 \doteq \begin{bmatrix} R & I \\ M^T & 0 \end{bmatrix}, \quad \Pi_2 \doteq \begin{bmatrix} I & S \\ 0 & N^T \end{bmatrix}.
$$

Then, it holds that

$$
P_{cl}\Pi_1 = \Pi_2;
$$

$$
\Pi_1^T P_{cl} A_{cl} \Pi_1 = \Pi_2^T A_{cl}\Pi_1 = \begin{bmatrix} AR + B_2\widehat{C} & A + B_2\widehat{D}C_2 \\ \widehat{A} & SA + \widehat{B}C_2 \end{bmatrix};
$$

$$
\Pi_1^T P_{cl} B_{cl} = \Pi_2^T B_{cl} = \begin{bmatrix} B_1 + B_2\widehat{D}D_{21} \\ SB_1 + \widehat{B}D_{21} \end{bmatrix};
$$

$$
C_{cl}\Pi_1 = \begin{bmatrix} C_1 R + D_{12}\widehat{C} & C_1 + D_{12}\widehat{D}C_2 \end{bmatrix};
$$

$$
\Pi_1^T P_{cl}\Pi_1 = \Pi_1^T \Pi_2 = \begin{bmatrix} R & I \\ I & S \end{bmatrix}.
$$

Applying the bounded real lemma to the closed-loop system Ω_T in (4.5), and performing the congruence transformation with $\mathrm{bdiag}(\Pi_1, I, I)$, we obtain a result which provides alternative solvability conditions in the modified controller variables. This is stated in the next theorem, see [236] for proof.

Theorem 4.2 (Alternative $\mathcal{H}_\infty$ LMI solution). *Consider the extended plant (4.1) with assumptions (4.2). Then, there exists a controller matrix Ω_K such that the closed loop is internally stable and $\|T_{w_\Delta, z_\Delta}\|_\infty < \gamma$ if and only if there exist symmetric matrices $R, S \in \mathbb{S}^{n_s}$ and matrices $\widehat{A}, \widehat{B}, \widehat{C}, \widehat{D}$ such that the following system of LMIs is feasible*[3]

$$
\begin{bmatrix}
AR + RA^T + B_2\widehat{C} + \widehat{C}^T B_2^T & \widehat{A}^T + A + B_2\widehat{D}C_2 & B_1 + B_2\widehat{D}D_{21} & * \\
* & A^T S + SA + \widehat{B}C_2 + C_2^T\widehat{B}^T & SB_1 + \widehat{B}D_{21} & * \\
* & * & -\gamma I & * \\
C_1 R + D_{12}\widehat{C} & C_1 + D_{12}\widehat{D}C_2 & D_{11} + D_{12}\widehat{D}D_{21} & -\gamma I
\end{bmatrix} \prec 0;
\tag{4.20}
$$

$$
\begin{bmatrix} R & I \\ I & S \end{bmatrix} \succ 0.
\tag{4.21}
$$

---

[3] The asterisks indicate elements whose values are easily inferred by symmetry.

We now discuss this result and show how a γ-suboptimal $\mathcal{H}_\infty$ controller can be obtained. For any matrices $\widehat{A}, \widehat{B}, \widehat{C}, \widehat{D}, R, S$ satisfying (4.20) and (4.21), we can recover a γ-suboptimal $\mathcal{H}_\infty$ controller by inverting the change of variables of Lemma 4.2 as follows:

1. Compute via SVD two square and invertible matrices $M, N \in \mathbb{R}^{n_s, n_s}$ such that $MN^T = I - RS$. We remark that this is always possible for full-order controller design due to the constraint (4.21);
2. The controller matrices are uniquely determined as

$$
\begin{aligned}
D_K &= \widehat{D}; \\
C_K &= (\widehat{C} - D_K C_2 R)M^{-T}; \\
B_K &= N^{-1}(\widehat{B} - S B_2 D_K); \\
A_K &= N^{-1}\left(\widehat{A} - N B_K C_2 R - S B_2 C_K M^T - S(A + B_2 D_K C_2)R\right)M^{-T}
\end{aligned}
\tag{4.22}
$$

where $M^{-T} = (M^{-1})^T = (M^T)^{-1}$.

4.1.3 μ synthesis

In this section, we consider the problem of designing a controller such that the interconnection in Figure 4.2 is well posed and robustly stable for structured uncertainty of the form

$$
\widetilde{\mathbb{D}} \doteq \{\Delta \in \mathcal{RH}_\infty^{r,c} : \Delta = \mathrm{bdiag}(\Delta_1, \ldots, \Delta_b), \Delta_i \in \mathcal{RH}_\infty^{r_i, r_i}\}
$$

with the bound $\|\Delta\|_\infty < \rho$. From the small μ theorem applied to the closed-loop system $\mathcal{F}_l(G, K)$, the problem amounts to determining a controller $K(s)$ such that

$$
\sup_{\omega \in \mathbb{R}} \mu_{\mathbb{D}}\left(\mathcal{F}_l(G(j\omega), K(j\omega))\right) \le \frac{1}{\rho}
$$

where $\mathbb{D}$ is the *purely complex* uncertainty structure

$$
\mathbb{D} = \{\Delta \in \mathbb{C}^{r,c} : \Delta = \mathrm{bdiag}(\Delta_1, \ldots, \Delta_b), \Delta_i \in \mathbb{C}^{r_i, r_i}\}.
$$

Unfortunately, this problem cannot be solved efficiently in general. Indeed, even evaluating $\mu_{\mathbb{D}}$ for *fixed* K is computationally difficult, as discussed in Section 3.7.3. However, in practice, the problem can be tackled using a sub-optimal iterative approach, generally denoted as D–K iteration, see for instance [21, 68, 295].

D–K iteration for μ synthesis.

1. Let $\widehat{K}(s)$ be any stabilizing controller for the system (usually, $\widehat{K}(s)$ is computed solving a standard $\mathcal{H}_\infty$ problem), and let $\{\omega_1, \ldots, \omega_N\}$ be a suitable frequency grid;

2. Fix $K(s) = \widehat{K}(s)$. Determine the sequence of scaling matrices of the form

$$\widetilde{D}(\omega_i) \doteq \mathrm{bdiag}\left(\widetilde{d}_1(\omega_i)I_{r_1}, \ldots, \widetilde{d}_{b-1}(\omega_i)I_{r_{b-1}}, I_{r_b}\right), \; i = 1, \ldots, N$$

with $\widetilde{d}_k(\omega_i) > 0$ for $k = 1, \ldots, b-1$, such that

$$\bar{\sigma}\left(\widetilde{D}(\omega_i)\mathcal{F}_l\left(G(j\omega_i), \widehat{K}(j\omega_i)\right)\widetilde{D}^{-1}(\omega_i)\right), \; i = 1, \ldots, N$$

is minimized;

3. Find scalar transfer functions $d_k(s) \in \mathcal{RH}_\infty$ such that $d_k^{-1}(s) \in \mathcal{RH}_\infty$ and $|d_k(j\omega_i)| \simeq \widetilde{d}_k(j\omega_i)$, for $i = 1, \ldots, N$ and $k = 1, \ldots, b-1$, and construct the transfer matrix

$$D(s) \doteq \mathrm{bdiag}\left(d_1(s)I_{r_1}, \ldots, d_{b-1}(s)I_{r_{b-1}}, I_{r_b}\right);$$

4. Consider the configuration in Figure 4.3, solve the $\mathcal{H}_\infty$ minimization problem

$$\widehat{K} = \arg\min \|D(s)\mathcal{F}_l\left(G(s), K(s)\right)D^{-1}(s)\|_\infty$$

and repeat the iterations from step 2.

The D–K iteration is terminated when either an $\mathcal{H}_\infty$ norm smaller than $1/\rho$ is achieved in step 4, or no significant improvement is obtained with respect to the previous iteration.

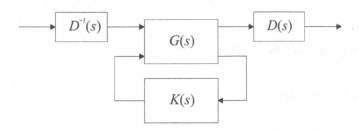

Fig. 4.3. Configuration for μ synthesis via D–K iteration.

Remark 4.3 (Convergence of D–K iteration). The D–K iteration is not guaranteed to converge to a global optimum (nor to a local one), but can be a useful tool for many practical design problems.

For fixed K, the D-subproblem (step 2) is a convex optimization problem that may be efficiently solved. Similarly, for fixed D, the K-subproblem (step 4) is a standard $\mathcal{H}_\infty$ design, which can be directly solved by means of the Riccati or LMI approach discussed in the previous sections. However, the problem is not jointly convex in D and K. The rational approximations required in step 3 may be performed using interpolation theory, see e.g [291], but this approach usually results in high-order transfer functions. This is clearly not desirable, since it increases the order of the scaled plant, and therefore of the controller. For this reason, a frequently used method is based on graphical matching by means of low-order transfer functions.

We finally remark that the D–K iteration may still be applied for more general uncertainty structures, involving for instance repeated scalar blocks, provided that suitable scalings are used in the evaluation of the upper bound of $\mu_{\mathbb{D}}$, see e.g. Section 3.7.3.   ◁

4.2 $\mathcal{H}_2$ design

In this section, we discuss the problem of designing a controller K such that its interconnection with the *nominal* model G, see Figure 4.4, provides closed-loop stability and attains a level γ of $\mathcal{H}_2$ performance on the $w \to z$ channel.

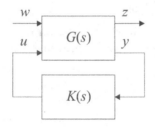

Fig. 4.4. Interconnection for nominal $\mathcal{H}_2$ design.

Let G be represented by

$$
\begin{aligned}
\dot{x} &= Ax + B_1 w + B_2 u; \\
z &= C_1 x + D_{11} w + D_{12} u; \\
y &= C_2 x + D_{21} w
\end{aligned}
\tag{4.23}
$$

where $A \in \mathbb{R}^{n_s,n_s}, B_1 \in \mathbb{R}^{n_s,q}, B_2 \in \mathbb{R}^{n_s,n_i}, C_1 \in \mathbb{R}^{p,n_s}, D_{11} \in \mathbb{R}^{p,q}, D_{12} \in \mathbb{R}^{p,n_i}, C_2 \in \mathbb{R}^{n_o,n_s}, D_{21} \in \mathbb{R}^{n_o,q}$. In this section, we make the following standing assumptions

$$
\begin{aligned}
&(A, B_2) \quad \text{stabilizable;} \\
&(A, C_2) \quad \text{detectable.}
\end{aligned}
\tag{4.24}
$$

Following the same derivation as in the beginning of Section 4.1, we obtain the closed-loop system $T_{w,z}$ described by the block matrix

$$\Omega_T = \left[\begin{array}{c|c} A_{cl} & B_{cl} \\ \hline C_{cl} & D_{cl} \end{array}\right]$$

given in (4.5). Then, for fixed $\gamma > 0$ the $\mathcal{H}_2$ performance design problem is formulated.

Problem 4.2 (γ-suboptimal $\mathcal{H}_2$ design). *For fixed $\gamma > 0$, find a controller K such that the closed loop between G and K is well posed and internally stable, and $\|T_{w,z}\|_2^2 < \gamma$.*

To obtain the $\mathcal{H}_2$ synthesis conditions we proceed as follows: first, we apply Lemma 3.1 to the closed-loop matrices $A_{cl}, B_{cl}, C_{cl}, D_{cl}$. Then, we introduce the change of controller variables of Lemma 4.2, and perform the congruence transformation with $\mathrm{bdiag}(\Pi_1, I)$, obtaining the result summarized in the next theorem, see [236].

Theorem 4.3 ($\mathcal{H}_2$ LMI solution). *Consider the extended plant (4.23) with assumptions (4.24). Then, there exists a controller matrix Ω_K such that the closed loop is internally stable and $\|T_{w,z}\|_2^2 < \gamma$ if and only if there exist symmetric matrices $R, S \in \mathbb{S}^{n_s}$ and matrices $Q, \widehat{A}, \widehat{B}, \widehat{C}, \widehat{D}$ such that the following system of LMIs is feasible*

$$\begin{bmatrix} AR + RA^T + B_2\widehat{C} + \widehat{C}^T B_2^T & \widehat{A}^T + A + B_2\widehat{D}C_2 & B_1 + B_2\widehat{D}D_{21} \\ * & A^T S + SA + \widehat{B}C_2 + C_2^T\widehat{B}^T & SB_1 + \widehat{B}D_{21} \\ * & * & -I \end{bmatrix} \prec 0;$$

$$\begin{bmatrix} R & I & (C_1 R + D_{12}\widehat{C})^T \\ I & S & (C_1 + D_{12}\widehat{D}C_2)^T \\ (C_1 R + D_{12}\widehat{C}) & (C_1 + D_{12}\widehat{D}C_2) & Q \end{bmatrix} \succ 0;$$

$$D_{11} + D_{12}\widehat{D}D_{21} = 0;$$
$$\mathrm{Tr}\, Q < \gamma.$$

For any matrices $\widehat{A}, \widehat{B}, \widehat{C}, \widehat{D}, R, S$ satisfying the LMI solvability conditions of this theorem, we recover a γ-suboptimal $\mathcal{H}_2$ controller by inverting the change of variables in (4.19) using (4.22).

Under the standard regularity assumptions introduced in Section 4.1.1, the solution of the $\mathcal{H}_2$ problem can be alternatively stated in terms of two *decoupled* AREs, as summarized in the next lemma, see e.g. [295].

Lemma 4.3 (ARE based $\mathcal{H}_2$ solution). *Consider the extended plant (4.23) satisfying the following regularity assumptions:*

1. $D_{11} = 0$;
2. $R_1 \doteq D_{12}^T D_{12} \succ 0$ and $R_2 \doteq D_{21}D_{21}^T \succ 0$;
3. $\begin{bmatrix} A - j\omega I & B_2 \\ C_1 & D_{12} \end{bmatrix}$ *has full column rank for all ω;*
4. $\begin{bmatrix} A - j\omega I & B_1 \\ C_2 & D_{21} \end{bmatrix}$ *has full row rank for all ω.*

Define

$$A_x \doteq A - B_2 R_1^{-1} D_{12}^T C_1, \qquad A_y \doteq A - B_1 D_{21}^T R_2^{-1} C_2$$

and let $X, Y \succeq 0$ be stabilizing solutions of the two AREs

$$XA_x + A_x^T X - XB_2 R_1^{-1} B_2^T X + C_1^T (I - D_{12} R_1^{-1} D_{12}^T) C_1 = 0; \qquad (4.25)$$
$$YA_y^T + A_y Y - YC_2^T R_2^{-1} C_2 Y + B_1 (I - D_{21}^T R_2^{-1} D_{21}) B_1^T = 0. \qquad (4.26)$$

Then, the controller

$$B_K = (B_1 D_{21}^T + YC_2^T) R_2^{-1}; \qquad (4.27)$$
$$C_K = -R_1^{-1} (D_{12}^T C_1 + B_2^T X); \qquad (4.28)$$
$$A_K = A + B_2 C_K - B_K C_2; \qquad (4.29)$$
$$D_K = 0 \qquad (4.30)$$

stabilizes the closed-loop system and minimizes the $\mathcal{H}_2$ norm of the $w \to z$ channel. Moreover, with this controller we have

$$\|T_{w,z}\|_2^2 = \operatorname{Tr} B_1^T X B_1 + \operatorname{Tr} C_K Y C_K^T R_1.$$

Remark 4.4 (LQG control). The $\mathcal{H}_2$ control problem is equivalent to classical linear quadratic Gaussian (LQG) control. The connection is established assuming that the exogenous signal $w(t)$ is white Gaussian noise with $\mathrm{E}(\mathbf{w}(t)) = 0$, $\mathrm{E}(\mathbf{w}(t)\mathbf{w}^T(t + \tau)) = I\delta(\tau)$, where $\delta(\cdot)$ is the Dirac impulse function. In this case, the objective is to determine a control law that stabilizes the closed loop and minimizes the average output power

$$\mathrm{E}\left(\lim_{T\to\infty} \frac{1}{T} \int_0^T \|z\|^2 \mathrm{d}t\right) = \|\mathbf{z}\|_{\mathrm{rms}}^2 = \|T_{w,z}\|_2^2.$$

$\triangleleft$

An important special case that falls out directly from the $\mathcal{H}_2$ problem is the classical *linear quadratic regulator* (LQR) problem, which is discussed next.

4.2.1 Linear quadratic regulator

In the traditional linear quadratic regulator problem, see for instance [11, 165], we consider the system description

$$\dot{x} = Ax + B_2 u \qquad (4.31)$$

with initial state $x(0) = x_0 \in \mathbb{R}^{n_s}$ given but arbitrary. We look for a state feedback control law of the form $u = -Kx$, where $K \in \mathbb{R}^{n_i, n_s}$ such that the closed loop is stable and the following cost is minimized

$$\mathcal{C} = \int_0^\infty \begin{bmatrix} x(t) \\ u(t) \end{bmatrix}^T Q \begin{bmatrix} x(t) \\ u(t) \end{bmatrix} dt \qquad (4.32)$$

where

$$Q \doteq \begin{bmatrix} Q_{xx} & Q_{xu} \\ Q_{xu}^T & Q_{uu} \end{bmatrix} \succeq 0. \qquad (4.33)$$

Since Q is positive semidefinite, it can be factored as

$$Q = \begin{bmatrix} C_1^T \\ D_{12}^T \end{bmatrix} [C_1 \; D_{12}].$$

Therefore, the above cost $\mathcal{C}$ is equivalent to $\|z\|_2^2$, where z is the output associated with (4.31)

$$z = C_1 x + D_{12} u. \qquad (4.34)$$

We now state a result analogous to Lemma 4.3 for the LQR problem.

Lemma 4.4 (Extended LQR). *Consider the plant (4.31), (4.34) satisfying the following hypotheses:*

1. *(A, B_2) is stabilizable;*
2. *$\begin{bmatrix} A - j\omega I & B_2 \\ C_1 & D_{12} \end{bmatrix}$ has full column rank for all ω;*
3. *$\begin{bmatrix} D_{12} & D_{12}^\perp \end{bmatrix}^T \begin{bmatrix} D_{12} & D_{12}^\perp \end{bmatrix} = \begin{bmatrix} Q_{uu} & 0 \\ 0 & I \end{bmatrix} \succ 0.$*

Then, there exists a unique optimal control minimizing the cost function given in (4.32)

$$K = Q_{uu}^{-1}(B_2^T X + D_{12}^T C_1)$$

where $X \succeq 0$ is the stabilizing solution of the ARE

$$(A - Q_{uu}^{-1} D_{12}^T C_1)^T X + X(A - Q_{uu}^{-1} D_{12}^T C_1) - X B_2 Q_{uu}^{-1} B_2^T X \qquad (4.35)$$
$$+ C_1^T D_{12}^\perp D_{12}^{\perp T} C_1 = 0.$$

Moreover, the optimal cost is $\mathcal{C} = x_0^T X x_0$.

Remark 4.5 (Standard LQR). We notice that in the standard LQR it is often assumed that $Q_{xu} = 0$, i.e. $C_1^T D_{12} = 0$. This assumption is made without loss of generality, since there always exists a coordinate transformation that brings a general LQR problem in standard form. In this situation, the cost function is

$$\mathcal{C} = \int_0^\infty (x^T(t) Q_{xx} x(t) + u^T(t) Q_{uu} u(t)) \, dt \qquad (4.36)$$

with $Q_{xx} \succeq 0$ and $Q_{uu} \succ 0$. Then, the ARE in (4.35) is simplified as

$$A^T X + X A - X B_2 Q_{uu}^{-1} B_2^T X + Q_{xx} = 0. \qquad (4.37)$$

Furthermore, the unique optimal control law is given by

$$u(t) = -Q_{uu}^{-1} B_2^T X x(t) \qquad (4.38)$$

where $X \succ 0$ is the corresponding stabilizing solution of (4.37).   ◁

In the next section we study the related problems of quadratic stabilizability and guaranteed-cost linear quadratic control.

4.2.2 Quadratic stabilizability and guaranteed-cost control

We next consider the case when the system (4.31) described in state space form is affected by uncertainty. In particular, we study

$$\dot{x} = A(\Delta)x + B_2 u, \quad x(0) = x_0 \in \mathbb{R}^{n_s} \qquad (4.39)$$

where Δ belongs to the uncertainty set $\mathcal{B}_{\mathbb{D}}(\rho)$ defined in (3.28). This system is said to be *quadratically stable* if there exists $P \succ 0$ such that the Lyapunov inequality

$$A(\Delta)P + PA^T(\Delta) \prec 0 \qquad (4.40)$$

is satisfied for all $\Delta \in \mathcal{B}_{\mathbb{D}}(\rho)$. Then, we address the problem of finding a control law $u = -Kx$, $K \in \mathbb{R}^{n_i, n_s}$, such that the closed-loop is quadratically stable. That is, we seek $P \succ 0$ and K such that the inequality

$$(A(\Delta) + B_2 K) P + P (A(\Delta) + B_2 K)^T \prec 0$$

is satisfied for all $\Delta \in \mathcal{B}_{\mathbb{D}}(\rho)$. This matrix inequality is not jointly linear in P and K. However, introducing the new matrix variable $Y = KP$ as in [49] we obtain the following LMI in the variables P, Y

$$A(\Delta)P + PA^T(\Delta) + B_2 Y + Y^T B_2^T \prec 0. \qquad (4.41)$$

Notice that $Y \in \mathbb{R}^{n_i, n_s}$ is not necessarily symmetric. We say that the system (4.39) is *quadratically stabilizable* if there exist matrices $P \succ 0$ and Y such that the above LMI holds for all $\Delta \in \mathcal{B}_{\mathbb{D}}(\rho)$. Relations between quadratic stabilization and $\mathcal{H}_\infty$ design can be found in a number of papers, including [155].

Remark 4.6 (Quadratic stability/stabilizability for interval matrices). In general, the matrix inequalities (4.40) or (4.41) are to be solved for all $\Delta \in \mathcal{B}_{\mathbb{D}}(\rho)$, i.e. we have to satisfy an infinite number of LMIs simultaneously. In many important cases, owing to convexity properties, this problem can be reduced to the solution of a *finite* number of LMIs. For example, let $\mathcal{A}$ be an $n_s \times n_s$ interval matrix family, defined as in (3.46). In this case, the problem of quadratic stability, i.e. finding a *common solution* $P \succ 0$ satisfying (4.40) for all $[A]_{i,k}$ in the intervals $[a_{ik}^-, a_{ik}^+]$, $i, k = 1, \ldots, n_s$, is equivalent to finding $P \succ 0$ that satisfies

$$A^\ell P + P(A^\ell)^T \prec 0, \quad \ell = 1, \ldots, 2^{n_s^2}$$

where A^ℓ is a vertex matrix obtained setting all the entries of the interval matrix to either lower a_{ik}^- or upper a_{ik}^+ bounds, see e.g. [136]. Extreme points results of this kind also hold for more general classes than interval matrices, see e.g. [24]. A major computational issue, however, is that the number of LMIs which should be simultaneously solved is exponential in n_s.   ◁

Quadratic stability and LQR control may be linked together in the so-called *guaranteed-cost* LQR control described below. Consider the system (4.39) and the standard quadratic cost

$$\mathcal{C}(\Delta) = \int_0^\infty \left(x^T(t) Q_{xx} x(t) + u^T(t) Q_{uu} u(t) \right) dt$$

with $Q_{xx} \succeq 0$ and $Q_{uu} \succ 0$. Then, for $\gamma > 0$ the objective is to find a state feedback law of the form

$$u(t) = -Q_{uu}^{-1} B_2^T P^{-1} x(t) \tag{4.42}$$

where the design matrix variable $P \succ 0$ is chosen so that the closed-loop system is quadratically stable and the cost

$$\mathcal{C}(\Delta) \le \gamma^{-1} x_0^T P^{-1} x_0 \tag{4.43}$$

is guaranteed for all $\Delta \in \mathcal{B}_\mathbb{D}(\rho)$.

The guaranteed-cost control problem can be easily formulated in terms of LMIs, see e.g. [49]. We present here an alternative characterization involving a quadratic matrix inequality (QMI).

Lemma 4.5. *Let $P \succ 0$ be a solution that simultaneously satisfies the QMIs*

$$A(\Delta)P + PA^T(\Delta) - 2B_2 Q_{uu}^{-1} B_2^T + \gamma(B_2 Q_{uu}^{-1} B_2^T + PQ_{xx}P) \prec 0 \quad (4.44)$$

for all $\Delta \in \mathcal{B}_\mathbb{D}(\rho)$. Then, the control law

$$u(t) = -Q_{uu}^{-1} B_2^T P^{-1} x(t)$$

quadratically stabilizes the system (4.31) and the cost

$$\mathcal{C}(\Delta) \le \gamma^{-1} x_0^T P^{-1} x_0$$

is guaranteed for all $\Delta \in \mathcal{B}_\mathbb{D}(\rho)$.

The proof of this result is standard and it is not reported here; see e.g. [162, 215] for statements and proofs of similar results.

Remark 4.7 (Special cases of guaranteed-cost control). We notice that if $\gamma \rightarrow$ 0, then the constraint (4.43) on the cost $\mathcal{C}(\Delta)$ vanishes and the guaranteed-cost control reduces to a special case of quadratic stabilizability with the specific choice of a control law of the form (4.42). On the other hand, by setting $\gamma = 2$ and $Q_{xx} = 0$ in equation (4.44), we recover the quadratic stability setup.   ◁

4.3 Discussion

The brief presentation of robustness analysis and control techniques given in this and the previous chapter is necessarily incomplete. The $\mathcal{H}_\infty$ theory alone has been the subject of a variety of studies, each addressing different viewpoints. Quoting Lanzon, Anderson and Bombois (see [169] and its previous versions), "Some control theorists may say that interpolation theory [291] is the essence of $\mathcal{H}_\infty$ control, whereas others may assert that unitary dilation [93] is the fundamental underlying idea of $\mathcal{H}_\infty$ control. Also, J-spectral factorization [157] is a well-known framework of $\mathcal{H}_\infty$ control. A substantial number of researchers may take differential game theory [30] as the most salient feature of $\mathcal{H}_\infty$ control and others may assert that the bounded real lemma is the most fundamental building block with its connections to LMI techniques. All these opinions contain some truth since some techniques may expose certain features of the problem that are hidden (to some degree) when using different viewpoints." The approach chosen in this chapter mainly focuses on LMI techniques, since this appears computationally attractive and quite flexible. For instance, multiobjective design with mixed $\mathcal{H}_2/\mathcal{H}_\infty$ performance specifications may be easily implemented in the LMI framework [235, 236, 259]. Also, robustness analysis based on *integral quadratic constraints* (IQCs) relies on LMI optimization [145, 185], as well as most of the recent results based on parameter-dependent Lyapunov functions [103]. In this chapter, we also presented well-known results related to *nominal* $\mathcal{H}_2$ control design. The issue of robust (or guaranteed-cost) $\mathcal{H}_2$ control is somewhat controversial and it is not included in this introductory exposition. The interested reader is referred for instance to the survey [212] and to [258, 290].

5. Some Limits of the Robustness Paradigm

In this chapter we discuss some limitations of the classical robustness paradigm. In particular, we study complexity issues, conservatism and discontinuity problems. As outlined in Chapters 3 and 4, stability and performance of control systems affected by bounded perturbations have been studied in depth in recent years. The attention of researchers and control engineers has been concentrated on specific descriptions of the uncertainty structures and related computational tools. Following the notation established in Section 3.6, we denote by Δ the uncertainty affecting a linear time-invariant system $M(s)$. In particular, Δ is generally assumed to belong to the structured set

$$\mathbb{D} = \{\Delta \in \mathbb{F}^{r,c} : \Delta = \text{bdiag}(q_1 I_{m_1}, \ldots, q_\ell I_{m_\ell}, \Delta_1, \ldots, \Delta_b)\}$$

where $q_1, \ldots, q_\ell$ represent real or complex uncertain parameters, possibly repeated with multiplicity $m_1, \ldots, m_\ell$, respectively, and $\Delta_1, \ldots, \Delta_b$ represent full blocks of appropriate dimensions. The structured matrix Δ is assumed to be bounded by a quantity ρ. That is, Δ belongs to the structured norm bounded set

$$\mathcal{B}_{\mathbb{D}}(\rho) \doteq \{\Delta \in \mathbb{D} : \|q\|_p \leq \rho, \bar{\sigma}(\Delta_i) \leq \rho, i = 1, \ldots, b\}.$$

Then, we consider the family of uncertain systems obtained when the uncertainty Δ varies over $\mathcal{B}_{\mathbb{D}}(\rho)$ and we say that a certain property, e.g. stability or performance, is *robustly satisfied* if it is satisfied for all members of the family. As discussed in Section 3.7, the largest value of ρ so that stability or performance holds for all Δ is called the robustness margin or stability radius

$$r_{\mathbb{D}}(M) = \sup\{\rho : \text{stability (or performance) holds for all } \Delta \in \mathcal{B}_{\mathbb{D}}(\rho)\}.$$

The main objective of robustness analysis is to develop efficient algorithms for computing $r_{\mathbb{D}}(M)$ for various uncertainty structures $\mathbb{D}$. In Chapter 4, we presented some classical results for special classes of $\mathbb{D}$. In general, however, the problem of computing $r_{\mathbb{D}}(M)$ is known to be difficult from the computational point of view. This issue is addressed in the next section.

5.1 Computational complexity

The first critical issue of the classical robustness paradigm is computational complexity. In particular, various robust control problems have been shown to fall into the category of the so-called "intractable" problems, which are practically unsolvable if the number of variables becomes sufficiently large. These problems are generally denoted as *NP-hard*. In this section, we present introductory material on computational complexity and then discuss some robust control problems which fall into the category of NP-hard problems. The reader interested in more advanced material regarding decidability concepts and NP-completeness may refer to standard references such as [4, 118], and for specific results on continuous computational complexity to [270]. The material presented here is based on the overview [42] on computational complexity results in systems and control.

5.1.1 Decidable and undecidable problems

First, we study decision problems, i.e. problems for which the answer is either "yes" or "no." An example of a decision problem is checking whether a given matrix with integer entries is nonsingular. This question can be answered, for instance, by computing the determinant of the matrix and verifying if it is zero or not. This nonsingularity problem is known to be *decidable*, that is there exists an algorithm which always halts with the right "yes/no" answer. Unfortunately, there are many *undecidable* problems for which there is no algorithm that always halts with the right answer. We now provide an example of a decidable problem, see for instance [42].

Example 5.1 (Decidable problem).
    Consider a set of m multivariate polynomials $p_1(x), \ldots, p_m(x)$ with rational coefficients in n real variables $x = [x_1 \cdots x_n]^T$. The problem is to decide whether a solution x exists satisfying m polynomial equalities and inequalities of the form

$$
\begin{aligned}
p_i(x_1, \ldots, x_n) &= 0 & i &= 1, \ldots, m_i; \\
p_k(x_1, \ldots, x_n) &> 0 & k &= 1, \ldots, m_k; \\
p_\ell(x_1, \ldots, x_n) &\geq 0 & \ell &= 1, \ldots, m_\ell
\end{aligned}
\tag{5.1}
$$

where $m_i + m_k + m_\ell = m$. This problem is decidable.   ⋆

A variation on this example is the following: to decide if there exist $x_1, \ldots, x_q \in \mathbb{R}$ such that (5.1) is satisfied for all $x_{q+1}, \ldots, x_n \in \mathbb{R}$. This problem is also decidable since it can be reduced to the previous one with the so-called Tarski–Seidenberg *quantifier elimination* (QE) method, see [242, 261]. On the other hand, there are many problems which are undecidable. Probably, the most famous one is the Hilbert's tenth problem on Diophantine equations, which is now explained.

Example 5.2 (Undecidable problem).
Given an integer coefficient polynomial $p(x)$ in the variables $x_1, \ldots, x_n$, the goal is to decide if there exists an integer solution. This problem has been shown to be undecidable in [183], see also [77] for an elementary exposition.

$\star$

5.1.2 Time complexity

Assuming that an algorithm halts, we define its *running time* as the sum of the "costs" of each instruction. In the so-called *random access machine* (RAM) model, each arithmetic operation involves a single instruction which is assumed to have a unit cost. More realistically, in the *bit* model, the cost of an arithmetic operation is given by the sum of the bit length of the integers involved in the operation, i.e. the running time of the algorithm depends on the size of the problem instance.

Since the running time may be different for different instances of the same size, we define the running time $T(n)$ as the worst-case running time over all instances of size n. Formally, we say that an algorithm runs in *polynomial time* if there exists an integer k such that its worst-case running time is[1]

$$T(n) = O(n^k).$$

We define P as the class of all decision problems having polynomial-time algorithms. In practice, the class P consists of all problems that are efficiently solvable. Notice that the notion of time complexity is associated with a specific algorithm and not with the problem itself. In other words, for the same problem, algorithms with different complexity may be derived.

An alternative definition of polynomial-time algorithms is related to the notion of "average" running time. An interesting example in this direction is the simplex method for solving linear programming problems. The complexity of the simplex method has been shown to be not polynomial-time in the worst case. However, it is known to have polynomial average running time. We will not discuss average complexity issues further, but we refer to [240]. An example of a problem which is solvable in polynomial time is stability of a continuous-time system, which is presented next.

Example 5.3 (Polynomial-time test for Hurwitz stability).
Given a monic polynomial of order n

$$p(s) = a_0 + a_1 s + a_2 s^2 + \cdots + s^n$$

we would like to ascertain if $p(s)$ is Hurwitz. This question can be easily answered using the Routh table. In particular, in [210] it is shown that the

---

[1] The notation $O(\cdot)$ means the following: for functions $f, g : \mathbb{R}^+ \to \mathbb{R}^+$, we write $f(n) = O(g(n))$ if there exist positive numbers n_0 and c such that $f(n) \leq cg(n)$ for all $n \geq n_0$.

number of arithmetic operations (additions, multiplications and divisions) required by the Routh test is given by

$$\frac{n^2 - 1}{4} \text{ for } n \text{ odd and } \frac{n^2}{4} \text{ for } n \text{ even.}$$

Therefore, we conclude that Hurwitz stability has a polynomial-time solution and, moreover, that the number of operations required is $O(n^2)$. In [210], a detailed complexity analysis of tabular and determinant methods for continuous and discrete-time stability is also reported.   ⋆

5.1.3 NP-completeness and NP-hardness

Unfortunately, for many decidable problems of interest, no polynomial-time algorithm is known. Many of these problems belong to a specific class of *nondeterministic polynomial time*, denoted as NP. A decision problem is said to belong to NP if the validity of a "yes" instance can be verified in polynomial time. Clearly, the class NP includes the class P. An example of a problem in the class NP is the so-called *satisfiability* problem (SAT) recalled next.

Example 5.4 (SAT problem).
       Consider the case of n binary variables $x_1, \ldots, x_n \in \{0, 1\}$ and a set of m linear equality and (strict) inequality constraints

$$\begin{aligned} c_i(x_1, \ldots, x_n) &= 0 \quad i = 1, \ldots, m_i; \\ c_k(x_1, \ldots, x_n) &> 0 \quad k = 1, \ldots, m_k \end{aligned} \tag{5.2}$$

with $m_i + m_k = m$. Notice that the sums and products in the constraints (5.2) can be interpreted as Boolean operations. The problem is to determine if a binary solution exists. Clearly, if we have a solution, then the "yes" instance can be verified in polynomial time, since it suffices to check if the conditions (5.2) are satisfied.   ⋆

It has been shown, [71], that SAT is the hardest problem in the class NP. That is, every problem in the class NP can be reduced to SAT in polynomial time. More precisely, let $\mathcal{R}$ be any problem in NP. Then, given an instance $\mathcal{I}$ of $\mathcal{R}$, there exists a polynomial-time algorithm which provides an "equivalent" instance $\mathcal{I}'$ of SAT. The "equivalence" statement means that $\mathcal{I}$ is a "yes" instance of $\mathcal{R}$ if and only if $\mathcal{I}'$ is a "yes" instance of SAT. This means that, if a polynomial-time algorithm for SAT were found then every problem in NP would be solvable in polynomial time. For this reason, the SAT problem belongs to the class of *NP-complete* problems. In words, a problem is NP-complete if it is at least as difficult as any other problem in NP. NP-completeness has been established for many problems, see e.g. [118].
       To elaborate further, suppose that a polynomial-time algorithm for SAT is known. In this case, any problem in NP could be solved in polynomial time,

by first reducing it to SAT and then using the polynomial-time algorithm for SAT. The immediate consequence of this fact would be that the classes NP and P coincide. Unfortunately, it is not known if a polynomial-time algorithm for SAT exists and, therefore, it is not known if P=NP, even though it is widely believed that P $\neq$ NP. The question of whether P is equal to NP is actually one of the main outstanding problems in theoretical computer science.

On the other hand, if a problem $\mathcal{R}$ is at least as hard as SAT (i.e. the SAT problem can be mapped into $\mathcal{R}$ in polynomial time), then we say that $\mathcal{R}$ is *NP-hard*. Therefore, a decision problem is NP-complete if and only if it is NP-hard and belongs to NP. In other words, if we could find an algorithm which runs in polynomial time for solving an NP-hard problem, we would have a polynomial-time solution for SAT. For this reason, NP-hardness is generally interpreted as "the problem is intractable."

We notice that there exist classes of problems which are harder than NP. For instance, consider the class EXP $\supset$ NP $\supset$ P of problems that can be solved in exponential time, i.e. with $O(2^{n^k})$ operations, for some integer k, where n is the problem size. Then, the hardest problems in this class are called EXP-complete. These problems are provably exponential time, i.e. P$\neq$EXP, see [213].

In recent years, a number of systems and control problems have been shown to be NP-hard. In the next subsection, we describe some of them.

5.1.4 Some NP-hard problems in systems and control

In Section 3.7, we defined the structured singular value $\mu_{\mathbb{D}}$ of a matrix $M \in \mathbb{F}^{c,r}$ with respect to the uncertainty structure $\mathbb{D}$

$$\mu_{\mathbb{D}}(M) = \frac{1}{\min\{\bar{\sigma}(\Delta) : \det(I - M\Delta) = 0, \Delta \in \mathbb{D}\}}$$

unless no $\Delta \in \mathbb{D}$ makes $I - M\Delta$ singular, in which case $\mu_{\mathbb{D}}(M) = 0$. As previously discussed, the computation of $\mu_{\mathbb{D}}$ for general uncertainty structures is a difficult problem. More precisely, it has been shown that the problem of deciding whether $\mu_{\mathbb{D}}(M) \geq 1$ is NP-hard in the following cases:

1. *Real μ problem.* M is a real matrix and

   $$\mathbb{D} = \{\Delta : \Delta = \mathrm{bdiag}(q_1 I_{m_1}, \ldots, q_\ell I_{m_\ell}), q_i \in \mathbb{R}\};$$

2. *Mixed μ problem.* M is a complex matrix and

   $$\mathbb{D} = \{\Delta : \Delta = \mathrm{bdiag}(q_1 I_{m_1}, \ldots, q_\ell I_{m_\ell}), q_i \in \mathbb{F}\};$$

3. *Purely complex μ problem.* M is a complex matrix and

   $$\mathbb{D} = \{\Delta : \Delta = \mathrm{bdiag}(q_1 I_{m_1}, \ldots, q_\ell I_{m_\ell}), q_i \in \mathbb{C}\}.$$

The first two results are proven in [50], and the third is stated in [268]. Subsequent specific results regarding the complexity of approximating μ are not discussed here but are reported in [42].

We now turn our attention to the class of $n \times n$ interval matrices defined in (3.46), where each entry $[A]_{i,k}$ of the matrix is bounded in the interval $[a_{ik}^-, a_{ik}^+]$, for all $i, k = 1, \ldots, n$. For this class, we are interested in establishing the complexity of various problems, including Hurwitz stability and nonsingularity. Clearly, stability can be detected by studying the characteristic polynomial. In the special case of interval matrices in controllability (observability) canonical form, with perturbations entering only in the last column (row), the characteristic polynomial is an interval polynomial with coefficients lying in independent intervals. Then, stability can be easily established through the Kharitonov theorem, see Section 3.8, which requires checking the stability of four specific polynomials. As discussed in Example 5.3, this test can be performed in $O(n^2)$ operations. However, similar results cannot be derived for stability of general interval matrices, and this problem is NP-hard. Similar negative complexity results have been established for different problems. More precisely, for the class of interval matrices, the following problems have been shown to be NP-hard:

1. *Stability.* Decide if all matrices in this class are Hurwitz;

2. *Nonsingularity.* Decide if all matrices in this class are nonsingular;

3. *Largest singular value.* Decide if all matrices in this class have spectral norm less than one;

4. *Positive definiteness.* Decide if all symmetric matrices in this class are positive definite.

These results have been established independently in [196, 216]. Related results for the nonsingularity problem are given in [74, 82].

Next, we turn our attention to the counterpart of stability of a family of matrices, namely the "existence" problem. It has been shown that establishing if there exists a Hurwitz matrix in the class of interval matrices is NP-hard. This problem is closely related to the well-known static output feedback problem: given state space matrices A, B and C, we are interested in determining if there exists a static feedback K such that $A + BKC$ is Hurwitz. Static output feedback has been shown to be NP-hard when lower and upper bounds on the entries of K are given, see e.g. [42]. This problem has also been reformulated in terms of decision methods, see [10]. More generally, reformulations of "difficult" control problems in terms of multivariate polynomial inequalities for the use of quantifier elimination methods have been addressed in various papers, see e.g. [89, 90]. However, such reformulations do not lead to a simplification, since QE problems have exponential complexity.

In recent years several computational improvements, aiming at a reduction of the number of operations, have been presented, see e.g. [44].

To conclude this section on complexity in systems and control, we mention that branch-and-bound techniques are often used to solve problems which are apparently intractable, see e.g. [20, 201]. These techniques seem to work reasonably well in practice, because the algorithms may run in polynomial time "on average", even though they are exponential time in the worst-case.

5.2 Conservatism of robustness margin

For real parametric uncertainty entering affinely into a control system, it is well known that the robustness margin can be computed exactly (modulo round-off errors). However, in real-world problems, the plant may be affected by nonlinear uncertainty. In many cases, this nonlinear uncertainty can be embedded into an affine structure by replacing the original family by a "larger" one. This process has the advantage of handling more general robustness problems, but it has the obvious drawback of giving only an approximate but guaranteed solution. Clearly, the quality of the approximation depends on the specific problem under consideration. This issue is analyzed in the next example.

Example 5.5 (Parameter overbounding and relaxation).
    To illustrate the overbounding methodology, consider the DC electric motor of Example 3.4 with two uncertain parameters $q_1 = R$ and $q_2 = J$. In this case, the closed-loop polynomial has the following multiaffine form

$$p(s,q) = K_m + (K_m^2 + Bq_1)s + (q_1q_2 + BL)s^2 + Lq_2s^3.$$

To overbound $p(s,q)$ with affine uncertainty, we set $q_3 = q_1q_2$. Given bounds $[q_1^-, q_1^+]$ and $[q_2^-, q_2^+]$ for q_1 and q_2, the range of variation $[q_3^-, q_3^+]$ for q_3 can be immediately computed

$$q_3^- = \min\{q_1^-q_2^-, q_1^-q_2^+, q_1^+q_2^-, q_1^+q_2^+\};$$
$$q_3^+ = \max\{q_1^-q_2^-, q_1^-q_2^+, q_1^+q_2^-, q_1^+q_2^+\}.$$

Clearly, the new uncertain polynomial

$$\widetilde{p}(s,q) = K_m + (K_m^2 + Bq_1)s + (q_3 + BL)s^2 + Lq_2s^3$$

has three uncertain parameters q_1, q_2, q_3 entering affinely into $\widetilde{p}(s,q)$. Since $q_3 = q_1q_2$, this new parameter is not independent of q_1 and q_2, and not all values of $[q_3^-, q_3^+]$ are physically realizable. However, since we neglect this constraint and assume that the parameters q_i are independent, this relaxation technique leads to an overbounding of $p(s,q)$ with $\widetilde{p}(s,q)$. In other words, if stability is guaranteed for $\widetilde{p}(s,q)$, then it is also guaranteed for $p(s,q)$, but

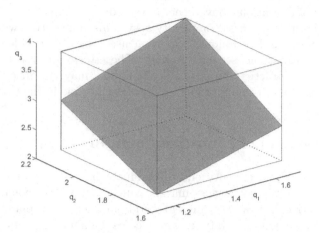

Fig. 5.1. Overbounding of the multiaffine uncertainty of Example 5.5.

the converse may not not true. Figure 5.1 illustrates the overbounding procedure for $q_1^- = 1.2, q_1^+ = 1.7, q_2^- = 1.7, q_2^+ = 2.2, q_3^- = 2.04$ and $q_3^+ = 3.74$.
In this figure, the gray square represents the set of all physically realizable
uncertainties, and the three-dimensional box shows its overbounding. The
possible conservatism of the approach is evident. $\star$

To generalize the discussion in this example, we restate the overbounding
problem for parametric uncertainty as follows: given an uncertain polynomial
$p(s, q)$ with nonlinear uncertainty structure and a hyperrectangle $\mathcal{B}_q$, find a
new uncertain polynomial $\widetilde{p}(s, q)$ with affine uncertainty structure and a new
hyperrectangle $\widetilde{\mathcal{B}}_q$. In general, since this procedure is not unique, there is no
systematic methodology to construct an "optimal" overbounding. The most
natural way may be to compute an interval overbounding for each coefficient
of the polynomial, i.e. an interval polynomial overbounding. To illustrate,
letting $a_i(q)$, $i = 0, 1, \ldots, n - 1$, denote the coefficients of $p(s, q)$, the lower
and upper bounds are given by

$$a_i^- = \min_{q \in \mathcal{B}_q} a_i(q); \quad a_i^+ = \max_{q \in \mathcal{B}_q} a_i(q).$$

If $a_i(q)$ are affine or multiaffine functions, these minimizations and maximiza-
tions can be easily performed. In fact, we have that

$$a_i^- = \min_{q \in \mathcal{B}_q} a_i(q) = \min_{k=1,\ldots,2^\ell} a_i(q^k);$$

$$a_i^+ = \max_{q \in \mathcal{B}_q} a_i(q) = \max_{k=1,\ldots,2^\ell} a_i(q^k)$$

where $q^1, \ldots, q^{2^\ell}$ are the vertices of the hyperrectangle $\mathcal{B}_q$. However, this is
not true for more general uncertainty structures, for example if $a_i(q)$ are poly-

nomial functions of q. In this case, a tight interval polynomial overbounding may be difficult to construct.

A technique frequently used to avoid the possible conservatism introduced by overbounding and relaxation techniques is the so-called *gridding approach*. Suppose we wish to establish if a polynomial $p(s, q)$ is Hurwitz when its coefficients are, for example, polynomial functions of $q \in \mathcal{B}_q$. In this case, we can take N equispaced points

$$q_i = q_i^- + k\frac{q_i^+ - q_i^-}{N - 1}, \quad k = 0, 1, \ldots, N - 1$$

for each parameter q_i, $i = 1, \ldots, \ell$, and check stability for every point in the grid. This procedure gives an "indication" of the robust stability of the system, which becomes more and more precise as the number of grid point increases. Clearly, the answer obtained by checking stability at the grid points does not provide any guarantee for the entire set $\mathcal{B}_q$. More importantly, the total number of grid points is $N_{\mathrm{grid}} = N^\ell$, which is exponential. This exponential growth is often referred to as the *curse of dimensionality*.[2] This fact clearly illustrates that the two issues of computational complexity and conservatism are indeed closely related and represent one of the main trade-offs in robustness analysis and design.

A similar situation arises in uncertain systems with nonparametric uncertainty described in M–Δ form. For example, for systems affected by more than one full real block or more than three full complex blocks, the computation of $\mu_\mathbb{D}$ (and therefore of the robustness margin) can be generally performed only at the expense of some conservatism. In fact, in these cases, only upper and lower bounds of $\mu_\mathbb{D}$ can be computed, see Section 3.7.3, but it may be difficult to estimate the degree of conservatism introduced. This issue is further addressed in the next example.

Example 5.6 (Conservatism in robustness margin computation).
We consider an example concerning a five-mass spring–damper model with parametric uncertainty on the stiffness and damping parameters and dynamic uncertainty due to unmodeled dynamics analyzed in Section 19.2.

This flexible structure may be modeled as an M–Δ configuration, with $M(s) = C(sI - A)^{-1}B$. The matrix Δ is structured and consists of two repeated real parameters q_1, q_2 and one transfer matrix $\Delta_1 \in \mathcal{RH}_\infty^{4,4}$, i.e.

$$\Delta \in \widetilde{\mathbb{D}} = \{\Delta : \Delta = \mathrm{bdiag}(q_1 I_5, q_2 I_5, \Delta_1)\}.$$

For this M–Δ system, $\mu_\mathbb{D}(M(j\omega))$ cannot be computed exactly, and the derivation of its upper and lower bounds requires the use of branch-and-bound techniques, which are computationally very expensive. Another approach, similar to the overbounding method previously studied, is to neglect

---

[2] The frequently used terminology "curse of dimensionality" has been coined by Bellman in 1957 in the classical book on dynamic programming [33].

the structure of Δ "lumping together" the uncertainty in one single block. In this case, an upper bound of the stability radius is immediately given by the small gain theorem studied in Section 3.7, which requires the computation of the $\mathcal{H}_\infty$ norm of $M(s)$. That is, we obtain the bound

$$r_\mathbb{D}(M(s)) \geq r_{\mathrm{LTI}}(M(s)) = \frac{1}{\|M(s)\|_\infty}.$$

$\star$

More generally, a wide class of robust synthesis problems can be recast as *bilinear matrix inequalities* (BMIs), see e.g. [121], but this reformulation does not solve the problem of efficiently deriving a controller. In fact, algorithms for finding global optimal solutions to BMI problems, based for instance on branch-and-bound techniques, have in general very high computational complexity. On the other hand, also in this case, methods based on relaxations and overbounding have been proposed, at the expense of introducing a certain degree of conservatism.

5.3 Discontinuity of robustness margin

There is another drawback inherent to the robustness paradigm that may arise even in the simple case when affine parametric uncertainty enters into the system. This is due to the fact that the robustness margin need not be a continuous function of the problem data. To show this phenomenon, we revisit a classical example [26] regarding the robustness margin of a system affected by linear parametric uncertainty.

Example 5.7 (Discontinuity of robustness margin).
    Consider a SISO plant of the form

$$G_\kappa(s, q) = \frac{N_\kappa(s, q)}{D_\kappa(s, q)}$$

where the numerator $N_\kappa(s, q)$ and denominator $D_\kappa(s, q)$ polynomials are given by

$$N_\kappa(s, q) = 4\kappa^2 + 10\kappa^2 q_1;$$
$$D_\kappa(s, q) = \kappa^2 + (20 + 8\kappa + 20\kappa q_1 - 20q_2)s + (44 + 2\kappa + 10q_1 - 40q_2)s^2$$
$$+ (20 - 20q_2)s^3 + s^4.$$

The uncertain parameters vector $q = [q_1 \ q_2]^T$ varies within the hyperrectangle of radius ρ

$$\mathcal{B}_q(\rho) = \{q \in \mathbb{R}^2 : \|q\|_\infty \leq \rho\}.$$

Taking a unit feedback $K(s) = 1$, we study robustness of the closed-loop polynomial

$$p_\kappa(s,q) = N_\kappa(s,q) + D_\kappa(s,q)$$

for "large" variations of the parameters q_1 and q_2 within $\mathcal{B}_q(\rho)$ and infinitesimal perturbations of κ. From this point of view, κ is considered as *problem data*, and not an uncertain parameter. Then, this example studies continuity properties of the robustness margin $r_q(\kappa)$ for small variations of κ.

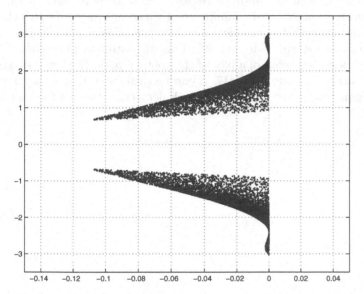

Fig. 5.2. Subset of the roots of $p_{\bar{\kappa}}(s,q)$, $q \in \mathcal{B}_q(\rho)$, $\rho = 0.417$, for Example 5.7.

More precisely, we study two cases. First, we consider the case where κ is fixed and set to $\kappa = \bar{\kappa} = 3 + 2\sqrt{2}$. Second, we consider a small perturbation $\epsilon > 0$ around κ, i.e.

$$\kappa(\epsilon) = \bar{\kappa} - \epsilon = 3 + 2\sqrt{2} - \epsilon.$$

Clearly, $\kappa(\epsilon) \to \bar{\kappa}$ as $\epsilon \to 0^+$. Then, we derive the closed-loop polynomials corresponding to these two cases, obtaining

$$p_{\bar{\kappa}}(s,q) = (5\bar{\kappa}^2 + 10\bar{\kappa}^2 q_1) + (20 + 8\bar{\kappa} + 20\bar{\kappa} q_1 - 20q_2)s \qquad (5.3)$$
$$+ (44 + 2\bar{\kappa} + 10q_1 - 40q_2)s^2 + (20 - 20q_2)s^3 + s^4$$

and

$$p_{\kappa(\epsilon)}(s,q) = p_{\bar{\kappa}}(s,q) - 2\bar{\kappa}\epsilon(2 + 5q_1).$$

Subsequently, it can be verified that

$$0.417 \approx 1 - \frac{\bar{\kappa}}{10} = \lim_{\epsilon \to 0^+} r_q(\kappa(\epsilon)) > r_q(\bar{\kappa}) = \frac{7 - \bar{\kappa}}{5} \approx 0.234.$$

Various robustness interpretations may be given, and we redirect the interested reader to the original example in [26] and to further discussions in [24]. Here, we only mention that this example illustrates the phenomenon called the "false sense of security" of the robustness margin. That is, the closed-loop polynomial $p_{\kappa(\epsilon)}(s, q)$ is destabilized by an uncertainty approximately of size 0.234, and the robustness margin corresponding to the polynomial $p_{\bar{\kappa}}(s, q)$ is given by 0.417. In other words, infinitesimal variations around $\bar{\kappa}$ (which can be caused, for example, by numerical round-off), may lead to an overestimate of the robustness margin of about 78%. This discontinuity phenomenon can also be illustrated by means of a plot of the roots of $p_{\bar{\kappa}}(s, q)$ when q ranges in a hyperrectangle of radius $\rho = 0.417$. It can be easily seen that some roots lie on the imaginary axis, see Figure 5.2, which depicts a subset of the roots.   ⋆

6. Probabilistic Methods for Robustness

In this chapter, we introduce a *probabilistic approach* for robustness analysis and design of uncertain systems. As pointed out in Chapter 5, many pessimistic results on the complexity-theoretic barriers of classical robust control have stimulated research in the direction of finding alternative solutions. One of these solutions, which is the main objective of the research pursued in this book, is first to shift the meaning of robustness from its usual deterministic sense to a probabilistic one. In this respect, we claim that a certain system property is "almost" robustly satisfied if it holds for "most" of the instances of the uncertainty. In other words, we accept the risk of this property being violated by a set of uncertainties having small probability measure. Such systems may be viewed as being *practically robust*.

6.1 Performance function for robustness

In the robustness analysis framework discussed in Chapter 3, the main objective is to guarantee that a certain system property is attained for all uncertainties Δ bounded within a specified set $\mathcal{B}_{\mathbb{D}}(\rho)$. To this end, it is useful to define a *performance function* (which is assumed to be measurable)

$$J(\Delta) : \mathbb{D} \to \mathbb{R}$$

where $\mathbb{D}$ is the uncertainty structured set defined in (3.27), and an associated performance level γ. In general, the function $J(\Delta)$ can take into account the simultaneous attainment of various performance requirements.

In the framework of robust synthesis studied in Chapter 4, the performance function depends also on some "design" parameters $\theta \in \Theta$ (e.g. the parameters of the controller), and takes the form

$$J(\Delta, \theta) : \mathbb{D} \times \Theta \to \mathbb{R}.$$

Two examples showing specific performance functions are now given.

Example 6.1 (Robust stability).
    Consider the feedback interconnection shown in Figure 6.1, where $M(s)$ is a given transfer function and the uncertainty Δ belongs to the structured

set $\widetilde{\mathbb{D}}$ defined in (3.25). We are interested in the internal stability of this interconnection when the uncertainty $\Delta \in \widetilde{\mathbb{D}}$ is such that $\|\Delta\|_\infty < \rho$. Consider a state space realization of the transfer function $M(s)$

$$M(s) = C(sI - A)^{-1}B + D,$$

where $A \in \mathbb{R}^{n_s,n_s}$ is stable, B, C, D are real matrices of suitable dimensions and $\Delta \in \mathcal{B}_{\mathbb{D}}(\rho)$. We assume that the well-posedness condition on D holds, i.e.

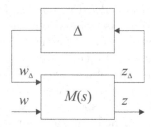

Fig. 6.1. M–Δ configuration of Example 6.1.

$(I - D\Delta)$ is nonsingular for all Δ. Then, from Lemma 3.4, internal stability of the system is equivalent to the matrix $A + B\Delta(I - D\Delta)^{-1}C$ being stable for all Δ belonging to the set $\mathbb{D}$ defined in (3.27) and with the bound $\bar{\sigma}(\Delta) < \rho$. Then, we choose the performance function $J(\Delta) : \mathbb{D} \to \mathbb{R}$ for robust stability of the M–Δ interconnection as

$$J(\Delta) = \begin{cases} 0 \text{ if } A + B\Delta(I - D\Delta)^{-1}C \text{ is stable;} \\ 1 \text{ otherwise.} \end{cases} \tag{6.1}$$

Setting the performance level for instance to $\gamma = 1/2$, robust stability is equivalent to checking if

$$J(\Delta) \leq \gamma$$

for all $\Delta \in \mathbb{D}$, $\bar{\sigma}(\Delta) < \rho$. If this check is satisfied for all Δ, we say that robust stability is guaranteed.   ⋆

Example 6.2 ($\mathcal{H}_\infty$ performance).
    Consider again the feedback interconnection in Figure 6.1. We now study the $\mathcal{H}_\infty$ norm of the transfer matrix between the disturbances w and the error z, defined in (3.29) as the upper linear fractional transformation $\mathcal{F}_u(M, \Delta)$. Given a level γ, we are interested in checking if the $\mathcal{H}_\infty$ performance is smaller than γ for all uncertainties Δ belonging to the structured bounded set $\mathcal{B}_{\mathbb{D}}(\rho)$ defined in (3.28). Assuming that $\mathcal{F}_u(M, \Delta)$ is stable and strictly proper for all Δ, we set the performance function to

$$J(\Delta) = \|\mathcal{F}_u\,(M,\Delta)\,\|_\infty.$$

Then, if $J(\Delta) \le \gamma$ for all Δ, we conclude that the performance level γ is robustly achieved.   ⋆

This example can be generalized to the robust satisfaction of a generic *control system property*, as now stated.

Problem 6.1 (Robust performance verification). *Let Δ be bounded in the set $\mathcal{B}_\mathbb{D}(\rho)$ defined in (3.28). Given a performance function $J(\Delta) : \mathbb{D} \to \mathbb{R}$ and associated level γ, check whether*

$$J(\Delta) \le \gamma$$

for all $\Delta \in \mathcal{B}_\mathbb{D}(\rho)$.

A different robustness problem is related to the computation of the so-called worst-case performance. That is, we are interested in evaluating the optimal guaranteed level of performance γ_{wc} such that $J(\Delta) \le \gamma_{\mathrm{wc}}$ for all $\Delta \in \mathcal{B}_\mathbb{D}(\rho)$. This amounts to evaluating the supremum of $J(\Delta)$ when Δ ranges in the set $\mathcal{B}_\mathbb{D}(\rho)$. This is shown in the next example.

Example 6.3 (Worst-case $\mathcal{H}_\infty$ performance).
We now revisit Example 6.2. Suppose we are interested in determining the smallest level γ such that the $\mathcal{H}_\infty$ norm of the transfer matrix between the disturbances w and the error z is less than γ. As in Example 6.2, we set the performance function to

$$J(\Delta) = \|\mathcal{F}_u\,(M,\Delta)\,\|_\infty.$$

Then, the worst-case $\mathcal{H}_\infty$ performance is given by the supremum of $J(\Delta)$ computed with respect to Δ.   ⋆

This example is now generalized to formally state the worst-case robust performance problem.

Problem 6.2 (Worst-case robust performance). *Let Δ be bounded in the set $\mathcal{B}_\mathbb{D}(\rho)$ defined in (3.28). Given a performance function $J(\Delta) : \mathbb{D} \to \mathbb{R}$, compute*

$$\gamma_{\mathrm{wc}} \doteq \sup_{\Delta \in \mathcal{B}_\mathbb{D}(\rho)} J(\Delta).$$

A related problem, also discussed in Chapter 3, is to compute the largest uncertainty radius $r_\mathbb{D} = r_\mathbb{D}(M)$ such that robust performance is guaranteed for all uncertainties $\Delta \in \mathbb{D}$ with $\|\Delta\| < r_\mathbb{D}$.

Example 6.4 (Robust stability radius).
    Consider the setting of Example 6.1 regarding robust stability of the $M-\Delta$ interconnection. Suppose we are interested in evaluating the maximum radius $r_{\mathbb{D}}$ such that the system is robustly stable for all $\Delta \in \mathcal{B}_{\mathbb{D}}(r_{\mathbb{D}})$. From Theorem 3.5, it follows that

$$r_{\mathbb{D}}(M) = \frac{1}{\sup_{\omega \in \mathbb{R}} \mu_{\mathbb{D}}(M(j\omega))}$$

where $\mu_{\mathbb{D}}(M(j\omega))$ is the structured singular value defined in Section 3.7.2. Choosing the performance function $J(\Delta)$ as in (6.1) and setting $\gamma = 1/2$, this problem can be rewritten as

$$r_{\mathbb{D}} = \sup\{\rho : J(\Delta) \leq \gamma \text{ for all } \Delta \in \mathcal{B}_{\mathbb{D}}(\rho)\}$$

or equivalently as

$$r_{\mathbb{D}} = \inf\{\rho : J(\Delta) > \gamma \text{ for some } \Delta \in \mathcal{B}_{\mathbb{D}}(\rho)\}.$$

We recall that the computation of $\mu_{\mathbb{D}}(M(j\omega))$, and consequently $r_{\mathbb{D}}(M)$, is NP-hard and only upper and lower bounds of it are in general available. $\star$

    In the following, we formally define the robust performance radius, which is a generalization of Definition 3.6.

Problem 6.3 (Robust performance radius). *Given a performance function $J(\Delta) : \mathbb{D} \to \mathbb{R}$ and associated level γ, compute*

$$r_{\mathbb{D}} \doteq \inf\{\rho : J(\Delta) > \gamma \text{ for some } \Delta \in \mathcal{B}_{\mathbb{D}}(\rho)\}$$

where the set $\mathcal{B}_{\mathbb{D}}(\rho)$ is defined in (3.28).

Remark 6.1 (Relationships between the robustness problems). The three robustness problems previously stated are closely related. In particular, if Problem 6.1 is solvable for fixed γ and ρ, then Problem 6.2 can be solved via a one-dimensional γ-iteration and Problem 6.3 can be solved via a one-dimensional ρ-iteration. $\triangleleft$

6.2 The good and the bad sets

We now study the performance problems previously discussed by introducing two sets, denoted as the *good set* and the *bad set*. These are subsets of $\mathcal{B}_{\mathbb{D}}(\rho)$ and represent, respectively, the collection of all Δ which satisfy or violate the control system property under attention. These sets are constructed so that their union coincides with the uncertainty set $\mathcal{B}_{\mathbb{D}}(\rho)$ and their intersection is empty. Formally, we define

$$\mathcal{B}_G \doteq \{\Delta \in \mathcal{B}_\mathbb{D}(\rho) : J(\Delta) \leq \gamma\};$$
$$\mathcal{B}_B \doteq \{\Delta \in \mathcal{B}_\mathbb{D}(\rho) : J(\Delta) > \gamma\}. \qquad (6.2)$$

In the case of purely parametric uncertainty, we usually consider the uncertainty set $\mathcal{B}_q(\rho)$ defined in (3.45) instead of $\mathcal{B}_\mathbb{D}(\rho)$. Hence, the sets $\mathcal{B}_G$ and $\mathcal{B}_B$ take the form

$$\mathcal{B}_G = \{\Delta \in \mathcal{B}_q(\rho) : J(\Delta) \leq \gamma\};$$
$$\mathcal{B}_B = \{\Delta \in \mathcal{B}_q(\rho) : J(\Delta) > \gamma\}.$$

Robustness of a control system is therefore equivalent to the case when the good set coincides with $\mathcal{B}_\mathbb{D}(\rho)$ and the bad set is empty. We now present two examples showing the computation of $\mathcal{B}_G$ and $\mathcal{B}_B$.

Example 6.5 (Continuous-time stability of a fourth-order system).
   In this example, we consider stability of a fourth-order continuous-time system affected by a vector of two real uncertain parameters. In particular, we study a closed-loop monic polynomial of the form

$$p(s,q) = 2 + q_1 + q_2 + (5 + q_1 + 3q_2)s + (6 + 3q_2)s^2 + (4 + q_2)s^3 + s^4.$$

In this case the structured set $\mathbb{D}$ coincides with $\mathbb{R}^2$ and the bounding set is the hyperrectangle

$$\mathcal{B}_q(\rho) = \{q \in \mathbb{R}^2 : \|q\|_\infty \leq \rho\}$$

with $\rho = 1.8$. Then, we introduce the performance function

$$J(q) = \begin{cases} 0 \text{ if } p(s,q) \text{ is Hurwitz}; \\ 1 \text{ otherwise} \end{cases} \qquad (6.3)$$

and set $\gamma = 1/2$. The good set coincides with the set of Hurwitz stable polynomials

$$\mathcal{B}_G = \{q \in \mathcal{B}_q(\rho) : p(s,q) \neq 0 \text{ for all } \operatorname{Re}(s) \geq 0\}.$$

In order to obtain a closed-form characterization of the set $\mathcal{B}_G$, we construct the Routh table, shown in Table 6.1.

Table 6.1. Routh table for Example 6.5.

s^4	1	$6 + 3q_2$	$2 + q_1 + q_2$
s^3	$4 + q_2$	$5 + q_1 + 3q_2$	
s^2	$a_1(q)/(4 + q_2)$	$2 + q_1 + q_2$	
s	$a_2(q)/a_1(q)$		
1	$2 + q_1 + q_2$		

Table 6.1 leads to the system of polynomial inequalities in q_1 and q_2

$$\begin{cases} 4 + q_2 > 0; \\ a_1(q) = 19 - q_1 + 15q_2 + 3q_2^2 > 0; \\ a_2(q) = (9 + q_1 + 4q_2)(7 - q_1 + 8q_2 + 2q_2^2) > 0; \\ 2 + q_1 + q_2 > 0. \end{cases}$$

These inequalities lead to the curves delimiting the Hurtwitz region in parameter space shown in Figure 6.2.   ⋆

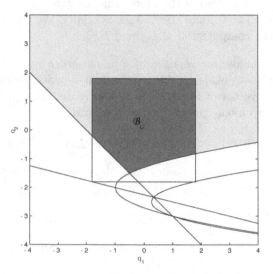

Fig. 6.2. Set of stable polynomials $\mathcal{B}_G$ for Example 6.5. The light-gray region indicates the Hurwitz region in parameter space.

Example 6.6 (Robust $\mathcal{H}_2$ performance).
    Consider a continuous-time system expressed in state space form

$$\dot{x} = A(q)x + Bw;$$
$$y = Cx$$

with

$$A(q) = \begin{bmatrix} -2 + q_1 & q_1 q_2 \\ 0 & -4 + q_2 \end{bmatrix}, \quad B = \begin{bmatrix} 1 \\ 1 \end{bmatrix}, \quad C = \begin{bmatrix} 1 & 0 \end{bmatrix}.$$

The uncertainty is confined in the hyperrectangle

$$\mathcal{B}_q(\rho) = \{q \in \mathbb{R}^2 : \|q\|_\infty \leq \rho\}$$

with $\rho = 1$. We are interested in checking if the squared $\mathcal{H}_2$ norm of the transfer function $G(s, q) = C(sI - A(q))^{-1}B$ between the disturbance w and the output z is less then $\gamma = 0.3$. Since the matrix $A(q)$ is upper triangular, it is easy to check that the system is stable for all values of the uncertainty, and therefore $G(s, q) \in \mathcal{H}_2$ for all $q \in \mathcal{B}_q(\rho)$. Then, letting $J(q) = \|G(s, q)\|_2^2$

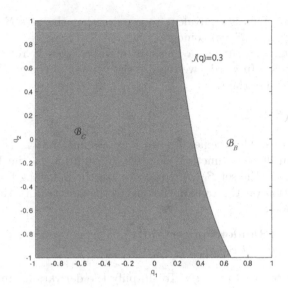

Fig. 6.3. Good and bad sets for Example 6.6.

and $\gamma = 0.3$, we define the good set as

$$\mathcal{B}_G = \{q \in \mathcal{B}_q(\rho) : J(q) \le \gamma\} = \{q \in \mathcal{B}_q(\rho) : \|G(s, q)\|_2^2 \le 0.3\}.$$

We can compute $\|G(s, q)\|_2^2 = \operatorname{Tr} C W_c C^T$, where the controllability Gramian $W_c \succeq 0$ is the solution of the Lyapunov equation

$$A(q)W_c + W_c A^T(q) + BB^T = 0.$$

For this simple case, straightforward but lengthy manipulations lead to

$$J(q) = -\frac{1}{2} \frac{q_1^2 q_2^2 - 2q_1 q_2(-4 + q_2) + (-2 + q_1)(-4 + q_2) + (-4 + q_2)^2}{(-4 + q_2)(-2 + q_1)^2 + (-2 + q_1)(-4 + q_2)^2}. \quad (6.4)$$

The level curve $J(q) = 0.3$ and the good set are depicted in Figure 6.3. $\star$

6.3 Probabilistic robustness analysis

In classical robustness analysis, one of the main objectives, discussed in Problem 6.1, is to check if a given system property is satisfied for all possible values of the uncertainty. That is, for a given performance level γ, we would like to guarantee that

$$J(\Delta) \leq \gamma$$

for all $\Delta \in \mathcal{B}_{\mathbb{D}}(\rho)$. This is equivalent to require that the sets $\mathcal{B}_G = \{\Delta \in \mathcal{B}_{\mathbb{D}}(\rho) : J(\Delta) \leq \gamma\}$ and $\mathcal{B}_{\mathbb{D}}(\rho)$ coincide.

In the probabilistic setting, a measure of robustness is related to the volume of the set $\mathcal{B}_G$. In words, we require the volume of the good set to be "sufficiently large", i.e. the ratio

$$\text{Vol}(\mathcal{B}_G)/\text{Vol}(\mathcal{B}_{\mathbb{D}}(\rho)) \tag{6.5}$$

to be "close" to one. More generally, we may assume that Δ is a random matrix with given density function. That is, in addition to the knowledge that Δ is bounded in the set $\mathcal{B}_{\mathbb{D}}(\rho)$, we also make the following fundamental standing assumption on the random nature of the uncertainty that is used throughout the book.

Assumption 6.1 (Random uncertainty). *The uncertainty $\boldsymbol{\Delta}$ is a random matrix with support $\mathcal{B}_{\mathbb{D}}(\rho)$.*

Remark 6.2 (Existence of the pdf). To simplify the derivations and improve readability, the results in this book are derived under the hypothesis that the probability density $f_{\boldsymbol{\Delta}}(\Delta)$ of $\boldsymbol{\Delta}$ exists. However, most of the results hold under the less restrictive assumption that only the distribution of $\boldsymbol{\Delta}$ is given.

$\lhd$

We recall that the performance function $J(\Delta) : \mathcal{B}_{\mathbb{D}}(\rho) \to \mathbb{R}$ is assumed to be measurable over its support. With these assumptions, probabilistic robustness of a control system is stated in terms of the probability that the desired performance is satisfied. In other words, if Δ is a random matrix, then the volume in (6.5) becomes a "weighted" volume, where the weight is the given probability density function $f_{\boldsymbol{\Delta}}(\Delta)$. That is, the key quantity to be computed is the probability of performance satisfaction

$$\text{PR}\left\{J(\boldsymbol{\Delta}) \leq \gamma\right\} = \int_{\mathcal{B}_G} f_{\boldsymbol{\Delta}}(\Delta)\mathrm{d}\Delta.$$

Clearly, if $f_{\boldsymbol{\Delta}}(\Delta)$ is the uniform density over $\mathcal{B}_{\mathbb{D}}(\rho)$, then this probability is indeed the (normalized) volume of the good set

$$\text{PR}\left\{J(\boldsymbol{\Delta}) \leq \gamma\right\} = \frac{\text{Vol}(\mathcal{B}_G)}{\text{Vol}(\mathcal{B}_{\mathbb{D}}(\rho))}.$$

We are now ready to formulate the probabilistic counterpart of Problem 6.1 previously defined.

Problem 6.4 (Probabilistic performance verification). *Given a performance function $J(\Delta)$ with associated level γ and a density function $f_\Delta(\Delta)$ with support $\mathcal{B}_\mathbb{D}(\rho)$, compute the probability of performance*

$$p(\gamma) \doteq \text{PR}\{J(\mathbf{\Delta}) \le \gamma\}. \tag{6.6}$$

The probability of performance $p(\gamma)$ measures the probability that a level of performance γ is achieved when $\mathbf{\Delta} \sim f_\Delta(\Delta)$. We remark that this probability is in general difficult to compute either analytically or numerically, since it basically requires the evaluation of a multidimensional integral. However, in some special cases it can be evaluated in closed form, as shown in the following examples.

Example 6.7 (Probability of stability).
    We now revisit Example 6.5 regarding stability of a fourth-order system with closed-loop polynomial

$$p(s, q) = 2 + q_1 + q_2 + (5 + q_1 + 3q_2)s + (6 + 3q_2)s^2 + (4 + q_2)s^3 + s^4.$$

We now assume that q is a random vector with uniform distribution in the set $\mathcal{B}_q(\rho)$ with $\rho = 1.8$, i.e. $\mathbf{q} \sim \mathcal{U}_{\mathcal{B}_q(1.8)}$. Then, we set $J(q)$ as in (6.3) and $\gamma = 1/2$. In this case, the volume of the good set can be computed by integrating the equations defining the Hurwitz region derived in Example 6.5, obtaining

$$\text{Vol}(\mathcal{B}_G) = 10.026.$$

The probability of stability is then immediately given by

$$\text{PR}_\mathbf{q}\{p(s, \mathbf{q}) \text{ Hurwitz}\} = p(\gamma) = \frac{\text{Vol}(\mathcal{B}_G)}{\text{Vol}(\mathcal{B}_q(1.8))} = \frac{10.026}{12.96} = 0.7736.$$

$\star$

Example 6.8 (Probability of $\mathcal{H}_2$ performance).
    We revisit Example 6.6 regarding robust $\mathcal{H}_2$ performance of a continuous-time system expressed in state space form. We now assume that the uncertainty is random with

$$\mathbf{q} \sim \mathcal{U}_{\mathcal{B}_q}$$

and we aim at computing the probability of performance

$$p(\gamma) = \text{PR}_\mathbf{q}\{J(\mathbf{q}) \le \gamma\} = \text{PR}_\mathbf{q}\{\|G(s, \mathbf{q})\|_2^2 \le 0.3\}.$$

Using the expression of $J(q)$ obtained in (6.4), we compute the probability of performance integrating in closed form the level function $J(q) = 0.3$, obtaining the value $p(\gamma) = 0.6791.$ $\star$

Example 6.9 (Probability of stability versus guaranteed stability).

This example is due to Truxal [272], and it has been subsequently reconsidered in [2]. We study stability of a third-order continuous-time system affected by a vector of uncertainties q bounded in the set

$$\mathcal{B}_q = \{q \in \mathbb{R}^2 : 0.3 \le q_1 \le 2.5; 0 \le q_2 \le 1.7\}.$$

The closed-loop polynomial is a bilinear function on the uncertainty and is given by

$$p(s,q) = 1 + \alpha^2 + 6q_1 + 6q_2 + 2q_1q_2 + (q_1 + q_2 + 3)s + (q_1 + q_2 + 1)s^2 + s^3$$

where α varies in the interval $[0, 0.7]$. It can be easily verified that the set of unstable polynomials, the bad set, is a disk in parameter space

$$\mathcal{B}_B = \{q \in \mathcal{B}_q : (q_1 - 1)^2 + (q_2 - 1)^2 \le \alpha^2\}$$

with volume $\text{Vol}(\mathcal{B}_B) = \pi\alpha^2$. The sets $\mathcal{B}_G$ and $\mathcal{B}_B$ are displayed in Figure 6.4 for $\alpha = 0.5$.

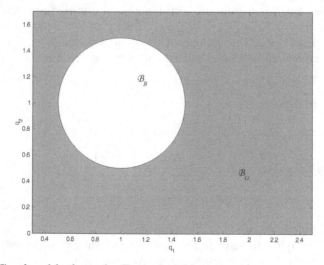

Fig. 6.4. Good and bad sets for Example 6.9.

Suppose now that the uncertainty is random with uniform density function over the set $\mathcal{B}_q$. Then, the probability of stability is

$$\text{PR}_\mathbf{q}\{p(s,\mathbf{q}) \text{ Hurwitz}\} = 1 - \frac{\pi\alpha^2}{3.74}.$$

We notice that by taking $\alpha = 0$ the set of unstable polynomials becomes a singleton centered at $q_1 = 1$ and $q_2 = 1$. In this case, the probability of stability is equal to one.   ★

Example 6.9 shows some interesting features of the probabilistic approach. In fact, taking the parameter α equal to zero, we obtain that the system is stable with probability one. Of course, the system in *not* robustly stable, even though stability is violated only in a set of measure zero. This clearly shows the differences between the two settings.

Example 6.10 (Probability of Schur stability).
In this example, we study Schur stability[1] of a discrete-time system affected by a vector q of parametric uncertainty bounded in the hyperrectangle $\mathcal{B}_q$. The closed-loop polynomial is a monic interval polynomial of order n

$$p(z, q) = q_0 + q_1 z + q_2 z^2 + \cdots + q_{n-1} z^{n-1} + z^n.$$

The set of coefficients $q \in \mathbb{R}^n$ leading to Schur stable polynomials (Schur region) is defined as

$$\mathcal{Q}_n \doteq \{q \in \mathbb{R}^n : p(z, q) \text{ Schur}\}.$$

Consider now the case when $\mathbf{q}$ is a random vector with uniform density over $\mathcal{B}_q$ and introduce the performance function

$$J(q) = \begin{cases} 0 \text{ if } p(z, q) \text{ is Schur;} \\ 1 \text{ otherwise.} \end{cases} \tag{6.7}$$

Setting $\gamma = 1/2$, the probability that the discrete-time system is stable is

$$p(\gamma) = \mathrm{PR}_{\mathbf{q}} \{J(\mathbf{q}) \leq \gamma\} = \mathrm{PR}_{\mathbf{q}} \{p(z, \mathbf{q}) \text{ Schur}\}.$$

That is, we define

$$\mathcal{B}_G = \{q \in \mathcal{B}_q : p(z, q) \text{ Schur}\} = \mathcal{B}_q \cap \mathcal{Q}_n.$$

The volume of Schur stable polynomials

$$\mathrm{Vol}(\mathcal{Q}_n) = \int_{\mathcal{Q}_n} dq$$

can be explicitly computed by the recursive formulae given in [107]

$$\mathrm{Vol}(\mathcal{Q}_{n+1}) = \frac{\mathrm{Vol}(\mathcal{Q}_n)^2}{\mathrm{Vol}(\mathcal{Q}_{n-1})} \text{ for } n \text{ odd;} \tag{6.8}$$

$$\mathrm{Vol}(\mathcal{Q}_{n+1}) = \frac{n \mathrm{Vol}(\mathcal{Q}_n) \mathrm{Vol}(\mathcal{Q}_{n-1})}{(n+1) \mathrm{Vol}(\mathcal{Q}_{n-2})} \text{ for } n \text{ even} \tag{6.9}$$

where $\mathrm{Vol}(\mathcal{Q}_1) = 2$, $\mathrm{Vol}(\mathcal{Q}_2) = 4$ and $\mathrm{Vol}(\mathcal{Q}_3) = 16/3$.

Next, we remark that a polytope in coefficient space which is guaranteed to contain the Schur region $\mathcal{Q}_n$ can be computed using the classical necessary conditions for Schur stability, see e.g. [179]

---

[1] A polynomial $p(z)$ is Schur stable if $p(z) \neq 0$ for all $|z| \geq 1$.

$$0 < p(1, q) < 2^n;$$
$$0 < (-1)^n p(-1, q) < 2^n;$$
$$|q_0| < 1.$$

(6.10)

Then, if this polytope is contained in the set $\mathcal{B}_q$, we conclude that condition $\mathcal{Q}_n \subseteq \mathcal{B}_q$ is satisfied. In this case, the good set coincides with $\mathcal{Q}_n$ and the probability of stability is immediately given in closed form as

$$p(\gamma) = \frac{\text{Vol}(\mathcal{Q}_n)}{\text{Vol}(\mathcal{B}_q)}.$$

(6.11)

To illustrate, consider a fourth-order polynomial $p(z, q)$, whose coefficients lie in the hyperrectangle

$$\mathcal{B}_q = \{q \in \mathbb{R}^4 : -1 \le q_0 \le 3; -5 \le q_1 \le 5; -3 \le q_2 \le 6; -4 \le q_3 \le 4\}.$$

Using conditions (6.10), it can be checked that $\mathcal{Q}_n \subseteq \mathcal{B}_q$ holds. Then, the probability of stability is given by

$$p(\gamma) = \frac{\text{Vol}(\mathcal{Q}_4)}{\text{Vol}(\mathcal{B}_q)} = \frac{64/9}{2880} = 0.0025.$$

However, if $\mathcal{Q}_n \not\subseteq \mathcal{B}_q$, equation (6.11) is no longer valid and other methods should be devised for computing $p(\gamma)$ exactly.   ⋆

We now consider the second problem presented Section 6.1, related to the computation of the worst-case performance, and introduce its probabilistic counterpart.

Problem 6.5 (Probabilistic worst-case performance). *Given a performance function $J(\Delta)$, a density function $f_{\boldsymbol{\Delta}}(\Delta)$ with support $\mathcal{B}_{\mathbb{D}}(\rho)$ and a probability level $p^* \in [0, 1]$, compute $\bar{\gamma}$ such that*

$$\bar{\gamma} \le \gamma_{\text{wc}} = \sup_{\Delta \in \mathcal{B}_{\mathbb{D}}(\rho)} J(\Delta)$$

and

$$\text{PR}\{J(\boldsymbol{\Delta}) \le \bar{\gamma}\} \ge p^*.$$

(6.12)

Remark 6.3 (Good and bad sets interpretation of Problem 6.5). Equation (6.12) can be interpreted in terms of good and bad sets. That is, if we define $\mathcal{B}_G = \{\Delta \in \mathcal{B}_{\mathbb{D}}(\rho) : J(\Delta) \le \bar{\gamma}\}$, we write this equation as

$$\text{PR}\{\boldsymbol{\Delta} \in \mathcal{B}_G\} \ge p^*.$$

In terms of a bad set, defining $\mathcal{B}_B = \{\Delta \in \mathcal{B}_{\mathbb{D}}(\rho) : J(\Delta) > \bar{\gamma}\}$ and letting $\epsilon = 1 - p^*$, we have

$$\mathrm{PR}\left\{\mathbf{\Delta} \in \mathcal{B}_B\right\} < \epsilon.$$

In other words, we accept an ϵ-risk that the performance is greater than the desired one, i.e.

$$\mathrm{PR}\left\{J(\mathbf{\Delta}) > \bar{\gamma}\right\} < \epsilon.$$

For uniform densities, this amounts to requiring that

$$\mathrm{Vol}(\mathcal{B}_B) < \epsilon \mathrm{Vol}(\mathcal{B}_{\mathbb{D}}(\rho)).$$

$\lhd$

Computation of the probabilistic worst-case performance is illustrated in the next example.

Example 6.11 (Probabilistic worst-case $\mathcal{H}_2$ performance).
Consider the state space representation of Example 6.6 and assume we are interested in computing a level of performance $\bar{\gamma}$ such that

$$\mathrm{PR}_{\mathbf{q}}\left\{J(\mathbf{q}) \leq \bar{\gamma}\right\} \geq 0.99.$$

To this aim, we repeat the procedure outlined in Example 6.8, and compute the corresponding probability of performance for different values of γ. In Figure 6.5 we report the level curves $J(q) = \gamma$, which were used to compute $p(\gamma)$. Using this procedure, we obtained that the level $\gamma = 0.65$ guarantees a probability $\mathrm{PR}_{\mathbf{q}}\left\{J(\mathbf{q}) \leq 0.65\right\} = 0.9931$, which is greater than the desired level $p^* = 0.99$.

$\star$

We now turn our attention to the third problem introduced in Section 6.1 and state its probabilistic version.

Problem 6.6 (Probabilistic performance radius). *Given a performance function $J(\mathbf{\Delta})$ with associated level γ and a probability level $p^* \in [0,1]$, compute $\bar{r}(p^*)$ as*

$$\bar{r}(p^*) \doteq \inf\left\{\rho : \mathrm{PR}\left\{J(\mathbf{\Delta}) \leq \gamma\right\} < p^*, \mathbf{\Delta} \in \mathcal{B}_{\mathbb{D}}(\rho), \mathbf{\Delta} \sim f_{\mathbf{\Delta}}(\Delta)\right\}.$$

We now further elaborate on this problem: for given radius ρ, we assume that $\mathbf{\Delta}$ is a random matrix with support $\mathcal{B}_{\mathbb{D}}(\rho)$ and with given density function $f_{\mathbf{\Delta}}(\Delta)$, also depending on ρ. Then, we define the *performance degradation function* as

$$\mathrm{degrad}(\rho) \doteq \mathrm{PR}\left\{J(\mathbf{\Delta}) \leq \gamma\right\}, \quad \mathbf{\Delta} \in \mathcal{B}_{\mathbb{D}}(\rho), \mathbf{\Delta} \sim f_{\mathbf{\Delta}}(\Delta). \tag{6.13}$$

Then, we conclude that the probabilistic performance radius becomes

$$\bar{r}(p^*) \doteq \inf\left\{\rho : \mathrm{degrad}(\rho) < p^*\right\}.$$

We now present an example showing the computation of the performance degradation function.

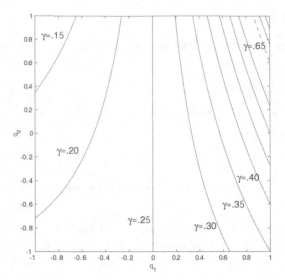

Fig. 6.5. Level curves $J(q) = \gamma$ for different values of γ for Example 6.11. The dashed line corresponds to $\gamma = 0.65$.

Example 6.12 (Probability degradation function).

We again revisit Example 6.5, regarding stability of a fourth-order system. We now assume that q is a random vector with uniform distribution in the set $\mathcal{B}_q(\rho)$. Setting $J(q)$ and γ as before, we evaluate the probability of stability when ρ varies in the interval $[0.5, 3]$, as depicted in Figure 6.6.

As in the previous case, the volume of the good set can be computed in closed form by integrating the equations defining the Hurwitz region

$$\text{Vol}(\mathcal{B}_G(\rho)) = \begin{cases} 4\rho^2 & \text{if } \rho \leq 1; \\ \frac{2}{3}\rho^3 - 3\rho^2 - (\alpha(\rho) - 13)\rho - \alpha(\rho) - \frac{16}{3} & \text{if } 1 < \rho \leq 1.5; \\ \frac{3}{2}\rho^2 + 4\rho + \frac{7}{24} - \frac{9}{2}\alpha(\rho)^3 & \text{if } 1.5 < \rho \leq 3 \end{cases}$$

where $\alpha(\rho) = \frac{1}{3}\sqrt{2\rho + 2}$. Then, the probability degradation function is immediately obtained as

$$\text{degrad}(\rho) = \frac{\text{Vol}(\mathcal{B}_G(\rho))}{\text{Vol}(\mathcal{B}_q(\rho))}$$

which is shown in Figure 6.7. Fixing a level of probability $p^* = 0.95$, from this figure we obtain that the probabilistic stability radius is given by

$$\bar{r}(0.95) = 1.2514.$$

$\star$

Finally, we remark that, in this example, the probability degradation function is monotonically decreasing.

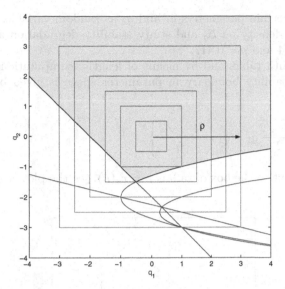

Fig. 6.6. The set $\mathcal{B}_q(\rho)$, $\rho \in [0.5, 3]$ for Example 6.12. The light-gray region indicates the Hurwitz region.

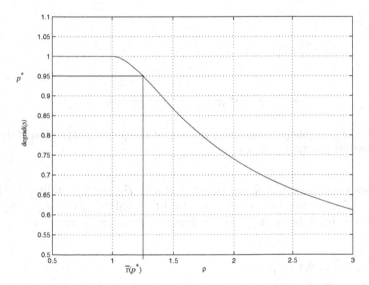

Fig. 6.7. Probability degradation function for robust stability for Example 6.12.

Example 6.13 (Probabilistic stability radius).

We revisit Example 5.7 regarding discontinuity of the robustness margin and study Hurwitz stability for $\bar{\kappa} = 3 + 2\sqrt{2}$ of the closed-loop polynomial $p_{\bar{\kappa}}(s, q)$ in (5.3). In Example 5.7, the stability radius of the polynomial is computed, obtaining $r_q(\bar{\kappa}) = 0.234$. Following a probabilistic approach, we now take the uncertain parameters q_1 and q_2 as random variables with uniform probability density on $\mathcal{B}_q$ and study stability degradation as ρ varies between $\rho = 0.234$ and $\rho = 0.417$.

Using the Routh table, and by means of lengthy computations, we conclude that the stability boundary in parameter space is given by the line segment

$$\left\{ q \in \mathcal{B}_q : \frac{1+\sqrt{2}}{2}q_1 + q_2 = \frac{4-\sqrt{2}}{5} \right\}.$$

This line is tangent to the box of radius $\rho = r_q(\bar{\kappa})$, see the plot of Figure 6.8.

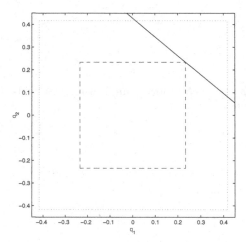

Fig. 6.8. Set of unstable polynomials in the box of radius $\rho = 0.417$ (dotted line), and box of radius $\rho = r_q(\bar{\kappa})$ (dashed line).

It follows that the set of unstable polynomials $\mathcal{B}_B$ corresponding to this line is a set of measure zero. Therefore, the ratio $\mathrm{Vol}(\mathcal{B}_B)/\mathrm{Vol}(\mathcal{B}_q)$ is equal to zero and the probability degradation function remains equal to one up to $\rho = 0.417$, i.e.

$$\mathrm{degrad}(\rho) = 1, \text{ for all } \rho \in [0, 0.417].$$

Notice that for $\kappa = \bar{\kappa} - \epsilon$ the diagonal line segment in Figure 6.8 disappears, giving rise to a discontinuity in the (deterministic) stability radius. This discontinuity is not present in its probabilistic counterpart.   ⋆

Remark 6.4 (Relationships between probabilistic robustness problems). Similar to the robustness problems studied in Section 6.1 (see Remark 6.1), the three probabilistic versions stated in this section are closely related. In particular, if Problem 6.4 is solvable for fixed γ and ρ, then Problem 6.5 can be solved via a one-dimensional γ iteration and Problem 6.6 can be solved via a one-dimensional ρ iteration. However, the probabilistic interpretation of these problems and the specific results obtained may be different. $\lhd$

Remark 6.5 (Closed-form computation). The examples reported in this chapter show the closed-form computation of the various probabilistic figures under study. Obviously, these exact computations are limited to special cases. The solution of these problems requires the evaluation of multidimensional integrals, which is in general a very difficult task. In subsequent chapters we develop randomized algorithms based on uncertainty sampling to obtain estimates of the required probabilities, up to a certain level of confidence. $\lhd$

6.4 Distribution-free robustness

In the probabilistic robustness setting described in this chapter, the density function of the uncertainty Δ is assumed to be known. If this is not the case, then clearly the probability of performance depends on the specific choice of $f_\Delta(\Delta)$. For example, in the previous section we have shown that the probability $p(\gamma) = \mathrm{PR}\{J(\Delta) \leq \gamma\}$ coincides with the ratio of volumes $\mathrm{Vol}(\mathcal{B}_G)/\mathrm{Vol}(\mathcal{B}_\mathbb{D}(\rho))$ if the pdf is uniform. For a different pdf, the probability $p(\gamma)$ may be dramatically different. In other words, without some reasoning attached to the selection of the probability density, the probability of performance obtained may be meaningless. Generally, the probability density function may be estimated directly from available data or prior information, but if this prior information is not available, then the selection of the distribution should be performed with great care.

To address this problem further we now consider an example.

Example 6.14 (Probabilistic stability for various distributions).
We now continue Example 6.9. Consider now the case when $\alpha = 0.1$ and the density function $f_\mathbf{q}(q)$ is a truncated Gaussian pdf with expected value $\mathrm{E}(\mathbf{q}) = [1\,1]^T$, covariance matrix $\mathrm{Cov}(\mathbf{q}) = \mathrm{diag}\,([\sigma^2\,\sigma^2])$ and support $\mathcal{B}_q = \{q \in \mathbb{R}^2 : 0.3 \leq q_1 \leq 2.5; 0 \leq q_2 \leq 1.7\}$. That is, we write

$$f_\mathbf{q}(q) = \begin{cases} Ce^{-\frac{(q_1-1)^2+(q_2-1)^2}{2\sigma^2}} & \text{if } q \in \mathcal{B}_q; \\ 0 & \text{otherwise} \end{cases}$$

where C is a normalizing constant obtained by imposing $\int_{\mathcal{B}_q} f_\mathbf{q}(q)dq = 1$. In this example we compute in closed form the probability of stability by solving explicitly the multiple integral required to compute the probability

$$\mathrm{PR}_\mathbf{q}\{p(s,\mathbf{q}) \text{ Hurwitz}\} = 1 - \int_{\mathcal{B}_B} f_\mathbf{q}(q)dq = 1 - 2\pi\sigma^2\left(1 - e^{-\frac{\alpha^2}{2\sigma^2}}\right)$$

Table 6.2. Probability of stability and constant C for decreasing values of σ.

σ	C	$\mathrm{PR}\,\{p(s,\mathbf{q})\ \text{Hurwitz}\}$
2	0.2955	0.9907
0.4	1.0199	0.9664
0.2	3.9789	0.8824
0.1	15.916	0.6065
0.01	1591.6	$\to 0$

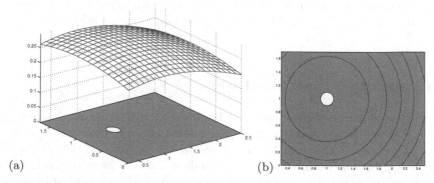

(a)   (b)

Fig. 6.9. (a) Gaussian pdf centered in $[1\,1]^T$ with $\sigma = 2$. (b) Level curves of the distribution. The disk of radius 0.1 represents the subset of unstable parameters $\mathcal{B}_B$.

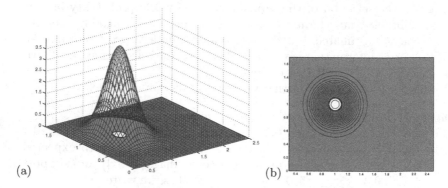

(a)   (b)

Fig. 6.10. (a) Gaussian pdf centered in $[1\,1]^T$ with $\sigma = 0.2$. (b) Level curves of the distribution. The disk of radius 0.1 represents the subset of unstable parameters $\mathcal{B}_B$.

where $\mathcal{B}_B = \{q \in \mathcal{B}_q : (q_1 - 1)^2 + (q_2 - 1)^2 \le \alpha^2\}$. Table 6.2 shows how different values of σ lead to very different values of the probability of stability. This behavior is also shown in Figures 6.9 and 6.10. When the value of σ decreases, the Gaussian pdf shrinks around the bad set. In the limit case $\sigma \to 0$, the probability of stability approaches zero, whereas for $\sigma \to \infty$ the Gaussian distribution tends to the uniform distribution and the probability of stability approaches the ratio $\mathrm{Vol}(\mathcal{B}_G)/\mathrm{Vol}(\mathcal{B}_q(\rho))$. The conclusion is that if the density function is chosen without any specific guideline, then the probability of stability may vary arbitrarily between the extreme values zero and one.   ⋆

Motivated by these considerations, in [27] the problem of distribution-free robustness is studied. In other words, the objective of this line of research is to determine the worst-case distribution in a certain class of probability measures. More precisely, let $\mathbf{q}$ be an ℓ-dimensional real vector of independent random variables with support

$$\mathcal{B}_q = \{\mathbf{q} \in \mathbb{R}^\ell : \|\mathbf{q}\|_\infty \le 1\}$$

and $\mathcal{B}_G \subset \mathbb{R}^\ell$ be a closed, convex and centrally symmetric set. Then, in [27] it is proven that

$$\min_{f_{\mathbf{q}} \in \mathcal{F}} \int_{\mathcal{B}_G} f_{\mathbf{q}}(q)\mathrm{d}q = \int_{\mathcal{B}_G} \mathcal{U}_{\mathcal{B}_q}\mathrm{d}q$$

where $\mathcal{U}_{\mathcal{B}_q}$ is the uniform probability density function with support $\mathcal{B}_q$ and $\mathcal{F}$ is the set of probability density functions satisfying two conditions:

1. The cdf $F_{\mathbf{q}}(q)$ is absolutely continuous, so that the density function

$$f_{\mathbf{q}}(q) = \prod_{i=1}^{\ell} f_{\mathbf{q}_i}(q_i)$$

   is well defined;
2. The marginal density functions $f_{\mathbf{q}_i}(q_i)$ are non increasing and centrally symmetric.

This result is generally denoted as the *uniformity principle*. In [27], applications to robustness analysis of affine polynomial families are also shown, taking $\mathcal{B}_G$ as the so-called value set. However, the fact that $\mathcal{B}_G$ needs to be convex and centrally symmetric may be a critical requirement that is generally not satisfied for the good and bad sets. This convexity assumption has been partially removed for unirectangular sets [168]. Subsequent research along this direction has been performed in various papers. The interested reader may also refer to the survey paper [167], which is focused on this particular line of research.

7. Monte Carlo Methods

In this chapter we discuss Monte Carlo (MC) and quasi-Monte Carlo (QMC) methods. The Monte Carlo method has a very long history, which officially began in 1949 with the seminal paper of Metropolis and Ulam [189]. We refer e.g. to [119] for historical remarks and introductory material. The quasi-Monte Carlo method is more recent and may be regarded as a deterministic version of MC, see for instance [203, 205].

7.1 Probability and expected value estimation

In this section, we discuss randomized techniques for probability and expected value estimation. In Chapter 6 we introduced probabilistic versions of classical robustness problems, which are based on the computation of the probability of performance. That is, for a given performance level γ, the objective is to estimate the probability

$$p(\gamma) = \text{PR}_{\boldsymbol{\Delta}} \left\{ J(\boldsymbol{\Delta}) \leq \gamma \right\} = \int_{\mathcal{B}_G} f_{\boldsymbol{\Delta}}(\Delta) \mathrm{d}\Delta.$$

The evaluation of this probability requires the solution of multiple integrals. Hence, its exact computation is very difficult in general, and only in special cases can $p(\gamma)$ be obtained in closed form, see the examples given in the previous chapters.

A classical tool for the numerical evaluation of multiple integrals is the Monte Carlo method. To estimate the probability $p(\gamma)$ with this approach, we generate N independent identically distributed (iid) random samples within the set $\mathcal{B}_{\mathbb{D}}(\rho)$

$$\boldsymbol{\Delta}^{(1...N)} \doteq \left\{ \boldsymbol{\Delta}^{(1)}, \ldots, \boldsymbol{\Delta}^{(N)} \right\} \tag{7.1}$$

according to the given density function $f_{\boldsymbol{\Delta}}(\Delta)$, where $\boldsymbol{\Delta}^{(1...N)}$ is called a *multisample* of $\boldsymbol{\Delta}$ of cardinality N. Then, we evaluate

$$J(\boldsymbol{\Delta}^{(1)}), \ldots, J(\boldsymbol{\Delta}^{(N)}).$$

A *Monte Carlo estimate* of $p(\gamma)$ is given by

$$\widehat{\mathbf{p}}_N(\gamma) = \frac{\mathbf{N}_G}{N}$$

where $\mathbf{N}_G$ is the number of "good" samples such that $J(\mathbf{\Delta}^{(i)}) \leq \gamma$. More formally, we define the indicator function associated with the good set

$$\mathbb{I}_{\mathcal{B}_G}(\Delta) \doteq \begin{cases} 1 \text{ if } \Delta \in \mathcal{B}_G; \\ 0 \text{ otherwise} \end{cases}$$

where $\mathcal{B}_G = \{\Delta \in \mathcal{B}_\mathbb{D}(\rho) : J(\Delta) \leq \gamma\}$ is defined in (6.2). Then, we write

$$\widehat{\mathbf{p}}_N(\gamma) = \frac{1}{N} \sum_{i=1}^{N} \mathbb{I}_{\mathcal{B}_G}(\mathbf{\Delta}^{(i)}). \tag{7.2}$$

That is, the probability $p(\gamma)$ is estimated by means of the empirical mean of the good set indicator function (see further discussions in Remark 7.1). The estimate $\widehat{\mathbf{p}}_N(\gamma)$ is a random variable usually referred to as the *empirical probability*. The weak and strong *laws of large numbers* presented next guarantee asymptotic convergence in probability and with probability one, respectively, of the empirical probability $\widehat{\mathbf{p}}_N(\gamma)$ to $p(\gamma)$.

Theorem 7.1 (Laws of large numbers for empirical probability). *For any $\epsilon \in (0,1)$, the* weak law of large numbers *states that*

$$\lim_{N \to \infty} \mathrm{PR}\{|p(\gamma) - \widehat{\mathbf{p}}_N(\gamma)| > \epsilon\} = 0. \tag{7.3}$$

The strong law of large numbers *guarantees that*

$$\lim_{N \to \infty} \frac{1}{N} \sum_{i=1}^{N} \mathbb{I}_{\mathcal{B}_G}(\mathbf{\Delta}^{(i)}) = \mathrm{PR}\{J(\mathbf{\Delta}) \leq \gamma\} \tag{7.4}$$

with probability one (a.e.).

The weak law (7.3) is classical, see e.g. [214], and follows directly from the Bernoulli bound presented in Chapter 9. The strong law (7.4) is a consequence of the Borel–Cantelli lemma, which gives a sufficient condition for a.e. convergence, see e.g. [284] for a proof and a discussion on the subject.

The MC approach can be readily used in the more general situation where the estimation of the expected value is of concern. That is, given a performance function $J(\Delta)$ and a density function $f_\mathbf{\Delta}(\Delta)$ with support $\mathcal{B}_\mathbb{D}(\rho)$, we aim at estimating

$$\mathrm{E}_\mathbf{\Delta}(J(\mathbf{\Delta})) = \int_{\mathcal{B}_\mathbb{D}(\rho)} J(\Delta) f_\mathbf{\Delta}(\Delta) \mathrm{d}\Delta.$$

In this case, we take the multisample $\mathbf{\Delta}^{(1...N)}$ defined in (7.1) and compute the so-called *empirical mean*

$$\widehat{\mathbf{E}}_N(J(\boldsymbol{\Delta})) \doteq \frac{1}{N} \sum_{i=1}^{N} J(\boldsymbol{\Delta}^{(i)}). \tag{7.5}$$

Subsequently, when clear from the context, the empirical mean will be denoted by $\widehat{\mathbf{E}}_N$. The two laws of large numbers are now stated for empirical mean estimation.

Theorem 7.2 (Laws of large numbers for empirical mean). *For any* $\epsilon > 0$, *we have*

$$\lim_{N \to \infty} \mathrm{PR} \left\{ \left| \mathbf{E}_{\boldsymbol{\Delta}}(J(\boldsymbol{\Delta})) - \widehat{\mathbf{E}}_N(J(\boldsymbol{\Delta})) \right| > \epsilon \right\} = 0.$$

Moreover, the empirical mean converges a.e. to the expected value

$$\lim_{N \to \infty} \frac{1}{N} \sum_{i=1}^{N} J(\boldsymbol{\Delta}^{(i)}) = \int_{\mathcal{B}_{\mathbb{D}}(\rho)} J(\Delta) f_{\boldsymbol{\Delta}}(\Delta) \mathrm{d}\Delta.$$

Remark 7.1 (Probability versus expected value). The probability estimation problem can be seen as a special case of expected value estimation. Indeed, if we introduce a new performance function

$$\tilde{J}(\Delta) \doteq \mathbb{I}_{\mathcal{B}_G}(\Delta)$$

then it follows that

$$\mathbf{E}_{\boldsymbol{\Delta}}\left(\tilde{J}(\boldsymbol{\Delta})\right) = \int_{\mathcal{B}_{\mathbb{D}}(\rho)} \mathbb{I}_{\mathcal{B}_G}(\Delta) f_{\boldsymbol{\Delta}}(\Delta) \mathrm{d}\Delta = \int_{\mathcal{B}_G} f_{\boldsymbol{\Delta}}(\Delta) \mathrm{d}\Delta = \mathrm{PR}_{\boldsymbol{\Delta}} \left\{ J(\boldsymbol{\Delta}) \leq \gamma \right\}.$$

◁

The asymptotic convergence of $\widehat{\mathbf{E}}_N(J(\boldsymbol{\Delta}))$ to the expected value $\mathrm{E}(J(\boldsymbol{\Delta}))$ is guaranteed by the laws of large numbers. For finite sample size N, it is of great interest to compute the expected value of the squared difference between $\mathrm{E}(J(\boldsymbol{\Delta}))$ and the empirical mean $\widehat{\mathbf{E}}_N(J(\boldsymbol{\Delta}))$. More precisely, in the next theorem, which is a classical result in the Monte Carlo literature, see e.g. [203], we explicitly compute

$$\mathrm{Var}\left(\widehat{\mathbf{E}}_N\right) \doteq \mathrm{E}_{\boldsymbol{\Delta}^{(1\ldots N)}}\left(\left(\mathrm{E}(J(\boldsymbol{\Delta})) - \widehat{\mathbf{E}}_N\right)^2\right)$$

$$= \int_{\mathcal{B}_{\mathbb{D}}(\rho)} \cdots \int_{\mathcal{B}_{\mathbb{D}}(\rho)} \left(\mathrm{E}(J(\boldsymbol{\Delta})) - \frac{1}{N} \sum_{i=1}^{N} J(\Delta^{(i)})\right)^2 \prod_{k=1}^{N} f_{\boldsymbol{\Delta}}(\Delta^{(k)}) \mathrm{d}\Delta^{(k)}.$$

The variance $\mathrm{Var}\left(\widehat{\mathbf{E}}_N\right)$ is clearly a measure of the "goodness" of the approximation error of MC methods.

Theorem 7.3. *If the variance* $\mathrm{Var}(J(\boldsymbol{\Delta}))$ *is finite, then for any* $N \geq 1$*, we have*

$$\mathbf{E}_{\boldsymbol{\Delta}^{(1\ldots N)}}\left(\left(\mathrm{E}(J(\boldsymbol{\Delta})) - \widehat{\mathbf{E}}_N\right)^2\right) = \frac{\mathrm{Var}(J(\boldsymbol{\Delta}))}{N}. \tag{7.6}$$

Proof. First, we define $h(\Delta) \doteq \mathrm{E}(J(\boldsymbol{\Delta})) - J(\Delta)$. Then, we have

$$\int_{\mathcal{B}_{\mathbb{D}}(\rho)} h(\Delta) f_{\boldsymbol{\Delta}}(\Delta) \mathrm{d}\Delta = 0$$

and

$$\mathrm{E}(J(\boldsymbol{\Delta})) - \widehat{\mathbf{E}}_N = \frac{1}{N}\sum_{i=1}^{N} h(\boldsymbol{\Delta}^{(i)}).$$

Hence, we write

$$\mathbf{E}_{\boldsymbol{\Delta}^{(1\ldots N)}}\left(\left(\mathrm{E}(J(\boldsymbol{\Delta})) - \widehat{\mathbf{E}}_N\right)^2\right) =$$

$$\int_{\mathcal{B}_{\mathbb{D}}(\rho)}\int_{\mathcal{B}_{\mathbb{D}}(\rho)}\cdots\int_{\mathcal{B}_{\mathbb{D}}(\rho)} \left(\frac{1}{N}\sum_{i=1}^{N} h(\boldsymbol{\Delta}^{(i)})\right)^2 \prod_{k=1}^{N} f_{\boldsymbol{\Delta}}(\Delta^{(k)})\mathrm{d}\Delta^{(k)} =$$

$$\frac{1}{N^2}\sum_{i=1}^{N}\int_{\mathcal{B}_{\mathbb{D}}(\rho)}\int_{\mathcal{B}_{\mathbb{D}}(\rho)}\cdots\int_{\mathcal{B}_{\mathbb{D}}(\rho)} h(\Delta^{(i)})^2 \prod_{k=1}^{N} f_{\boldsymbol{\Delta}}(\Delta^{(k)})\mathrm{d}\Delta^{(k)} +$$

$$\frac{2}{N^2}\sum_{i=1}^{N}\sum_{k>i}^{N}\int_{\mathcal{B}_{\mathbb{D}}(\rho)}\int_{\mathcal{B}_{\mathbb{D}}(\rho)}\cdots\int_{\mathcal{B}_{\mathbb{D}}(\rho)} h(\Delta^{(i)})h(\Delta^{(k)}) \prod_{k=1}^{N} f_{\boldsymbol{\Delta}}(\Delta^{(k)})\mathrm{d}\Delta^{(k)} =$$

$$\frac{1}{N}\int_{\mathcal{B}_{\mathbb{D}}(\rho)} h(\Delta)^2 f_{\boldsymbol{\Delta}}(\Delta)\mathrm{d}\Delta = \frac{\mathrm{Var}(J(\boldsymbol{\Delta}))}{N}.$$

$\square$

Remark 7.2 (Breaking the curse of dimensionality). As a consequence of this theorem, we obtain that the average absolute value of the approximation error of the MC method is given by $\sigma_J/\sqrt{N}$, where $\sigma_J = \sqrt{\mathrm{Var}(J(\boldsymbol{\Delta}))}$ is the standard deviation. Assuming that the variance $\mathrm{Var}(J(\boldsymbol{\Delta}))$ is known, then the number of samples necessary to guarantee a given error can be established *a priori*. That is, we compute

$$N \geq \frac{\mathrm{Var}(J(\boldsymbol{\Delta}))}{\mathrm{Var}\left(\widehat{\mathbf{E}}_N\right)}.$$

Unfortunately, since $\mathrm{Var}(J(\boldsymbol{\Delta}))$ is generally unknown, equation (7.6) can only be used to conclude that the error is of the order $O(N^{-1/2})$. An important consequence of this discussion is that the mean square error of the Monte Carlo estimate is *independent* of the problem dimension. This is the reason why Monte Carlo methods are said to *break the curse of dimensionality*. This issue is further discussed in Section 8.4 when dealing with computational complexity of randomized algorithms.   ◁

Remark 7.3 (Random sample generation). One of the key issues regarding the application of MC techniques in systems and control is the availability of efficient algorithms for the generation of the multisample (7.1) according to a given density function over the support $\mathcal{B}_{\mathbb{D}}(\rho)$. This problem is fully addressed in Chapters 16 and 18, where algorithms for generating samples of random vectors and matrices in the structured set $\mathcal{B}_{\mathbb{D}}(\rho)$ are presented. These algorithms reduce the problem to the univariate generation of uniform samples in the interval $[0, 1]$, which is the standard random number generation problem discussed in Chapter 14.   ◁

The techniques and the convergence results presented in this section are at the basis of Monte Carlo methods for computation of multiple integrals, which are discussed next.

7.2 Monte Carlo methods for integration

Monte Carlo methods address the general problem of computing numerically the multiple integral

$$\int_{\mathcal{Y}} g(x)\mathrm{d}x \tag{7.7}$$

of a multivariate measurable function $g(x) : \mathbb{R}^n \to \mathbb{R}$ with domain $\mathcal{Y} \subset \mathbb{R}^n$. The main idea is to transform this integral into an expected value computation problem. This can be done by factorizing the function $g(x)$ into the product of two terms $J(x)$ and $f_{\mathbf{x}}(x)$ such that $f_{\mathbf{x}}(x)$ is a probability density function with support $\mathcal{Y}$, and

$$g(x) = J(x)f_{\mathbf{x}}(x). \tag{7.8}$$

With this choice, the integral (7.7) can be viewed as the expected value of $J(\mathbf{x})$ with respect to the density function $f_{\mathbf{x}}(x)$, i.e.

$$\int_{\mathcal{Y}} g(x)\mathrm{d}x = \int_{\mathcal{Y}} J(x)f_{\mathbf{x}}(x)\mathrm{d}x = \mathrm{E}_{\mathbf{x}}(J(\mathbf{x})). \tag{7.9}$$

An MC estimate of the integral (7.7) is then immediately obtained via the techniques described in the previous section. That is, we approximate the expected value in (7.9) with the empirical mean

$$\widehat{\mathbf{E}}_N(J(\mathbf{x})) = \frac{1}{N} \sum_{i=1}^{N} J(\mathbf{x}^{(i)})$$

where the multisample

$$\mathbf{x}^{(1...N)} \doteq \left\{ \mathbf{x}^{(1)}, \ldots, \mathbf{x}^{(N)} \right\} \tag{7.10}$$

of cardinality N is generated according to the pdf $f_\mathbf{x}(x)$ with support $\mathcal{Y}$. Recalling Theorem 7.3, we immediately obtain that the variance of the estimate is equal to

$$\mathrm{Var}\left(\widehat{\mathbf{E}}_N(J(\mathbf{x})) \right) = \frac{\mathrm{Var}(J(\mathbf{x}))}{N}.$$

Various techniques have been developed to reduce this error for finite and fixed sample size. The most straightforward approach is to reduce the variance $\mathrm{Var}(J(\mathbf{x}))$ by choosing the density function $f_\mathbf{x}(x)$ in (7.8) in a suitable way. This leads to the methods of stratified and importance sampling, see e.g. [227].

To conclude this section, in the following example we compare the complexity of integration based on the MC method with that obtained with a trapezoidal rule of integration.

Example 7.1 (Trapezoidal rule for integration).
The computation of multiple integrals can be performed using the multidimensional trapezoidal rule for integration of functions with continuous and bounded second partial derivatives. For example, consider the integral

$$\int_{\mathcal{Y}} g(x) \mathrm{d}x \tag{7.11}$$

where $g(x) : \mathbb{R}^n \to \mathbb{R}$ is a twice-differentiable function and the integration domain $\mathcal{Y}$ is the unit cube in $\mathbb{R}^n$. In this case, we construct a trapezoidal approximation of (7.11) based on N grid points for each x_i

$$\sum_{i_1=1}^{N} \cdots \sum_{i_n=1}^{N} w_{i_1} \cdots w_{i_n} g(\widetilde{x}(i_1, \ldots, i_n))$$

where

$$\widetilde{x}(i_1, \ldots, i_n) = \left[\left(\frac{i_1 - 1}{N-1} \right) \quad \cdots \quad \left(\frac{i_n - 1}{N-1} \right) \right]^T$$

and the weights are given by

$$w_1 = \frac{1}{2N-2}, w_2 = \frac{1}{N-1}, w_3 = \frac{1}{N-1}, \ldots, w_{N-1} = \frac{1}{N-1}, w_N = \frac{1}{2N-2}.$$

It is well known that the deterministic error given by the trapezoidal rule is of the order $O(N^{-2/n})$, see for instance [203]. However, in Theorem 7.3, we have shown that the average error of MC algorithms is of the order $O(N^{-1/2})$, where N is the number of random samples. Hence, comparing these errors we conclude that Monte Carlo methods improve upon classical deterministic algorithms based on the trapezoidal rule for integration for $n > 4$. Obviously, it should be noticed that the Monte Carlo mean square error given in Theorem 7.3 is of probabilistic, and not deterministic, nature.   ⋆

7.3 Monte Carlo methods for optimization

In this section we briefly discuss the application of Monte Carlo techniques in optimization problems, see [227] for a survey of the literature. In particular, consider a bounded multivariate function $g(x) : \mathcal{Y} \to \mathbb{R}$ and suppose we are interested in evaluating

$$g^* = g(x^*) = \sup_{x \in \mathcal{Y}} g(x). \tag{7.12}$$

A simple algorithm for estimating the optimal value of $g(x)$ has been proposed in [53] using the so-called *nonadaptive random search algorithm*. First, we draw N iid points $\mathbf{x}^{(i)}$, $i = 1, \ldots, N$ in $\mathcal{Y}$ according to a given density $f_{\mathbf{x}}(x)$. Then, an approximation of the maximum in (7.12) is given by the *empirical maximum*

$$\widehat{\mathbf{g}}_N = \max_{i=1,\ldots,N} g(\mathbf{x}^{(i)}).$$

The next theorem studies asymptotic convergence of the estimate $\widehat{\mathbf{g}}_N$ to g^*.

Theorem 7.4 (Laws of large numbers for empirical maximum). *Assume that the density function $f_{\mathbf{x}}(x)$ assigns a nonzero probability to every neighborhood of x^*, and $g(x)$ is continuous at x^*. Then, for any $\epsilon > 0$*

$$\lim_{N \to \infty} \mathrm{PR}\left\{g^* - \widehat{\mathbf{g}}_N > \epsilon\right\} = 0.$$

Moreover, the empirical maximum converges a.e. to the true maximum

$$\lim_{N \to \infty} \widehat{\mathbf{g}}_N = g^*.$$

More sophisticated Monte Carlo algorithms for global optimization have been widely studied in the literature. We recall here the *multistart random search*, which performs a series of gradient descents starting from random generating initial points, and the *simulated annealing* algorithm, for further details see for instance [294].

7.4 Quasi-Monte Carlo methods

The quasi-Monte Carlo method is a deterministic version of Monte Carlo with the primary goal of obtaining guaranteed (instead of probabilistic) errors. Some motivations for studying QMC methods may also come from the difficulty of generating "truly" random samples, see Chapter 14 for further discussions and the classical reference [203] for a complete treatment of the subject.

In the quasi-Monte Carlo method, *deterministic points* chosen according to some optimality criterion are used instead of random samples generated according to a given probability density function. For integration of a multivariate function, the most frequently used criterion is the so-called *discrepancy*, which is a measure of how the sample set is "evenly distributed" within the integration domain, which is usually taken as the unit cube in $\mathbb{R}^n$. The problem therefore is to find specific sequences of points which minimize the discrepancy, or upper bounds on it. There are many such sequences, including Sobol', Halton, and Niederreiter. For integrands with a sufficiently low "degree of regularity," these sequences guarantee a deterministic error bound for integration of $O(N^{-1}(\log N)^n)$. For fixed dimension n, and for large N, this error is therefore smaller than the mean square error of the MC estimate, which is $O(N^{-1/2})$. QMC methods are also used for optimization problems. In this case, the optimality criterion used is the *dispersion*, which is a measure of denseness, rather than equidistribution, as in the case of discrepancy. In this section, which is largely based on [203], we study both integration and optimization problems and the related sequences.

Various applications of QMC methods and numerical comparisons with MC methods have been developed, in particular in the areas of path integrals for mathematical finance, see for instance [271] and references therein, motion and path planning [51], and congestion control of communication networks [7].

7.4.1 Discrepancy and error bounds for integration

In this section we consider the integration of a (measurable) multivariate function $g(x) : \mathbb{R}^n \to \mathbb{R}$

$$\int_{[0,1]^n} g(x)\mathrm{d}x$$

with domain

$$[0,1]^n \doteq \{x \in \mathbb{R}^n : x_i \in [0,1], i = 1, \ldots, n\}.$$

This integral is approximated by the sum

$$\frac{1}{N} \sum_{i=1}^{N} g(x^{(i)})$$

where $x^{(i)} \in [0,1]^n$, $i = 1, \ldots, N$, are now *deterministic* vector points. The deterministic multisample of cardinality N

$$x^{(1 \ldots N)} \doteq \{x^{(1)}, \ldots, x^{(N)}\} \tag{7.13}$$

is usually called a *point set*. Intuitively, these points should be "evenly distributed," so that the irregularity of their distribution within the unit cube is minimized. This intuitive concept leads to the definition of discrepancy.

Definition 7.1 (Discrepancy). *Let S be a nonempty family of subsets of $[0,1]^n$. The discrepancy $D_N(S, x^{(1 \ldots N)})$ of a point set $x^{(1 \ldots N)}$ of cardinality N with respect to S is defined as*

$$D_N(S, x^{(1 \ldots N)}) \doteq \sup_{S \in \mathcal{S}} \left| \frac{\sum_{i=1}^N \mathbb{I}_S(x^{(i)})}{N} - \mathrm{Vol}(S) \right| \tag{7.14}$$

where $\mathbb{I}_S(x)$ is the indicator function of S and $\mathrm{Vol}(S)$ is its volume.

We remark that the discrepancy is a nonnegative quantity and it is upper bounded by one. Next, we define the star discrepancy $D_N^*(x^{(1 \ldots N)})$ and the extreme discrepancy $D_N^e(x^{(1 \ldots N)})$, obtained by considering specific choices of the family $\mathcal{S}$.

Definition 7.2 (Star discrepancy). *Let $\mathcal{S}^*$ be the family of all subintervals of the semi-open unit cube[1] $[0,1)^n$ of the form $\{x \in \mathbb{R}^n : x_i \in [0, v_i), i = 1, \ldots, n\}$. Then, the star discrepancy $D_N^*(x^{(1 \ldots N)})$ is defined as*

$$D_N^*(x^{(1 \ldots N)}) \doteq D_N(\mathcal{S}^*, x^{(1 \ldots N)}).$$

Definition 7.3 (Extreme discrepancy). *Let $\mathcal{S}^e$ be the family of all subintervals of the semi-open unit cube $[0,1)^n$ of the form $\{x \in \mathbb{R}^n : x_i \in [u_i, v_i), i = 1, \ldots, n\}$. Then, the extreme discrepancy $D_N^e(x^{(1 \ldots N)})$ is defined as*

$$D_N^e(x^{(1 \ldots N)}) = D_N(\mathcal{S}^e, x^{(1 \ldots N)}).$$

For any $x^{(1 \ldots N)}$ in the unit cube, it can be shown that the extreme and the star discrepancies are related as follows

$$D_N^*(x^{(1 \ldots N)}) \leq D_N^e(x^{(1 \ldots N)}) \leq 2^n D_N^*(x^{(1 \ldots N)}).$$

The definition of extreme discrepancy will be used later in Section 7.4.4 when studying the connections with dispersion.

By means of the star discrepancy, we can establish error bounds on the integration error

---

[1] The semi-open unit cube is defined as $[0,1)^n \doteq \{x \in \mathbb{R}^n : x_i \in [0,1), i = 1, \ldots, n\}$.

$$\left| \int_{[0,1]^n} g(x)\mathrm{d}x - \frac{1}{N} \sum_{i=1}^N g(x^{(i)}). \right| \tag{7.15}$$

A classical result in this direction, often called the *Koksma–Hlawka inequality* [134, 159], can be stated in terms of the total variation. Namely, for functions with continuous partial derivatives on $[0,1]^n$, the total variation of g in the sense of Vitali is defined as

$$V^{(n)}(g) = \int_0^1 \cdots \int_0^1 \left| \frac{\partial^n g}{\partial x_1 \cdots \partial x_n} \right| \mathrm{d}x_1 \cdots \mathrm{d}x_n.$$

Then, if g has bounded variation on $[0,1]^n$ in the sense of Vitali, and the restriction of g to each k-dimensional face of $[0,1]^n$, for $k = 1, \ldots, n$, is of bounded variation in the sense of Vitali, then g is said to be of bounded variation on $[0,1]^n$ in the sense of Hardy and Krause. This concept is an n-dimensional extension of the scalar variation in the interval $[0,1]$.

Theorem 7.5 (Koksma–Hlawka inequality). *Assume that $g : \mathbb{R}^n \to \mathbb{R}$ has bounded variation $V^{(n)}(g)$ on $[0,1]^n$ in the sense of Hardy and Krause. Then, for any $x^{(1\ldots N)}$ with $x^{(i)} \in [0,1)^n$, $i = 1, \ldots, N$, we have*

$$\left| \int_{[0,1]^n} g(x)\mathrm{d}x - \frac{1}{N} \sum_{i=1}^N g(x^{(i)}) \right| \le V^{(n)}(g) D_N^*(x^{(1\ldots N)}). \tag{7.16}$$

From this theorem it follows that, for given variation $V^{(n)}(g)$, the integration error in (7.16) is minimized if the point set $x^{(1\ldots N)}$ is selected so that the star discrepancy $D_N^*(x^{(1\ldots N)})$ is minimized.

Next, we present two results stated in [203] that give a precise characterization of the star and extreme discrepancies in the special case $n = 1$.

Theorem 7.6. *If $n = 1$ and $0 \le x^{(1)} \le \cdots \le x^{(N)} \le 1$, then*

$$D_N^*(x^{(1\ldots N)}) = \frac{1}{2N} + \max_{1 \le i \le N} \left| x^{(i)} - \frac{2i-1}{2N} \right|.$$

Theorem 7.7. *If $n = 1$ and $0 \le x^{(1)} \le \cdots \le x^{(N)} \le 1$, then*

$$D_N^e(x^{(1\ldots N)}) = \frac{1}{N} + \max_{1 \le i \le N} \left(\frac{i}{N} - x^{(i)} \right) - \min_{1 \le i \le N} \left(\frac{i}{N} - x^{(i)} \right).$$

From these results, it can be easily verified that the two inequalities

$$D_N^*(x^{(1\ldots N)}) \ge \frac{1}{2N} \text{ and } D_N^e(x^{(1\ldots N)}) \ge \frac{1}{N}$$

hold for any $N \ge 1$. We remark that equality is attained with the choice

$$x^{(i)} = \frac{2i-1}{2N}. \qquad (7.17)$$

In this case, both star and extreme discrepancy are minimized. That is, if one wants to place N points in the interval $[0,1]$ in order to minimize the error bound for integration given in (7.16), then the "optimal gridding" is the one given in (7.17), which corresponds to the N-panel midpoint integration rule, see e.g. [78]. Notice, however, that these facts may be used only when N is known a priori. A subsequent problem, addressed in the next section, is how to construct *recursive sequences* of points that guarantee low discrepancy.

7.4.2 One-dimensional low discrepancy sequences

We first study low discrepancy sequences for $n = 1$ on the semi-open interval $[0,1)$. For an integer $b \geq 2$, we define

$$\mathcal{Z}_b \doteq \{0, 1, \ldots, b-1\}.$$

Every integer $k \geq 0$ has a unique digit expansion in base b

$$k = \sum_{i=0}^{\infty} a_i(k) b^i \qquad (7.18)$$

where $a_i(k) \in \mathcal{Z}_b$ for all $i \geq 0$ and $a_i(k) = 0$ for all sufficiently large i. Then, for an integer $b \geq 2$, we define the *radical-reversal* function in base b as

$$\phi_b(k) = \sum_{i=0}^{\infty} a_i(k) b^{-i-1} \qquad (7.19)$$

for all integers $k \geq 0$ and where $a_i(k)$ is given by (7.18). We are now ready to define the van der Corput sequence [277].

Definition 7.4 (van der Corput sequence). *Let $n = 1$. For an integer $b \geq 2$, the van der Corput sequence in base b is the sequence*

$$x^{(1,\ldots)} \doteq x^{(1)}, x^{(2)}, \ldots$$

where $x^{(i)} = \phi_b(i-1)$ for all $i \geq 1$.

Example 7.2 (Binary van der Corput sequence).
    We study the van der Corput sequence in base $b = 2$. To illustrate, we compute the element $x^{(24)} = \phi_2(23)$. First, we write

$$23 = \sum_{i=0}^{\infty} a_i(23) 2^i.$$

With straightforward computations, we obtain

$$a_0(23) = 1, a_1(23) = 1, a_2(23) = 1, a_2(23) = 0, a_4(23) = 1$$

and $a_i(23) = 0$ for all $i \geq 5$. Then, we have

$$x^{(24)} = \phi_2(23) = \frac{a_0(23)}{2} + \frac{a_1(23)}{4} + \frac{a_2(23)}{8} + \frac{a_3(23)}{16} + \frac{a_4(23)}{32} =$$

$$\frac{1}{2} + \frac{1}{4} + \frac{1}{8} + \frac{1}{32} = 0.9063.$$

Similarly, we construct the other elements $x^{(1)} = \phi_2(0), x^{(2)} = \phi_2(1), \cdots$ obtaining the sequence

0,   0.5,   0.25,   0.75,   0.125,   0.625,   0.375,   0.875,   0.0625,
0.5625, 0.3125, 0.8125, 0.1875, 0.6875, 0.4375, 0.9375, 0.0313, 0.5313,
0.2813, 0.7813, 0.1563, 0.6563, 0.4063, 0.9063, ...

The first 24 points of this sequence are plotted in Figure 7.1.   ⋆

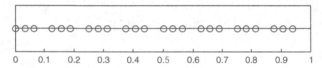

Fig. 7.1. First 24 points of the binary van der Corput sequence.

The discrepancy of the van der Corput sequence is of the order of magnitude of $O(N^{-1} \log N)$ for all $N \geq 2$ (see Theorem 7.8). It can be shown [203] that this is the best bound achievable by any sequence of points in $[0, 1]$. This result can be compared with Theorem 7.6, which states that the discrepancy of an N-point set in $[0, 1]$ can be made to be $O(N^{-1})$.

In the next section, we study low discrepancy sequences for $n > 1$.

7.4.3 Low discrepancy sequences for $n > 1$

The van der Corput sequence can be extended to any dimension n. This leads to the definition of the so-called Halton sequence [126].

Definition 7.5 (Halton sequence). *Let $b_1, \ldots, b_n$ be integers ≥ 2 and let ϕ_{b_i} be defined as in (7.19) for $b = b_i$. The Halton sequence in the bases $b_1, \ldots, b_n$ is the sequence*

$$x^{(1,\ldots)} \doteq x^{(1)}, x^{(2)}, \ldots$$

where

$$x^{(i)} = [\phi_{b_1}(i-1) \cdots \phi_{b_n}(i-1)]^T$$

for all $i \geq 0$.

Samples of a two-dimensional Halton sequence are shown in Figure 7.2. We now present a result which gives an upper bound on the star discrepancy of a Halton sequence.

Theorem 7.8. *For all $N \geq 1$, if $x^{(1...N)}$ are the first N points of the Halton sequence in the pairwise relatively prime bases $b_1, \ldots, b_n$, then*

$$D_N^*(x^{(1...N)}) < \frac{n}{N} + \frac{1}{N} \prod_{i=1}^{n} \left[\frac{b_i - 1}{2 \log b_i} \log N + \frac{b_i + 1}{2} \right]. \tag{7.20}$$

Combining this result with Theorem 7.5, we conclude that, for functions with finite variation $V(g)$, the integration error given in (7.15) is of the order $O(N^{-1}(\log N)^n)$. It can be easily verified that, asymptotically, this error is much smaller than $O(N^{-1/2})$, which is that associated with classical Monte Carlo. However, when n is large, the factor $(\log N)^n$ becomes huge, and it takes an impracticably large sample size N before the performance of QMC becomes superior to MC.

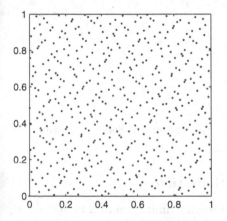
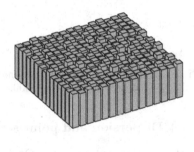

Fig. 7.2. Samples of two-dimensional Halton sequence for $N = 500$ and histogram of the relative frequency for $N = 10,000$ (each dimension is partitioned in 15 bins).

Many other low discrepancy sequences are studied in the quasi-Monte Carlo literature. We recall in particular the Sobol' [248], Faure [109] and Niederreiter [202] sequences. The basic idea underlying these methods is to suitably permute the elements of a Halton sequence. In particular, the Sobol' sequence uses only the basis 2, whereas in the Faure sequence the basis is the smallest prime number $b \geq n$. For illustrative purposes, $1,000$ points are generated in the unit box for the case of $n = 2$ for Halton, Faure, Sobol' and Niederreiter sequences. These points are shown in Figure 7.3.

Finally, we would like to mention that discrepancy is not the only optimality criterion used in QMC methods. For example, the so-called *dispersion* is generally used in the context of optimization.

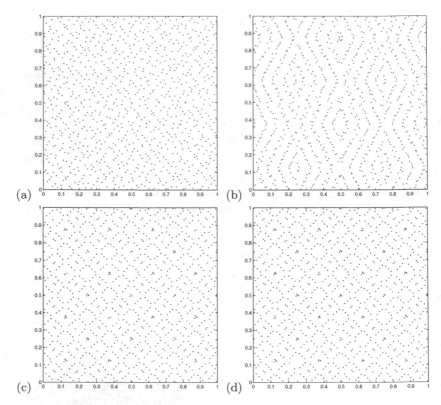

Fig. 7.3. Samples of two-dimensional sequences: (a) Halton, (b) Sobol', (c) Faure and (d) Niederreiter for $N = 1,000$.

7.4.4 Dispersion and point sets for optimization

In this section we study the QMC approach for maximization of a bounded multivariate function $g(x) : \mathbb{R}^n \to \mathbb{R}$ over the unit cube $[0,1]^n$

$$\sup_{x \in [0,1]^n} g(x).$$

We consider the QMC approximation

$$\max_{i=1,\dots,N} g(x^{(i)})$$

where $x^{(i)}$, $i = 1, \dots, N$, belong to a deterministic point set $x^{(1\dots N)}$. Clearly, the approximation error

$$\sup_{x \in [0,1]^n} g(x) - \max_{i=1,\dots,N} g(x^{(i)}) \tag{7.21}$$

is related to specific properties of the point set. Hence, we define the dispersion of a point set $x^{(1\dots N)}$ in the n-dimensional unit cube $[0,1]^n$.

Definition 7.6 (Dispersion). *The dispersion* $d_N(x^{(1...N)})$ *of a point set* $x^{(1...N)}$ *with cardinality* N *is defined as*

$$d_N(x^{(1...N)}) = \sup_{x \in [0,1]^n} \min_{1 \le i \le N} \|x - x^{(i)}\|_\infty. \tag{7.22}$$

The next theorem relates the approximation error (7.21) to the dispersion of the point set $x^{(1...N)}$. Let $\varpi(g, r)$ be the *modulus of continuity* of $g(x)$, which is defined as

$$\varpi(g, r) = \sup_{\substack{x, y \in [0,1]^n \\ \|x - y\|_\infty \le r}} |g(x) - g(y)|$$

for $r \ge 0$.

Theorem 7.9. *Let* $g(x) : \mathbb{R}^n \to \mathbb{R}$ *be a bounded function on the unit cube* $[0, 1]^n$. *For any point set* $x^{(1...N)}$, *we have*

$$\sup_{x \in [0,1]^n} g(x) - \max_{i=1,...,N} g(x^{(i)}) \le \varpi(g, d_N(x^{(1...N)})).$$

From this theorem, we see that point sets with low dispersion are required when dealing with optimization problems. The following result establishes a precise connection between the dispersion and the extreme discrepancy of a point set.

Theorem 7.10. *For any point set* $x^{(1...N)}$ *of cardinality* N *we have*

$$d_N(x^{(1...N)}) \le \left[D_N^e(x^{(1...N)})\right]^{1/n}.$$

Therefore, we conclude that low extreme discrepancy implies low dispersion, but the converse is not necessarily true.

We now turn our attention to the computation of the dispersion of a given point set. First, we present a result for the special case $n = 1$.

Theorem 7.11. *If* $n = 1$ *and* $0 \le x^{(1)} \le \cdots \le x^{(N)} \le 1$, *then*

$$d_N(x^{(1...N)}) =$$
$$\max \left(x^{(1)}, \frac{1}{2}(x^{(2)} - x^{(1)}), \ldots, \frac{1}{2}(x^{(N)} - x^{(N-1)}), 1 - x^{(N)} \right).$$

It can be easily verified that the same set of points given in (7.17)

$$x^{(i)} = \frac{2i - 1}{2N}$$

guarantees $d_N(x^{(1...N)}) = 1/(2N)$, which is the minimum value of dispersion for any point set $x^{(1...N)}$ in the unit cube.

For the n-dimensional case, we present a universal lower bound on the dispersion stated in [257], which gives a characterization of the minimum attainable dispersion for any point set.

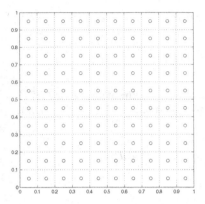

Fig. 7.4. Plot of 100 points chosen according to the Sukharev sampling criterion.

Theorem 7.12 (Sukharev inequality). *For any point set $x^{(1...N)}$ of cardinality N, we have*

$$d_N(x^{(1...N)}) \geq \frac{1}{2\lfloor N^{1/n} \rfloor}. \tag{7.23}$$

Remark 7.4 (Sukharev sampling criterion). The implications of this result are now briefly discussed. To simplify the discussion, suppose that $N^{1/n}$ is an integer, and suppose we are interested in generating N points in $[0,1]^n$ with "optimal" dispersion. Then, it can be shown that equality in (7.23) is attained if the points of $x^{(1...N)}$ are placed in a grid with discretization interval $N^{1/n}$ and the first point is shifted of $(N^{1/n})/2$ from the origin. This particular choice of point set is sometimes called the Sukharev sampling criterion. Figure 7.4 illustrates this criterion for $N = 100$ points in two dimensions. In addition, we notice that, if we solve equation (7.23) for N, we obtain

$$N \geq \left(2d_N(x^{(1...N)})\right)^{-n}.$$

Hence, the number of points N is an exponential function in n, regardless of how the sample set is generated.   ◁

8. Randomized Algorithms in Systems and Control

In Chapter 6 we introduced the concept of "probabilistic robustness" of a control system. It should be noticed that, except for some special cases (which include the examples discussed in Chapter 6), this probabilistic shift in the meaning of robustness does not automatically imply a simplification of the analysis or design problem. Indeed, assessing probabilistic robustness of a given system property may be even computationally harder than assessing robustness in the usual deterministic sense, since it requires the exact computation of a multidimensional probability integral. At this point *randomization* comes into play: the probability of robust performance is *estimated* by randomly sampling the uncertainties, and tail inequalities are used to bound the estimation error. Since the estimated probability is itself a random quantity, this method always entails a certain *risk of failure*, i.e. there exists a nonzero probability of making an erroneous estimation.

In this chapter, we present meta-algorithms for analysis and design of uncertain systems. These algorithms will be later specified in Chapters 11, 12 and 13 for various classes of problems. Crucial steps for implementing these algorithms are the determination of an appropriate sample size N, discussed in Chapters 9 and 10, and the construction of efficient algorithms for random sampling of the uncertainty $\mathbf{\Delta}$ within the structured set $\mathcal{B}_{\mathbb{D}}(\rho)$, see Chapters 14, 16 and 18.

8.1 Probabilistic robustness via randomized algorithms

As we shall see in the following, a *randomized algorithm* (RA) for probabilistic robustness analysis is an algorithm that, based on random extractions of uncertainty samples, returns an estimated probability of satisfaction of the performance specifications. The estimate provided by the randomized algorithm should be within an a priori specified *accuracy* $\epsilon \in (0,1)$ from the true value, with high *confidence* $1 - \delta$, $\delta \in (0,1)$. That is, the algorithm may indeed fail to return an approximately correct estimate, but with probability at most δ.

A similar situation also arises in the more difficult case of robust *synthesis*. The complication stems from the fact that the probability to be estimated is no longer fixed, but it is instead a function of some unknown "controller" or "design" parameters $\theta \in \Theta$. In this case, a randomized algorithm should return a controller, represented by a design vector θ, which guarantees the desired performance with an a priori specified (high) probability $p^* \in (0,1)$. As in the analysis case, this algorithm may fail with probability at most δ. An alternative synthesis paradigm is to seek a controller that (approximately) minimizes the average performance index. That is, the objective is to design a controller that performs well *on average*.

We highlight some points that are specific to the randomized approach that we intend to pursue. First, we aim to establish probabilistic statements with emphasis on finite sample bounds, as opposed to asymptotic results arising in other branches of probability theory. In particular, the *sample complexity* of a randomized algorithm is the minimum number of uncertainty samples N (sample size) that need to be drawn in order to achieve the desired levels of probabilistic robustness.

A conceptual distinction with respect to the common use of randomized algorithms in other fields, such as optimization, is that the randomization process is (whenever possible) applied only to the actual uncertainties present in the system, and not to other deterministic decision parameters. Therefore, randomness is not artificially introduced into the problem by the solution technique. Instead, the "natural" randomness due to the presence of stochastic uncertainty in the plant is exploited in the probabilistic solution. Finally, we point out that, in principle, control design under probabilistic constraints can be viewed as a *chance constrained* optimization problem, see for instance [223, 275]. However, chance constrained problems are hard to solve exactly, and therefore randomization can provide a viable approach for their solution.

In this exposition, we denote by *randomized algorithm* any algorithm that makes some random choices in the course of its execution. The outcome of such a decision process is hence a random variable, and the quality of this outcome is to be assessed via a probabilistic statement. In this chapter, we specialize this general definition to the context of uncertain systems and control, considering the specific features of randomized algorithms used for probabilistic performance analysis and synthesis.

8.2 Randomized algorithms for analysis

The two basic types of randomized algorithms for robust performance analysis presented next are related to the problems of "Probabilistic performance verification" and "Probabilistic worst-case performance" introduced in Chapter 6 (Problems 6.4 and 6.5). First, we specify the characteristics that an RA for probabilistic performance verification should comply with.

Definition 8.1 (RA for probabilistic performance verification). *Let* $\mathbf{\Delta} \in \mathbb{D}$ *be random with density* $f_{\mathbf{\Delta}}(\Delta)$, *and let* $\epsilon \in (0,1)$, $\delta \in (0,1)$ *be assigned probability levels. Given a performance function* $J(\Delta) : \mathbb{D} \to \mathbb{R}$ *and associated level* γ, *the RA should return with probability at least* $1 - \delta$ *an estimate* $\widehat{\mathbf{p}}_N(\gamma)$ *of the probability of performance*

$$p(\gamma) = \mathrm{PR}\left\{J(\mathbf{\Delta}) \le \gamma\right\}$$

that is within ϵ *from* $p(\gamma)$. *The estimate* $\widehat{\mathbf{p}}_N(\gamma)$ *should be constructed based on a finite number* N *of random samples of* $\mathbf{\Delta}$.

Notice that a simple RA for performance verification is directly constructed by means of the Monte Carlo method presented in Chapter 7. This is summarized in the following meta-algorithm.

Algorithm 8.1 (Probabilistic performance verification).
Given $\epsilon, \delta \in (0, 1)$ *and* γ, *this RA returns with probability at least* $1 - \delta$ *an estimate* $\widehat{\mathbf{p}}_N(\gamma)$ *such that*

$$|p(\gamma) - \widehat{\mathbf{p}}_N(\gamma)| < \epsilon.$$

1. Determine finite sample size $N = \overline{N}_{\mathrm{ch}}(\epsilon, \delta)$;
2. Draw N samples $\mathbf{\Delta}^{(1)}, \ldots, \mathbf{\Delta}^{(N)}$ according to $f_{\mathbf{\Delta}}$;
3. Return the empirical probability

$$\widehat{\mathbf{p}}_N(\gamma) = \frac{1}{N} \sum_{i=1}^{N} \mathbb{I}_{\mathcal{B}_G}(\mathbf{\Delta}^{(i)})$$

where $\mathbb{I}_{\mathcal{B}_G}(\cdot)$ is the indicator function of the set $\mathcal{B}_G = \{\Delta : J(\Delta) \le \gamma\}$.

We now comment on step 1 of this algorithm. In Chapter 7 we stated the laws of large numbers for empirical probability, which guarantee the asymptotic convergence $\widehat{\mathbf{p}}_N(\gamma) \to p(\gamma)$. However, to use Algorithm 8.1 we need estimates based on a *finite* number of samples. This topic is fully addressed in Chapter 9, in which finite size estimates are derived. In particular, if N in Algorithm 8.1 is chosen according to the well-known Chernoff bound (9.14), that is

$$\overline{N}_{\mathrm{ch}}(\epsilon, \delta) \doteq \left\lceil \frac{1}{2\epsilon^2} \log \frac{2}{\delta} \right\rceil$$

then this RA satisfies the requirements of Definition 8.1; see also Remark 9.3.

Similarly, an RA for probabilistic worst-case performance should determine a performance level $\widehat{\gamma}_N$ which is guaranteed for most of the uncertainty instances. This is summarized in the next definition.

Definition 8.2 (RA for probabilistic worst-case performance). *Let* $\Delta \in \mathcal{B}_{\mathbb{D}}(\rho)$ *be random with density* $f_{\Delta}(\Delta)$ *having support* $\mathcal{B}_{\mathbb{D}}(\rho)$, *and let* $p^* \in (0,1)$, $\delta \in (0,1)$ *be assigned probability levels. Given a performance function* $J(\Delta) : \mathbb{D} \to \mathbb{R}$, *the RA should return with probability at least* $1 - \delta$ *a performance level* $\widehat{\gamma}_N \leq \sup_{\Delta \in \mathcal{B}_{\mathbb{D}}(\rho)} J(\Delta)$ *such that*

$$\mathrm{PR}\left\{J(\Delta) \leq \widehat{\gamma}_N\right\} \geq p^*.$$

The performance level $\widehat{\gamma}_N$ *should be constructed based on a finite number* N *of random samples of* Δ.

An RA for probabilistic worst-case performance is described in the following meta-algorithm.

Algorithm 8.2 (Probabilistic worst-case performance).
Given $p^*, \delta \in (0,1)$, *this RA returns with probability at least* $1 - \delta$ *a level* $\widehat{\gamma}_N \leq \sup_{\Delta \in \mathcal{B}_{\mathbb{D}}(\rho)} J(\Delta)$ *such that*

$$\mathrm{PR}\left\{J(\Delta) \leq \widehat{\gamma}_N\right\} \geq p^*.$$

1. Determine finite sample size $N = \overline{N}_{\mathrm{wc}}(p^*, \delta)$;
2. Draw N samples $\Delta^{(1)}, \dots, \Delta^{(N)}$ according to f_{Δ};
3. Return the empirical maximum

$$\widehat{\gamma}_N = \max_{i=1,\dots,N} J(\Delta^{(i)}).$$

In Chapter 9, a finite sample size bound is also derived for this RA. In particular, Theorem 9.1 shows that if N in Algorithm 8.2 is chosen according to

$$\overline{N}_{\mathrm{wc}}(p^*, \delta) \doteq \left\lceil \frac{\log \frac{1}{\delta}}{\log \frac{1}{p^*}} \right\rceil$$

then this RA satisfies the requirements of Definition 8.2.

8.3 Randomized algorithms for synthesis

In the next sections, we introduce definitions of RAs for probabilistic controller synthesis and present the associated meta-algorithms. In the synthesis paradigm, the performance function measures the closed-loop performance of the system with "controller" θ, hence we write $J = J(\Delta, \theta) : \mathbb{D} \times \Theta \to \mathbb{R}$.

Two different philosophies are currently employed for control synthesis in the probabilistic context. The first philosophy aims at designing a controller that performs well *on average*, i.e. a controller that minimizes the average (with respect to the uncertainty Δ) performance index. An alternative synthesis paradigm is to seek a controller which satisfies the performance specification for most values of the uncertainties, i.e. that is robust in a probabilistic sense. The first approach is discussed in the next section and the probabilistic robust design approach is presented in Section 8.3.2.

8.3.1 Randomized algorithms for average synthesis

We introduce here the requirements that an RA for average performance synthesis should fulfill.

Definition 8.3 (RA for average performance synthesis). *Let $\Delta \in \mathbb{D}$ be random with density $f_{\Delta}(\Delta)$, and let $\epsilon \in (0,1)$, $\delta \in (0,1)$ be assigned probability levels. Consider a performance function $J(\Delta, \theta) : \mathbb{D} \times \Theta \to [0,1]$. Denote the* average *performance (with respect to Δ) of the controlled plant by*

$$\phi(\theta) \doteq \mathrm{E}_{\Delta}(J(\Delta, \theta))$$

and the optimal achievable average performance by $\phi^* \doteq \inf_{\theta \in \Theta} \phi(\theta)$. An RA for average synthesis should return with probability at least $1 - \delta$ a design vector $\widehat{\theta}_N \in \Theta$, such that

$$\phi(\widehat{\theta}_N) - \phi^* \leq \epsilon.$$

The controller parameter $\widehat{\theta}_N$ should be constructed based on a finite number N of random samples of Δ.

An RA for average synthesis satisfying the requirements of Definition 8.3 is derived using results from statistical learning theory discussed in Chapter 10. In this approach, the design algorithm follows two steps. In the first step, the expected value $\phi(\theta) = \mathrm{E}_{\Delta}(J(\Delta, \theta))$ is estimated empirically, and in the second step a minimization is performed on the estimate of $\phi(\theta)$ to obtain the "optimal" controller.

Algorithm 8.3 (Average performance synthesis).
Given $\epsilon, \delta \in (0,1)$, this RA returns with probability at least $1 - \delta$ a design vector $\widehat{\theta}_N$ such that

$$\phi(\widehat{\theta}_N) - \phi^* \leq \epsilon.$$

1. Determine finite sample size $N = \overline{N}_{\mathrm{av}}(\epsilon, \delta)$;
2. Draw N samples $\Delta^{(1)}, \ldots, \Delta^{(N)}$ according to f_{Δ};
3. Return the controller parameter

$$\widehat{\theta}_N = \arg \inf_{\theta \in \Theta} \frac{1}{N} \sum_{i=1}^{N} J(\Delta^{(i)}, \theta). \tag{8.1}$$

The sample complexity of this algorithm may be assessed using the uniform convergence bounds derived in Chapter 10. In particular, the sample size in Algorithm 8.3 can be derived using Corollary 10.2, obtaining

$$\overline{N}_{\mathrm{av}}(\epsilon,\delta) \doteq \left\lceil \frac{128}{\epsilon^2} \left(\log \frac{8}{\delta} + d \left(\log \frac{32e}{\epsilon} + \log \log \frac{32e}{\epsilon} \right) \right) \right\rceil \tag{8.2}$$

where d is an upper bound on the so-called P-dimension of the function family $\{J(\cdot,\theta), \theta \in \Theta\}$; see Definition 10.3.

Notice that, in principle, the minimization over $\theta \in \Theta$ in (8.1) can be performed by any numerical optimization method. However, since this constrained minimization problem is in general nonconvex, there are obvious difficulties in finding its global solution. Thus, a viable approach is to use a randomized algorithm also for this minimization. In this way, randomness is artificially introduced for the solution of the nonconvex optimization problem over the controller parameters. This global randomized algorithm, presented in Section 10.3, returns a so-called probable approximate minimizer of the averaged performance function.

Certainly, a notable feature of this synthesis approach based on "double randomization" is its complete generality. However, two main criticisms are in order. The first one relates to the "looseness" of the bound (8.2), which makes the required number of samples so large to be hardly useful in practice. Secondly, randomization over the controller parameters, which is in general necessary, leads to rather weak statements on the quality of the resulting solutions. This latter problem can be in principle avoided when the performance function $J(\Delta, \theta)$ is *convex* in θ for any fixed Δ.

In the next section, we discuss RAs for robust (as opposed to average) performance synthesis, which are based on sequential stochastic iterations.

8.3.2 Randomized algorithms for robust synthesis

A different synthesis philosophy follows from the discussion in the introduction of this chapter, and aims at designing controllers that satisfy the performance specification for "most" values of the uncertainties, i.e. controllers that are robust in a probabilistic sense. An RA that meets this design approach is specified next.

Definition 8.4 (RA for robust performance synthesis). *Let $\Delta \in \mathbb{D}$ be random with density $f_\Delta(\Delta)$, and let $p^* \in (0,1)$ and $\delta \in (0,1)$ be assigned probability levels. Given a performance function $J(\Delta, \theta) : \mathbb{D} \times \Theta \to \mathbb{R}$ and associated level γ the RA should return with probability at least $1 - \delta$ a design vector $\widehat{\boldsymbol{\theta}}_N \in \Theta$, such that*

$$\mathrm{PR} \left\{ J(\boldsymbol{\Delta}, \widehat{\boldsymbol{\theta}}_N) \leq \gamma \right\} \geq p^*. \tag{8.3}$$

The controller parameter $\widehat{\boldsymbol{\theta}}_N$ should be constructed based on a finite number N of random samples of $\boldsymbol{\Delta}$.

In Chapters 11, 12 and 13, we study randomized algorithms that fulfill the requirements of Definition 8.4, and do not require randomization over the controller parameters. The RAs of Chapters 11 and 12 belong to the general class of stochastic optimization algorithms, and are applicable whenever $J(\Delta, \theta)$ is *convex* in θ for any fixed Δ. In particular, they are based on an *update rule* $\psi_{\mathrm{upd}}(\Delta^{(i)}, \theta^{(k)})$ which is applied to the current design parameter $\theta^{(k)}$, when this parameter leads to a performance violation, to determine a new candidate parameter $\theta^{(k+1)}$. As discussed further in Chapter 11, this update rule can, for instance, be a simple gradient descent step, or a step of the ellipsoid method for convex optimization. An important role in these algorithms is played by the *performance violation function*, see Definition 11.1. A meta-algorithm for robust performance synthesis is now presented.

Algorithm 8.4 (Robust performance synthesis).
Given $p^*, \delta \in (0,1)$ and γ, this RA returns with probability at least $1 - \delta$ a design vector $\widehat{\theta}_N$ such that

$$\mathrm{PR}\left\{ J(\Delta, \widehat{\theta}_N) \le \gamma \right\} \ge p^*.$$

1. Initialization.
   ▷ Determine a sample size function $N(k) = \overline{N}_{\mathrm{ss}}(p^*, \delta, k)$;
   ▷ Set $k = 0$, $i = 0$, and choose $\theta^{(0)} \in \Theta$;
2. Feasibility loop.
   ▷ Set $\ell = 0$ and feas=true;
   ▷ While $\ell < N(k)$ and feas=true
      − Set $i = i + 1$, $\ell = \ell + 1$;
      − Draw $\Delta^{(i)}$ according to f_Δ;
      − If $J(\Delta^{(i)}, \theta^{(k)}) > \gamma$ set feas=false;
   ▷ End While;
3. Exit condition.
   ▷ If feas=true
      − Set $N = i$;
      − Return $\widehat{\theta}_N = \theta^{(k)}$ and Exit;
   ▷ End If;
4. Update.
   ▷ Update $\theta^{(k+1)} = \psi_{\mathrm{upd}}(\Delta^{(i)}, \theta^{(k)})$;
   ▷ Set $k = k + 1$ and goto 2.

We now make some comments on this algorithm. First notice that, for any candidate controller $\theta^{(k)}$, the inner feasibility loop of the algorithm (step 2) performs a randomized check of robust feasibility on the current solution. If this test is passed, i.e. the feasibility loop is run up to $N(k)$, then the algorithm returns the solution $\widehat{\boldsymbol{\theta}}_N = \boldsymbol{\theta}^{(k)}$. Otherwise, if $\boldsymbol{\Delta}^{(i)}$ is found such that the performance is violated (i.e. $J(\boldsymbol{\Delta}^{(i)}, \boldsymbol{\theta}^{(k)}) > \gamma$), then the current solution $\boldsymbol{\theta}^{(k)}$ is updated and the process is repeated.

An important observation is that the number of checks to be performed in the inner feasibility loop is not fixed but is a *function* of the number of updates k executed so far. In particular, if this function is chosen according to the bound given in [206], i.e.

$$\overline{N}_{\mathrm{ss}}(p^*, \delta, k) \doteq \left\lceil \frac{\log \frac{\pi^2 (k+1)^2}{6\delta}}{\log \frac{1}{1-\epsilon}} \right\rceil$$

then, with probability greater than $1 - \delta$, (8.3) holds if the feasibility test is passed. In Chapters 11 and 12 we give results for specific control problems guaranteeing finite convergence of the above algorithm under different update rules. In particular, upper bounds on the total number of samples N to be drawn are reported.

Chapter 13 deals instead with a scenario approach for design and studies a different randomized algorithm based on one-shot convex optimization. For this algorithm, which is not sequential, a finite sample size N may be determined a priori.

8.4 Computational complexity of randomized algorithms

The computational complexity of an RA is due to three main sources: the computational cost of generating random samples of $\boldsymbol{\Delta}$ according to the density $f_{\boldsymbol{\Delta}}(\Delta)$, the computational cost of evaluating the performance $J(\Delta)$ for fixed Δ, and the minimum number N of samples required to attain the desired probabilistic levels. A formal definition of an *efficient* randomized algorithm is presented next.

Definition 8.5 (Efficient RA). *An RA is said to be* efficient *if:*

1. *The random sample generation can be performed in polynomial time;*
2. *The performance function can be evaluated in polynomial time for fixed Δ;*
3. *The sample size is polynomial in the problem size and probabilistic levels.*

The random sample generation depends on the type and structure of the set in which randomization is performed. This issue is discussed in detail in Chapters 16 and 18. More precisely, it is shown that in many cases of practical interest in robustness, namely vectors and block-structured matrices with norm bound, uniform sample generation can be performed efficiently in polynomial time.

Concerning the second issue above, we remark that, in the majority of cases arising in control problems, the performance function $J(\Delta)$ can be efficiently evaluated for fixed Δ. For example, stability tests, as well as other performance tests based on the solution of linear matrix inequalities, can be performed in polynomial time.

The third source of complexity is indeed the most critical one for a randomized algorithm, and it is discussed in detail in the Chapters 9 and 10, for each specific case of interest. In particular, for analysis RAs, explicit bounds on the sample complexity are available. These bounds depend polynomially on the probabilistic levels, and are actually *independent* of the problem size (dimension and structure of Δ, and type of performance function $J(\Delta)$). For synthesis RAs, similar bounds are shown which are polynomial in the probabilistic levels and in the size and structure of the problem.

9. Probability Inequalities

In this chapter we address the issue of finite sample size in probability estimation. That is, we analyze the reliability of the probabilistic estimates introduced in Chapter 7, for *finite* values of N. This issue is crucial in the development of randomized algorithms for uncertain systems and control and makes a clear distinction with the *asymptotic* laws of large numbers preliminarily discussed in Chapter 7.

9.1 Probability inequalities

This section presents some standard material on probability inequalities, which is the backbone for the sample size bounds subsequently derived in this chapter. The first fundamental result is the Markov inequality.

Markov inequality. *Let* $\mathbf{x} \in [0, \infty)$ *be a nonnegative random variable with* $E(\mathbf{x}) < \infty$. *Then, for any* $\epsilon > 0$, *we have*

$$\mathrm{PR}\left\{\mathbf{x} \geq \epsilon\right\} \leq \frac{E(\mathbf{x})}{\epsilon}. \tag{9.1}$$

Proof. The proof of this result is immediate and follows from the chain of inequalities

$$E(\mathbf{x}) = \int_0^\infty x f_{\mathbf{x}}(x) \mathrm{d}x \geq \int_\epsilon^\infty x f_{\mathbf{x}}(x) \mathrm{d}x \geq \epsilon \int_\epsilon^\infty f_{\mathbf{x}}(x) \mathrm{d}x = \epsilon \mathrm{PR}\left\{\mathbf{x} \geq \epsilon\right\}.$$

$\square$

Obviously, the Markov inequality, as well as the other inequalities presented in this section, is meaningful only when the right-hand side of (9.1) is not greater than one. We now show that various classical inequalities can be derived from the Markov inequality. To this end, let a and $m > 0$ be two real numbers and observe that the random variable $|\mathbf{x} - a|^m$ is nonnegative. Then, applying the Markov inequality to this random variable, we obtain

$$\mathrm{PR}\left\{|\mathbf{x} - a|^m \geq \epsilon^m\right\} \leq \frac{E(|\mathbf{x} - a|^m)}{\epsilon^m}.$$

Taking $a = E(\mathbf{x})$, we immediately derive to the so-called Bienaymé inequality

$$\mathrm{PR}\{|\mathbf{x} - E(\mathbf{x})| \geq \epsilon\} \leq \frac{E(|\mathbf{x} - E(\mathbf{x})|^m)}{\epsilon^m}. \tag{9.2}$$

The well-known Chebychev inequality is a special case of Bienaymé inequality, obtained for $m = 2$.

Chebychev inequality. *Let* $\mathbf{x}$ *be a random variable with* $\mathrm{Var}(\mathbf{x}) < \infty$. *Then, for any* $\epsilon > 0$, *we have*

$$\mathrm{PR}\{|\mathbf{x} - E(\mathbf{x})| \geq \epsilon\} \leq \frac{\mathrm{Var}(\mathbf{x})}{\epsilon^2}. \tag{9.3}$$

We remark that, while in the Markov inequality $\mathbf{x}$ is a nonnegative random variable, in Bienaymé and Chebychev inequalities there is no sign restriction. However, these latter inequalities hold only when $\mathbf{x}$ has bounded variance or bounded moment of order m.

Remark 9.1 (Historical remarks). The problems concerning the computation of probability inequalities given moments of different order of a random variable have a long history and a rich literature. For example, moment problems have been analyzed in the early works of Stieltjes [255, 256]. An elementary introduction to Chebychev and Markov inequalities is given in [214]. The interested reader may also refer to the original paper of Chebychev [64] and to the thesis of his student Markov [180]. Multivariate generalizations of these inequalities are studied in [181, 269] and in [222], which also contains an historical overview of the topic. Additional related results on large deviation methods can be found in [81].   ◁

We now analyze other less well known inequalities. In all these inequalities we assume $\epsilon > 0$. The first one we study was derived by Uspensky [276]

$$\mathrm{PR}\{\mathbf{x} \geq (1 + \epsilon)E(\mathbf{x})\} \leq \frac{\mathrm{Var}(\mathbf{x})}{\mathrm{Var}(\mathbf{x}) + \epsilon^2 E(\mathbf{x})^2}.$$

This inequality always improves upon the so-called right-sided Chebychev inequality

$$\mathrm{PR}\{\mathbf{x} \geq (1 + \epsilon)E(\mathbf{x})\} \leq \frac{\mathrm{Var}(\mathbf{x})}{\epsilon^2 E(\mathbf{x})^2}.$$

For completeness, we also state the left-sided Chebychev inequality

$$\mathrm{PR}\{\mathbf{x} \leq (1 - \epsilon)E(\mathbf{x})\} \leq \frac{\mathrm{Var}(\mathbf{x})}{\mathrm{Var}(\mathbf{x}) + \epsilon^2 E(\mathbf{x})^2}.$$

Other inequalities are proved in [222] for the case when moments up to third order are given. These inequalities are "tight" in a certain sense and have been derived via a general convex optimization reformulation, see [38]. For instance, for a nonnegative random variable $\mathbf{x}$, it is shown that

$$\text{PR}\left\{|\mathbf{x} - \text{E}(\mathbf{x})| \geq \epsilon\text{E}(\mathbf{x})\right\} \leq \min\left(1, 1 + 27\frac{\alpha^2 + \beta^4 - \epsilon^2}{4 + 3(1 + 3\epsilon^2) + 2(1 + 3\epsilon^2)^{\frac{3}{2}}}\right)$$

where

$$\alpha = \frac{\text{E}(\mathbf{x})\text{E}(\mathbf{x}^3) - \text{E}(\mathbf{x}^2)}{\text{E}(\mathbf{x})^4} \quad \text{and} \quad \beta = \frac{\text{Var}(\mathbf{x})}{\text{E}(\mathbf{x})^2}.$$

In the next section we study applications of the previous inequalities (and in particular of the Markov inequality) to the problem of bounding the probability of deviation from the mean of sums of random variables.

9.2 Deviation inequalities for sums of random variables

We here focus our attention on inequalities for the tail probabilities of the *sum of random variables*. That is, we consider N independent random variables $\mathbf{x}_1, \ldots, \mathbf{x}_N$, define the new random variable

$$\mathbf{s}_N \doteq \sum_{i=1}^{N} \mathbf{x}_i$$

and aim to compute bounds on the probability $\text{PR}\left\{|\mathbf{s}_N - \text{E}(\mathbf{s}_N)| \geq \epsilon\right\}$.

A first simple inequality for sums of random variables may be directly derived from the Chebychev inequality, obtaining

$$\text{PR}\left\{|\mathbf{s}_N - \text{E}(\mathbf{s}_N)| \geq \epsilon\right\} \leq \frac{\text{Var}(\mathbf{s}_N)}{\epsilon^2} = \frac{\sum_{i=1}^{N}\text{Var}(\mathbf{x}_i)}{\epsilon^2}. \tag{9.4}$$

A tighter classical inequality, which holds for the case of bounded random variables, is due to Hoeffding [135]. Before stating and proving the Hoeffding inequality, we need to introduce a lemma.

Lemma 9.1. *Let $\mathbf{x} \in [a, b]$ be a random variable with $\text{E}(\mathbf{x}) = 0$. Then, for any $\lambda > 0$*

$$\text{E}(e^{\lambda\mathbf{x}}) \leq e^{\lambda^2(b-a)^2/8}. \tag{9.5}$$

Proof. Since $\mathbf{x} \in [a, b]$, we write it as a convex combination of a and b, namely $\mathbf{x} = \eta b + (1-\eta)a$, where $\eta = (\mathbf{x} - a)/(b - a)$. So, by convexity of $e^{\lambda\mathbf{x}}$ we have

$$e^{\lambda\mathbf{x}} \leq \frac{\mathbf{x} - a}{b - a}e^{\lambda b} + \frac{b - \mathbf{x}}{b - a}e^{\lambda a}.$$

Taking expectation of both sides and using the fact that $E(\mathbf{x}) = 0$, we get

$$E\left(e^{\lambda \mathbf{x}}\right) \leq -\frac{a}{b-a}e^{\lambda b} + \frac{b}{b-a}e^{\lambda a} = \left(1 - p + pe^{\lambda(b-a)}\right)e^{-p\lambda(b-a)}$$

where $p = -a/(b-a)$. Next, defining the function

$$L(u) \doteq -pu + \log(1 - p + pe^u)$$

we have, for $u = \lambda(b-a)$

$$E\left(e^{\lambda \mathbf{x}}\right) \leq e^{L(u)}.$$

The first and second derivatives of $L(u)$ are given by

$$L'(u) = -p + \frac{pe^u}{1 - p + pe^u};$$

$$L''(u) = \frac{p(1-p)e^u}{(1 - p + pe^u)^2} \leq \frac{1}{4} \quad \text{for all } u > 0.$$

Therefore, using Taylor expansion, we have that for some $\xi \in (0, u)$

$$L(u) = L(0) + uL'(0) + \frac{u^2}{2}L''(\xi) = \frac{u^2}{2}L''(\xi) \leq \frac{u^2}{8} = \frac{\lambda^2(b-a)^2}{8}$$

which proves the lemma.   □

We now state the Hoeffding inequality.

Hoeffding inequality. *Let* $\mathbf{x}_1, \ldots, \mathbf{x}_N$ *be independent bounded random variables with* $\mathbf{x}_i \in [a_i, b_i]$. *Then, for any* $\epsilon > 0$, *we have*

$$\mathrm{PR}\left\{\mathbf{s}_N - E(\mathbf{s}_N) \geq \epsilon\right\} \leq e^{-2\epsilon^2/\sum_{i=1}^N (b_i - a_i)^2} \tag{9.6}$$

and

$$\mathrm{PR}\left\{\mathbf{s}_N - E(\mathbf{s}_N) \leq -\epsilon\right\} \leq e^{-2\epsilon^2/\sum_{i=1}^N (b_i - a_i)^2}. \tag{9.7}$$

Proof. The inequality is derived using the Chernoff bounding method. That is, for any random variable $\mathbf{x}$, we write the Markov inequality for the random variable $e^{\lambda \mathbf{x}}$ with $\lambda > 0$, obtaining

$$\mathrm{PR}\left\{e^{\lambda \mathbf{x}} \geq \alpha\right\} \leq \frac{E\left(e^{\lambda \mathbf{x}}\right)}{\alpha}$$

for any $\alpha > 0$. Then, taking $\alpha = e^{\lambda \epsilon}$ and replacing $\mathbf{x}$ with $\mathbf{x} - E(\mathbf{x})$, we write

$$\mathrm{PR}\left\{\mathbf{x} - E(\mathbf{x}) \geq \epsilon\right\} \leq e^{-\lambda \epsilon}E\left(e^{\lambda(\mathbf{x} - E(\mathbf{x}))}\right). \tag{9.8}$$

Applying this bound to the random variable $\mathbf{s}_N$, due to the independence of the random variables $\mathbf{x}_i$, we obtain

$$\mathrm{PR}\left\{\mathbf{s}_N - \mathrm{E}(\mathbf{s}_N) \geq \epsilon\right\} \leq e^{-\lambda\epsilon}\mathrm{E}\left(e^{\lambda \sum_{i=1}^{N}(\mathbf{x}_i - \mathrm{E}(\mathbf{x}_i))}\right)$$

$$= e^{-\lambda\epsilon}\prod_{i=1}^{N}\mathrm{E}\left(e^{\lambda(\mathbf{x}_i - \mathrm{E}(\mathbf{x}_i))}\right). \tag{9.9}$$

To complete the proof, we apply the result of Lemma 9.1 in combination with (9.9), obtaining

$$\mathrm{PR}\left\{\mathbf{s}_N - \mathrm{E}(\mathbf{s}_N) \geq \epsilon\right\} \leq e^{-\lambda\epsilon}\prod_{i=1}^{N}\mathrm{E}\left(e^{\lambda(\mathbf{x}_i - \mathrm{E}(\mathbf{x}_i))}\right)$$

$$\leq e^{-\lambda\epsilon}\prod_{i=1}^{N}e^{\lambda^2(b_i - a_i)^2/8} = e^{-\lambda\epsilon}e^{\lambda^2\sum_{i=1}^{N}(b_i - a_i)^2/8}.$$

Inequality (9.6) is obtained by selecting λ such that the exponent is minimized

$$\lambda = \frac{4\epsilon}{\sum_{i=1}^{N}(b_i - a_i)^2}.$$

Inequality (9.7) follows from similar derivations. $\qquad\square$

The Hoeffding inequality takes a simpler form in the case when the random variables $\mathbf{x}_i$ are independent and bounded in the same interval $[a, b]$. In this case, combining (9.6) and (9.7), we derive the inequality presented next.

Two-sided Hoeffding inequality. *Let* $\mathbf{x}_1, \ldots, \mathbf{x}_N$ *be independent random variables such that* $\mathbf{x}_i \in [a, b]$. *Then, for any* $\epsilon > 0$, *we have*

$$\mathrm{PR}\left\{|\mathbf{s}_N - \mathrm{E}(\mathbf{s}_N)| \geq \epsilon\right\} \leq 2e^{-2\epsilon^2/(N(b-a)^2)}. \tag{9.10}$$

Finally, we state without proof another classical inequality due to Bernstein [37], see additional details and extensions in [35].

Bernstein inequality. *Let* $\mathbf{x}_1, \ldots, \mathbf{x}_N$ *be independent random variables with* $\mathbf{x}_i \in [-a, a]$, $\mathrm{E}(\mathbf{x}_i) = 0$ *and* $\mathrm{Var}(\mathbf{x}_i) < \infty$. *Then, for any* $\epsilon > 0$, *we have*

$$\mathrm{PR}\left\{\mathbf{s}_N \geq \epsilon\right\} \leq e^{-\epsilon^2/(2N\sigma^2 + 2a\epsilon/3)} \tag{9.11}$$

where $\sigma^2 = \frac{1}{N}\sum_{i=1}^{N}\mathrm{Var}(\mathbf{x}_i)$.

Remark 9.2 (Concentration inequalities). More general functions (other than sums) of independent random variables may also be bounded using so-called "concentration" inequalities, such as the Efron–Stein inequality [102] or Talagrand inequality, but this topic goes beyond the scope of this discussion. We address the reader to [47, 260] and the references therein for further details on these issues.

9.3 Sample size bounds for probability estimation

In this section we specialize the previous inequalities to derive bounds on the sample size for the randomized algorithms presented in Chapter 8. In fact, these inequalities are the key tools for determining the minimum number of samples needed to compute the reliability of the estimate

$$\widehat{\mathbf{p}}_N(\gamma) = \frac{1}{N} \sum_{i=1}^{N} \mathbb{I}_{\mathcal{B}_G}(\mathbf{\Delta}^{(i)})$$

of the probability of performance introduced in Chapter 6

$$p(\gamma) = \mathrm{PR}_{\mathbf{\Delta}} \{J(\mathbf{\Delta}) \leq \gamma\}.$$

This reliability is measured in terms of the "closeness" of $\widehat{\mathbf{p}}_N(\gamma)$ to the true probability $p(\gamma)$. That is, given $\epsilon \in (0,1)$, we wish to assure that the event

$$|\widehat{\mathbf{p}}_N(\gamma) - p(\gamma)| < \epsilon$$

holds with high probability. Since $\widehat{\mathbf{p}}_N(\gamma)$ is estimated via random sampling, it is itself a random variable which depends on the multisample $\mathbf{\Delta}^{(1...N)}$. Therefore, for given $\delta \in (0,1)$, we require that

$$\mathrm{PR}_{\mathbf{\Delta}^{(1...N)}} \{|\widehat{\mathbf{p}}_N(\gamma) - p(\gamma)| < \epsilon\} > 1 - \delta. \tag{9.12}$$

The problem is then finding the minimal N such that (9.12) is satisfied for fixed *accuracy* $\epsilon \in (0,1)$ and *confidence* $\delta \in (0,1)$.

The first bound on the sample size was derived in 1713 by Jacob Bernoulli [36] and is reported for historical reasons and for its simplicity.

Bernoulli bound. *For any $\epsilon \in (0,1)$ and $\delta \in (0,1)$, if*

$$N \geq \frac{1}{4\epsilon^2 \delta} \tag{9.13}$$

then, with probability greater than $1 - \delta$, we have $|\widehat{\mathbf{p}}_N(\gamma) - p(\gamma)| < \epsilon$.

Proof. This result is proved by means of the Chebychev inequality. The empirical probability $\widehat{\mathbf{p}}_N(\gamma)$ is a random variable binomially distributed, with expected value $\mathrm{E}(\widehat{\mathbf{p}}_N(\gamma)) = p(\gamma)$ and variance $\mathrm{Var}(\widehat{\mathbf{p}}_N(\gamma)) = p(\gamma)(1 - p(\gamma))/N$. Substituting these values in (9.3) for $\mathbf{x} = \widehat{\mathbf{p}}_N(\gamma)$, we obtain

$$\mathrm{PR}\{|\widehat{\mathbf{p}}_N(\gamma) - p(\gamma)| \geq \epsilon\} \leq \frac{p(\gamma)(1 - p(\gamma))}{N\epsilon^2} \leq \frac{1}{4N\epsilon^2}$$

for all $p(\gamma) \in (0,1)$. The bound (9.13) then follows immediately from this inequality.   □

A significant improvement on the previous bound is given by the classical Chernoff bound [67].

Chernoff bound. *For any $\epsilon \in (0,1)$ and $\delta \in (0,1)$, if*

$$N \geq \frac{1}{2\epsilon^2} \log \frac{2}{\delta} \tag{9.14}$$

then, with probability greater than $1 - \delta$, we have $|\widehat{\mathbf{p}}_N(\gamma) - p(\gamma)| < \epsilon$.

Proof. The Chernoff bound follows from direct application of the Hoeffding inequality to the random variables $\mathbf{x}_1, \ldots, \mathbf{x}_N$, defined as

$$\mathbf{x}_i = \mathbb{I}_{\mathcal{B}_G}(\mathbf{\Delta}^{(i)}) = \begin{cases} 1 & \text{if } \mathbf{\Delta}^{(i)} \in \mathcal{B}_G; \\ 0 & \text{otherwise} \end{cases}$$

for $i = 1, \ldots, N$. Since $\mathbf{x}_i \in [0,1]$, letting $\mathbf{s}_N = \sum_{i=1}^{N} \mathbf{x}_i$ and applying inequality (9.10) we get

$$\mathrm{PR}\{|\mathbf{s}_N - \mathrm{E}(\mathbf{s}_N)| \geq \epsilon\} \leq 2e^{-2\epsilon^2/N}.$$

Now, observing that $\widehat{\mathbf{p}}_N(\gamma) = \mathbf{s}_N/N$ and $\mathrm{E}(\widehat{\mathbf{p}}_N(\gamma)) = p(\gamma)$, we write

$$\mathrm{PR}\{|\widehat{\mathbf{p}}_N(\gamma) - p(\gamma)| \geq \epsilon\} \leq 2e^{-2N\epsilon^2}$$

from which the desired bound follows immediately. $\square$

Remark 9.3 (Chernoff bound and RAs). The Chernoff bound (9.14) is the key tool for the application of the randomized algorithm for probabilistic performance verification (Algorithm 8.1) presented in Chapter 8. $\triangleleft$

Remark 9.4 (Chernoff inequalities). In the proof of the Chernoff bound, applying the two-sided Hoeffding inequality, we obtained the Chernoff inequality

$$\mathrm{PR}\{|\widehat{\mathbf{p}}_N(\gamma) - p(\gamma)| \geq \epsilon\} \leq 2e^{-2N\epsilon^2} \text{ for } \epsilon \in (0,1).$$

Similarly, applying the one-sided Hoeffding inequalities, we obtain the one-sided Chernoff inequalities

$$\mathrm{PR}\{\widehat{\mathbf{p}}_N(\gamma) - p(\gamma) \geq \epsilon\} \leq e^{-2N\epsilon^2}; \tag{9.15}$$

$$\mathrm{PR}\{\widehat{\mathbf{p}}_N(\gamma) - p(\gamma) \leq -\epsilon\} \leq e^{-2N\epsilon^2} \tag{9.16}$$

for $\epsilon \in (0,1)$. The above inequalities are often denoted as *additive* Chernoff inequalities to distinguish them from the so-called *multiplicative* Chernoff inequalities, see e.g. [284]. These latter inequalities take the form

$$\mathrm{PR}\{\widehat{\mathbf{p}}_N(\gamma) - p(\gamma) \geq \epsilon p(\gamma)\} \leq e^{-p(\gamma)N\epsilon^2/3}; \tag{9.17}$$

$$\mathrm{PR}\{\widehat{\mathbf{p}}_N(\gamma) - p(\gamma) \leq -\epsilon p(\gamma)\} \leq e^{-p(\gamma)N\epsilon^2/2} \tag{9.18}$$

for $\epsilon \in (0,1)$. Contrary to those in the additive form, we observe that these inequalities are not symmetric. Notice further that, for sufficiently large values of $p(\gamma)$, the multiplicative inequalities are more conservative than the additive ones. In particular, comparing (9.15) and (9.17), it can be easily shown that for $p(\gamma) > 1/6$ the right-tail additive bound is superior. For $p(\gamma) > 1/4$, the same conclusion holds for the left-tail bound. To conclude this discussion, we remark that the so-called Chernoff–Okamoto inequality is less conservative than the previous ones, but it holds only for $p(\gamma) \leq 0.5$. Refer to [284] for additional discussions on these topics.   ◁

From the one-sided Chernoff inequality we immediately derive the one-sided Chernoff bound.

One-sided Chernoff bound. *For any $\epsilon \in (0,1)$ and $\delta \in (0,1)$, if*

$$N \geq \frac{1}{2\epsilon^2} \log \frac{1}{\delta} \tag{9.19}$$

then, with probability greater than $1 - \delta$, we have $\widehat{\mathbf{p}}_N(\gamma) - p(\gamma) < \epsilon$.

Example 9.1 (RA for probability of Schur stability).
    We revisit Example 6.10 regarding Schur stability of a discrete-time system affected by parametric uncertainty $q \in \mathcal{B}_q$. In particular, we consider the same fourth-order monic polynomial

$$p(z, q) = q_0 + q_1 z + q_2 z^2 + q_3 z^3 + z^4$$

but we now take a different hyperrectangle given by

$$\mathcal{B}_q = \{q \in \mathbb{R}^4 : \|q\|_\infty \leq 0.5\}.$$

In this case, conditions (6.10) do not guarantee that $\mathcal{Q}_n \subseteq \mathcal{B}_q$, and the formulae (6.8) and (6.9) cannot be used. Therefore, we estimate the probability of stability using Algorithm 8.1 of Chapter 8. In particular, we select the performance function (6.7) and set $\gamma = 1/2$. Then, we assign probability levels $\epsilon = 0.005$ and $\delta = 0.001$ and, by means of the Chernoff bound, we determine the sample size

$$\overline{N}(\epsilon, \delta) = 1.52 \times 10^5.$$

With this (or larger) sample size we guarantee that with probability at least $1 - \delta = 0.999$

$$|p(\gamma) - \widehat{\mathbf{p}}_N(\gamma)| \leq 0.005.$$

In particular, we applied the RA, using $N = 155,000$ iid samples of the random vector $\mathbf{q} \in \mathcal{B}_q$ uniformly distributed within $\mathcal{B}_q$, obtaining an estimated probability

$$\widehat{p}_N = 0.9906.$$

In words, we can say with 99.9% confidence that at least 98.56% of the uncertain polynomials are Schur stable.   ⋆

Example 9.2 (RA for probability of $\mathcal{H}_\infty$ performance).
In this example we consider a continuous-time SISO system affected by parametric uncertainty, originally presented in [24]. The transfer function of the system is

$$G(s,q) = \frac{0.5q_1 q_2 s + 10^{-5}q_1}{(10^{-5} + 0.005q_2)s^2 + (0.00102 + 0.5q_2)s + (2\times10^{-5} + 0.5q_1^2)}$$

$$(9.20)$$

where the uncertainty q is bounded in the hyperrectangle

$$\mathcal{B}_q = \{q \in \mathbb{R}^2 : 0.2 \le q_1 \le 0.6, 10^{-5} \le q_2 \le 3\times10^{-5}\}.$$

Assume further that $\mathbf{q}$ is random, with uniform distribution over the set $\mathcal{B}_q$. We are interested in evaluating the probability of attaining an $\mathcal{H}_\infty$ performance level no larger than $\gamma = 0.003$. We apply a randomized approach, setting the performance function to $J(\mathbf{q}) = \|G(s,\mathbf{q})\|_\infty$. Then, we assign probability levels $\epsilon = 0.01$ and $\delta = 0.001$ and apply Algorithm 8.1. By means of the Chernoff bound, we determine the sample size

$$\overline{N}(\epsilon, \delta) = 3.801\times10^4$$

which guarantees that with probability at least $1 - \delta = 0.999$

$$|p(\gamma) - \widehat{\mathbf{p}}_N(\gamma)| \le 0.01.$$

To apply the RA, we generated $N = 40,000$ iid samples of $\mathbf{q} \in \mathcal{B}_q$, uniformly distributed within $\mathcal{B}_q$ and obtained the empirical mean

$$\widehat{p}_N = 0.3482.$$

In practice, we may conclude with 99.9% confidence that the $\mathcal{H}_\infty$ performance level is below $\gamma = 0.003$ for at least 33.8% of the uncertain plants.   ⋆

We remark that the Chernoff bound largely improves upon the Bernoulli bound. In particular, whereas the sample size in Bernoulli depends on $1/\delta$, the Chernoff bound is a function of $\log(2/\delta)$. However, in both cases, the dependence on ϵ is unchanged and it is inversely proportional to ϵ^2. We conclude that confidence is "cheaper" than accuracy. Table 9.1 shows a comparison between these bounds for several values of ϵ and δ.

We observe that the bounds discussed in these sections can be computed *a priori* and are explicit. That is, given ϵ and δ one can find directly the minimum value of N without generating the samples $\Delta^{(1\cdots N)}$ and evaluating $J(\Delta^{(i)})$ for $i = 1, \ldots, N$. On the other hand, when computing the classical *confidence intervals*, see e.g. [70], the sample size obtained is not explicit. More precisely, for given $\delta \in (0,1)$, the lower and upper confidence intervals $\mathbf{p}_L$ and $\mathbf{p}_U$ are such that

Table 9.1. Comparison of the sample size obtained with Bernoulli and Chernoff bounds for different values of ϵ and δ.

ϵ	$1-\delta$	Bernoulli	Chernoff
0.05	0.95	2000	738
	0.99	1.00×10^4	1060
	0.995	2.00×10^4	1199
	0.999	1.00×10^5	1521
0.01	0.95	5.00×10^4	1.84×10^4
	0.99	2.50×10^5	2.65×10^4
	0.995	5.00×10^5	3.00×10^4
	0.999	2.50×10^6	3.80×10^4
0.005	0.95	2.00×10^5	7.38×10^4
	0.99	1.00×10^6	1.06×10^5
	0.995	2.00×10^6	1.20×10^5
	0.999	1.00×10^7	1.52×10^5
0.001	0.95	5.00×10^6	1.84×10^6
	0.99	2.50×10^7	2.65×10^6
	0.995	5.00×10^7	3.00×10^6
	0.999	2.50×10^8	3.80×10^6

$$\mathrm{PR}_{\boldsymbol{\Delta}(1\ldots N)}\left\{\mathbf{p}_L \le p(\gamma) \le \mathbf{p}_U\right\} > 1 - \delta.$$

The evaluation of this probability requires the solution with respect to $\mathbf{p}_L$ and $\mathbf{p}_U$ of equations of the type

$$\sum_{k=\mathbf{N}_G}^{N} \binom{N}{k} \mathbf{p}_L{}^k (1 - \mathbf{p}_L)^{N-k} = \delta_L; \qquad (9.21)$$

$$\sum_{k=0}^{\mathbf{N}_G} \binom{N}{k} \mathbf{p}_U{}^k (1 - \mathbf{p}_U)^{N-k} = \delta_U \qquad (9.22)$$

with $\delta_L + \delta_U = \delta$, where $\mathbf{N}_G$ is the number of samples such that $J(\boldsymbol{\Delta}^{(i)}) \le \gamma$. Clearly, the probabilities $\mathbf{p}_L$ and $\mathbf{p}_U$ are random variables which can be computed only *a posteriori*, once the value of $\mathbf{N}_G$ is known. Moreover, an explicit solution of the previous equations is not available, so that standard tables or numerical methods are generally used; see e.g. [225]. Figure 9.1 shows the confidence intervals for $\delta = 0.002$ and various values of N. The figure should be interpreted as follows: if, for instance, $N = 1,000$ and $\mathbf{N}_G = 700$, then the estimated probability is $\widehat{p}_N(\gamma) = N_G/N = 0.7$ and the values $p_U = 0.74$, $p_L = 0.65$ can be obtained from the plot.

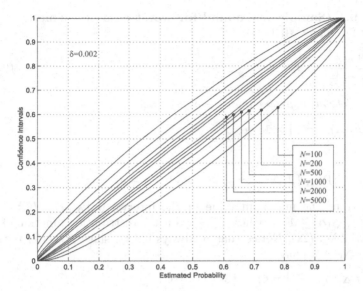

Fig. 9.1. Confidence intervals for $\delta = 0.002$.

9.4 Sample size bounds for estimation of extrema

The sample size bounds considered in the previous section apply to the estimation of probabilities, which are essentially expectations of random variables defined by means of indicator functions. In this section, we deal instead with sample bounds that apply to the estimation of extrema (as opposed to expectation) of a random variable. Computing the extremum of a function is directly related to the issue of assessing the worst-case performance of a system, as defined in Problem 6.2. In particular, we consider the problem of computing a probabilistic estimate of the worst-case performance

$$\gamma_{\mathrm{wc}} = \sup_{\Delta \in \mathcal{B}_{\mathbb{D}}(\rho)} J(\Delta).$$

To this end, we introduce a random sampling scheme, generating N iid samples of Δ according to $f_{\Delta}(\Delta)$ with support $\mathcal{B}_{\mathbb{D}}(\rho)$, and define the empirical maximum

$$\widehat{\gamma}_N = \max_{i=1,\dots,N} J(\Delta^{(i)}).$$

For this specific problem, a bound on the sample size, derived in [154] and [264], is now given.

Theorem 9.1 (Sample size bound for worst-case performance). *For any $\epsilon \in (0,1)$ and $\delta \in (0,1)$, if*

$$N \geq \frac{\log \frac{1}{\delta}}{\log \frac{1}{1-\epsilon}} \qquad (9.23)$$

then, with probability greater than $1 - \delta$, we have

$$\mathrm{PR}_{\boldsymbol{\Delta}} \left\{ J(\boldsymbol{\Delta}) \leq \widehat{\gamma}_N \right\} \geq 1 - \epsilon.$$

That is

$$\mathrm{PR}_{\boldsymbol{\Delta}^{(1...N)}} \left\{ \mathrm{PR}_{\boldsymbol{\Delta}} \left\{ J(\boldsymbol{\Delta}) \leq \widehat{\gamma}_N \right\} \geq 1 - \epsilon \right\} > 1 - \delta.$$

Proof. Let α be the minimum value in the interval $[\inf_{\boldsymbol{\Delta} \in \mathcal{B}_{\mathbb{D}}(\rho)} J(\boldsymbol{\Delta}), \gamma_{\mathrm{wc}}]$ such that $F_{J(\boldsymbol{\Delta})}(\alpha) \geq 1 - \epsilon$, where $F_{J(\boldsymbol{\Delta})}(\cdot)$ is the distribution function of the random variable $J(\boldsymbol{\Delta})$. Notice that α always exists, since $F_{J(\boldsymbol{\Delta})}(\cdot)$ is right continuous. Now, we have

$$\mathrm{PR}\left\{ F_{J(\boldsymbol{\Delta})}(\widehat{\gamma}_N) \geq 1 - \epsilon \right\} = \mathrm{PR}\left\{ \widehat{\gamma}_N \geq \alpha \right\}$$
$$= 1 - \mathrm{PR}\left\{ \widehat{\gamma}_N < \alpha \right\} = 1 - F_{J(\boldsymbol{\Delta})}(\alpha^-)^N$$

where $F_{J(\boldsymbol{\Delta})}(\alpha^-)$ is the limit of $F_{J(\boldsymbol{\Delta})}(\alpha)$ from the left. In addition, observe that $F_{J(\boldsymbol{\Delta})}(\alpha^-) \leq 1 - \epsilon$. Then

$$\mathrm{PR}\left\{ F_{J(\boldsymbol{\Delta})}(\widehat{\gamma}_N) \geq 1 - \epsilon \right\} \geq 1 - (1 - \epsilon)^N.$$

Next, notice that if (9.23) holds, then $(1 - \epsilon)^N \leq \delta$. Thus

$$\mathrm{PR}\left\{ F_{J(\boldsymbol{\Delta})}(\widehat{\gamma}_N) \geq 1 - \epsilon \right\} \geq 1 - \delta.$$

This completes the proof.   □

It is shown in [66] that the bound (9.23) is tight if the distribution function is continuous. This is a consequence of the fact that the bound on the sample size is minimized if and only if

$$\sup_{\{\gamma : F_{J(\boldsymbol{\Delta})}(\gamma) \leq 1 - \epsilon\}} F_{J(\boldsymbol{\Delta})}(\gamma) = 1 - \epsilon.$$

Comparing the bound (9.23) with the Chernoff bound, it can be observed that in the former case the bound grows as $1/\epsilon$, since $\log(1/(1 - \epsilon)) \approx \epsilon$, whereas in the latter case it grows as $1/\epsilon^2$. This fact leads to a major reduction in the number of samples needed, as shown in Figure 9.2, which compares the worst-case bound of Theorem 9.1 with the Chernoff bound for various values of ϵ and δ. This is not surprising, since in the worst-case bound the performance level is selected *a posteriori*, once the samples are generated and $\widehat{\gamma}_N$ is computed, while the Chernoff bound holds for any a priori specified value of γ. Note however that the two bounds apply to different problems. Namely, the Chernoff bound applies to estimation of expected values or probabilities, while the bound in Theorem 9.1 applies to the estimation of extrema.

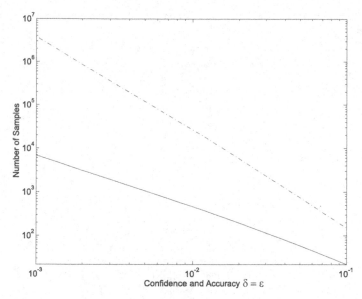

Fig. 9.2. Comparison between the Chernoff bound (dash-dotted line) and the worst-case bound (solid line).

Remark 9.5 (Worst-case bound interpretation). In general, there is no assurance that $\widehat{\gamma}_N$ is actually close to the maximum γ_{wc}. As shown in Figure 9.3, the bound previously discussed only guarantees (in probability) that the performance is less than $\widehat{\gamma}_N$ with high probability $p^* = 1 - \epsilon$. In other words, the set of points greater than the estimated value has a measure smaller than ϵ, and this is true with a probability at least $1 - \delta$. In turn, this implies that, if the function $J(\Delta)$ is sufficiently smooth, then the estimated and actual maximum may be close. For this reason, care should be exercised when the bound is applied for solving optimization problems with a randomized approach.   ◁

Example 9.3 (Randomized algorithm for worst-case $\mathcal{H}_\infty$ performance).
    We revisit Example 9.2 about $\mathcal{H}_\infty$ performance of the continuous-time SISO system (9.20). We are now interested in evaluating a performance level $\widehat{\gamma}_N$ which is guaranteed with high probability $p^* = 0.9999$. To this aim, we apply Algorithm 8.2. Setting a confidence level $\delta = 0.0001$, we determine the sample size by means of the bound given in Theorem 9.1 (with $\epsilon = 1 - p^*$), obtaining

$$\overline{N}(p^*, \delta) = 9.210 \times 10^4.$$

To apply the RA, we hence generated $N = 100,000$ iid samples uniformly distributed within $\mathcal{B}_q$ and obtained the empirical maximum

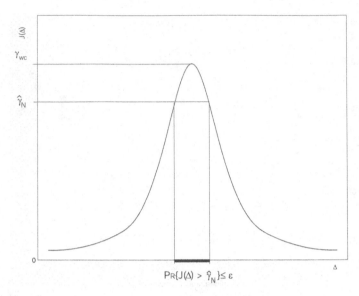

Fig. 9.3. Interpretation of the worst-case bound.

$$\widehat{\gamma}_N = 0.0087.$$

Then, with confidence greater than $1 - \delta = 0.9999$, we may conclude that the performance of the uncertain system is below $\widehat{\gamma}_N$ for at least 99.99% of the uncertain plants. ★

Remark 9.6 (Bounds for one level of probability). We notice that [19] investigates the minimal sample size problem with one level of probability, when the performance function J is in the class of all Lipschitz continuous functions $\mathcal{J}_L$. More precisely, the problem addressed in [19] is to compute the minimum number of samples required to satisfy

$$\text{PR}_{\mathbf{\Delta}^{(1...N)}} \left\{ \sup_{J \in \mathcal{J}_L} |\gamma_{\text{wc}} - \widehat{\gamma}_N| \le \epsilon \right\} \ge 1 - \delta.$$

In the same paper, it is also shown that the uniform distribution is "optimal" in the sense of minimizing the sample size, but the sample size obtained may be an exponential function of the number of variables and, therefore, computational complexity becomes a critical issue. Contrary to the bounds previously discussed, in this line of research the performance function is not fixed, rather it varies within the class $\mathcal{J}_L$. ◁

10. Statistical Learning Theory and Control Design

In the first part of this chapter, we give a summary of some key results of *statistical learning theory*. This theory provides a fundamental extension of the probability inequalities studied in Chapter 9 to the case when parameterized *families* of performance functions are considered, instead of a *fixed* function. For an advanced treatment of this topic, the interested reader may refer to [83, 97, 177, 279, 284]. In the second part of the chapter, we discuss applications of this theory to robust control design problems following the approach introduced in [282, 283].

10.1 Deviation inequalities for finite families

In Chapter 9 we studied several deviation-type inequalities for random variables. Here, we begin by reconsidering the two-sided Hoeffding inequality, applied to the general case of expected value estimation. Consider a performance function $J : \mathbb{D} \to [0, 1]$, and its empirical mean

$$\widehat{\mathbf{E}}_N(J(\mathbf{\Delta})) = \frac{1}{N} \sum_{i=1}^{N} J(\mathbf{\Delta}^{(i)})$$

computed using a multisample $\mathbf{\Delta}^{(1\ldots N)}$ of cardinality N. Then, it immediately follows from (9.10) that the probability of deviation of the empirical mean from the actual one is bounded as

$$\mathrm{PR}\left\{\left|\mathrm{E}(J(\mathbf{\Delta})) - \widehat{\mathbf{E}}_N(J(\mathbf{\Delta}))\right| \geq \epsilon\right\} \leq 2\mathrm{e}^{-2N\epsilon^2}. \tag{10.1}$$

It should be emphasized that this inequality holds for a *fixed* performance function J. On the other hand, if we wish to consider m performance functions simultaneously, we need to define a *finite class* of functions, consisting of m elements

$$\mathcal{J}_m \doteq \{J_1, \ldots, J_m\}$$

where $J_i : \mathbb{D} \to [0, 1]$, $i = 1, \ldots, m$. We now aim at bounding the probability of deviation of the empirical mean from the actual one, for all functions in the considered class. This worst-case probability

$$q(\mathcal{J}_m, N, \epsilon) \doteq \mathrm{PR} \left\{ \sup_{J \in \mathcal{J}_m} \left| \mathrm{E}(J(\mathbf{\Delta})) - \widehat{\mathbf{E}}_N(J(\mathbf{\Delta})) \right| > \epsilon \right\}$$

can be bounded by repeated application of the inequality (10.1), obtaining

$$q(\mathcal{J}_m, N, \epsilon) \le 2m e^{-2N\epsilon^2}. \tag{10.2}$$

Equivalently, we have

$$\mathrm{PR} \left\{ \exists J \in \mathcal{J}_m : \left| \mathrm{E}(J(\mathbf{\Delta})) - \widehat{\mathbf{E}}_N(J(\mathbf{\Delta})) \right| > \epsilon \right\} \le 2m e^{-2N\epsilon^2}.$$

Notice that this result is distribution free, i.e. the actual distribution of the data does not play a role in the upper bound.

A similar bound on the expected value of the maximal deviation

$$\mathrm{E}_{\mathbf{\Delta}^{(1\ldots N)}} \left(\sup_{J \in \mathcal{J}_m} \left| \mathrm{E}(J(\mathbf{\Delta})) - \widehat{\mathbf{E}}_N(J(\mathbf{\Delta})) \right| \right)$$

can be derived. This bound is reported in the following lemma, whose proof may be found for instance in [177].

Lemma 10.1. *Let $\mathcal{J}_m$ be a family of performance functions of finite cardinality m. Then*

$$\mathrm{E}_{\mathbf{\Delta}^{(1\ldots N)}} \left(\sup_{J \in \mathcal{J}_m} \left| \mathrm{E}(J(\mathbf{\Delta})) - \widehat{\mathbf{E}}_N(J(\mathbf{\Delta})) \right| \right) \le \sqrt{\frac{\log(2m)}{2N}}.$$

10.2 Vapnik–Chervonenkis theory

Notice that inequality (10.2) implies that, for any family of performance functions having finite cardinality, the worst-case deviation probability for the class, $q(\mathcal{J}_m, N, \epsilon)$, approaches zero asymptotically as N goes to infinity for all $\epsilon > 0$. This fundamental property goes under the name of *uniform convergence of empirical means* (UCEM). The UCEM property is therefore satisfied by any family of performance functions having finite cardinality. The problem that naturally arises is whether *infinite* families of performance functions may have the UCEM property. This problem, and the derivation of related inequalities, is the focal point of learning theory, which was initiated by Vapnik and Chervonenkis in their seminal paper [280]. In this section, we report some of the key results of this theory.

Consider a possibly infinite family $\mathcal{J}$ of measurable binary functions mapping a generic set $\mathcal{B} \subseteq \mathbb{R}^n$ into $\{0, 1\}$. Associated with this family of functions, we define the corresponding class $\mathcal{S}_\mathcal{J}$, whose elements are the sets

$$S_J \doteq \{\Delta \in \mathcal{B} : J(\Delta) = 1\}, \quad J \in \mathcal{J}.$$

In the following, we use families of functions or families of sets interchangeably, depending on which one is more convenient in the context. In fact, given a family of functions $\mathcal{J}$, we construct the corresponding family of sets $\mathcal{S}_J$ as shown above. Conversely, to a family of measurable sets $\mathcal{S}$ we associate the family of binary valued functions $\mathcal{J}$ whose elements are

$$J(\Delta) \doteq \mathbb{I}_S(\Delta), \quad S \in \mathcal{S}.$$

Notice that with these definitions we have

$$\mathrm{PR}_\Delta \{\Delta \in S_J\} = \mathrm{E}_\Delta (J(\Delta)).$$

We now define the "shatter coefficient" of a set of points $\Delta^{(1...N)}$.

Definition 10.1 (Shatter coefficient). *Let $\Delta^{(1...N)} = \{\Delta^{(1)}, \dots, \Delta^{(N)}\}$ be a set of points of cardinality N, and let*

$$\mathrm{N}_{\mathcal{J}}\left(\Delta^{(1...N)}\right) \doteq \mathrm{Card}\left(\Delta^{(1...N)} \cap S_J, S_J \in \mathcal{S}_J\right)$$

be the number of different subsets of $\Delta^{(1...N)}$ obtained intersecting $\Delta^{(1...N)}$ with the elements of $\mathcal{S}_J$. When $\mathrm{N}_{\mathcal{J}}\left(\Delta^{(1...N)}\right)$ equals the maximum number of possible different subsets of $\Delta^{(1...N)}$, which is 2^N, we say that $\mathcal{S}_J$ shatters the set $\Delta^{(1...N)}$. The shatter coefficient of the family $\mathcal{J}$, or equivalently of the family of sets $\mathcal{S}_J$, is defined as

$$\mathbb{S}_{\mathcal{J}}(N) \doteq \max_{\Delta^{(1...N)}} \mathrm{N}_{\mathcal{J}}\left(\Delta^{(1...N)}\right). \tag{10.3}$$

Thus, $\mathbb{S}_{\mathcal{J}}(N)$ is the maximum number of different subsets of any point set $\Delta^{(1...N)}$ of cardinality N that can be obtained by intersecting $\Delta^{(1...N)}$ with elements of $\mathcal{S}_J$.

Example 10.1 (Shatter coefficient of a family of half-spaces in $\mathbb{R}^2$).
    Consider the family $\mathcal{J}$ of binary valued functions $J(\Delta)$ mapping $\mathbb{R}^2$ into $\{0, 1\}$ of the form

$$J(\Delta) = \begin{cases} 1 & \text{if } \theta_1^T \Delta + \theta_2 \geq 0; \\ 0 & \text{otherwise} \end{cases}$$

with parameters $\theta_1 \in \mathbb{R}^2$, $\theta_2 \in \mathbb{R}$. We associate to the family $\mathcal{J}$ the family of sets $\mathcal{S}_J$ formed by all possible linear half-spaces of $\mathbb{R}^2$. Then, in Figure 10.1, we consider a point set $\Delta^{(1,2,3)} = \{\Delta^{(1)}, \Delta^{(2)}, \Delta^{(3)}\}$ of cardinality three and observe that

$$\mathrm{N}_{\mathcal{J}}\left(\Delta^{(1,2,3)}\right) = 8 = 2^3.$$

Therefore, we say that the family $\mathcal{S}_J$ shatters the point set, and $\mathbb{S}_{\mathcal{J}}(3) = 8$. Similarly, in Figure 10.2 we consider a point set $\Delta^{(1,\dots,4)}$ of cardinality four and observe that

$$\mathbb{N}_{\mathcal{J}}\left(\Delta^{(1,\dots,4)}\right) = 14.$$

Since $14 < 2^4$, we conclude that the set considered is *not* shattered by the family. It can indeed be shown that the shatter coefficient of the family is $\mathbb{S}_{\mathcal{J}}(4) = 14$, i.e. it is not possible to find *any* point set of cardinality four that can be shattered by $\mathcal{S}_J$.   ⋆

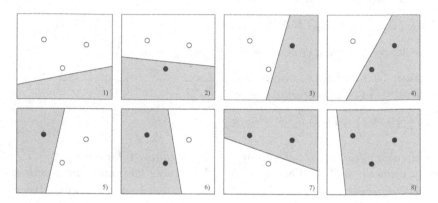

Fig. 10.1. A family of half-spaces shatters a point set of cardinality three.

The celebrated Vapnik–Chervonenkis inequality states that, for any $\epsilon > 0$

$$\mathrm{PR}\left\{\sup_{J \in \mathcal{J}}\left|\mathrm{E}(J(\boldsymbol{\Delta})) - \widehat{\mathbf{E}}_N(J(\boldsymbol{\Delta}))\right| > \epsilon\right\} \le 4\,\mathbb{S}_{\mathcal{J}}(2N)\,e^{-N\epsilon^2/8}. \qquad (10.4)$$

The Vapnik–Chervonenkis inequality therefore provides a bound on the uniform deviation of empirical means, in terms of the combinatorial parameter $\mathbb{S}_{\mathcal{J}}(N)$. This parameter can be interpreted as a measure of the "richness" of the class of functions $\mathcal{J}$. An important issue therefore is the determination of explicit upper bounds for $\mathbb{S}_{\mathcal{J}}(N)$. To this end, we now define the Vapnik–Chervonenkis (VC) dimension of a family of binary-valued functions.

Definition 10.2 (VC dimension). *The VC dimension* $\mathrm{VC}(\mathcal{J})$ *of the family* $\mathcal{J}$ *of binary-valued functions is defined as the largest integer* k *such that* $\mathbb{S}_{\mathcal{J}}(k) = 2^k$. *If* $\mathbb{S}_{\mathcal{J}}(k) = 2^k$ *for all* k, *then we say that* $\mathrm{VC}(\mathcal{J}) = \infty$.

In words, the VC dimension of $\mathcal{J}$ is the largest integer N such that there exists a set of cardinality N that is shattered by $\mathcal{J}$.

Example 10.2 (VC dimension of a family of half-spaces in $\mathbb{R}^2$).
    In Example 10.1 we considered a family of half-spaces in $\mathbb{R}^2$ and showed that the maximum cardinality of a point set shattered by $\mathcal{S}_J$ is three. We hence conclude that the VC dimension of this family is $\mathrm{VC}(\mathcal{J}) = 3$.   ⋆

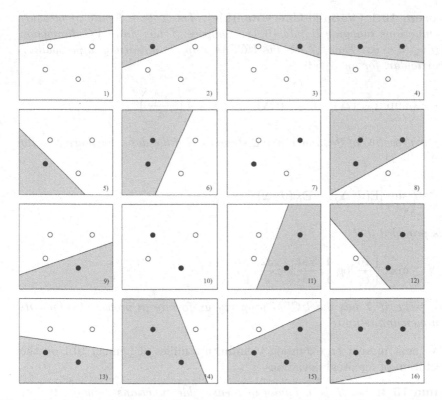

Fig. 10.2. A point set of cardinality four is not shattered by $\mathcal{S}_J$. In fact, the subsets in 7) and 10) cannot be obtained intersecting an element of $\mathcal{S}_J$ with $\Delta^{(1,\dots,4)}$.

The VC dimension is used to determine a bound on the shatter coefficient, by means of the so-called Sauer lemma, see [233].

Lemma 10.2 (Sauer). *Let $\mathcal{J}$ be a family of measurable functions mapping $\mathcal{B}$ into $\{0,1\}$ and let $\mathrm{vc}\,(\mathcal{J}) \leq d < \infty$. Then*

$$\mathbb{S}_{\mathcal{J}}(k) \leq \sum_{i=0}^{d} \binom{k}{i}.$$

Moreover, for all $k \geq d$, we have

$$\sum_{i=0}^{d} \binom{k}{i} \leq \left(\frac{ke}{d}\right)^d.$$

Combining this bound and the Vapnik–Chervonenkis inequality (10.4), we obtain the following fundamental result. The reader may refer to the original paper [280] and to [284] for different proofs.

Theorem 10.1 (Vapnik–Chervonenkis). *Let $\mathcal{J}$ be a family of measurable functions mapping $\mathcal{B}$ into $\{0,1\}$. Suppose $\mathcal{J}$ has finite VC dimension $\mathrm{vc}\,(\mathcal{J}) \le d < \infty$. Then, $\mathcal{J}$ has the UCEM property uniformly in probability. In particular, for any $\epsilon > 0$*

$$\mathrm{PR}\left\{ \sup_{J \in \mathcal{J}} \left| \mathrm{E}(J(\mathbf{\Delta})) - \widehat{\mathbf{E}}_N(J(\mathbf{\Delta})) \right| > \epsilon \right\} \le 4 \left(\frac{2eN}{d} \right)^d e^{-N\epsilon^2/8}$$

holds irrespective of the underlying distribution of the data. Furthermore, for any $\delta \in (0,1)$

$$\mathrm{PR}\left\{ \sup_{J \in \mathcal{J}} \left| \mathrm{E}(J(\mathbf{\Delta})) - \widehat{\mathbf{E}}_N(J(\mathbf{\Delta})) \right| > \epsilon \right\} \le \delta$$

holds provided that

$$N \ge \max \left\{ \frac{16}{\epsilon^2} \log \frac{4}{\delta}, \frac{32d}{\epsilon^2} \log \frac{32e}{\epsilon^2} \right\}.$$

Conversely, if $\mathcal{J}$ has the UCEM property uniformly in probability, then its VC dimension is finite.

We next present an extension to infinite families of Lemma 10.1 on the expectation of maximal deviations.

Lemma 10.3. *Let $\mathcal{J}$ be a family of measurable functions mapping $\mathcal{B}$ into $\{0,1\}$ and let $\mathrm{vc}\,(\mathcal{J}) \le d < \infty$. Then, for any $\epsilon > 0$*

$$\mathbf{E}_{\mathbf{\Delta}^{(1\ldots N)}} \left(\sup_{J \in \mathcal{J}} \left| \mathrm{E}(J(\mathbf{\Delta})) - \widehat{\mathbf{E}}_N(J(\mathbf{\Delta})) \right| \right) \le 2\sqrt{\frac{d \log(N+1) + \log 2}{N}}$$

holds irrespective of the underlying distribution of the data.

The Vapnik–Chervonenkis theory can be extended from binary valued functions to families of continuous valued functions taking values in the interval $[0,1]$. In this case, the analogous concept of the VC dimension is the Pollard or pseudo (P) dimension of the class of functions, which is now defined.

Definition 10.3 (P dimension). *Suppose $\mathcal{J}$ is a family of measurable functions mapping $\mathcal{B}$ into $[0,1]$. A point set $\Delta^{(1\ldots N)} = \{\Delta^{(1)}, \ldots, \Delta^{(N)}\}$ is said to be P-shattered by $\mathcal{J}$ if there exists a vector*

$$v = [v_1 \cdots v_N]^T, \; v_i \in [0,1], \; i = 1, \ldots, N$$

such that for every binary vector

$$b = [b_1 \cdots b_N]^T, \; b_i \in \{0,1\}, \; i = 1, \ldots, N$$

there exists a function $J \in \mathcal{J}$ such that

$$\begin{cases} J(\Delta^{(i)}) \geq v_i, \text{ if } b_i = 1; \\ J(\Delta^{(i)}) < v_i, \text{ if } b_i = 0 \end{cases}$$

for all $i = 1, \ldots, N$.

The P dimension of $\mathcal{J}$, denoted as P-DIM $(\mathcal{J})$, is the largest integer N such that there exists a set of cardinality N that is P-shattered by $\mathcal{J}$.

We now state a result, due to [217], which is the analogous of Theorem 10.1, for the case of continuous-valued functions with finite P dimension.

Theorem 10.2 (Pollard). *Let $\mathcal{J}$ be a family of measurable functions mapping $\mathcal{B}$ into $[0,1]$ and P-DIM $(\mathcal{J}) \leq d < \infty$. Then, $\mathcal{J}$ has the UCEM property uniformly in probability. In particular, for any $\epsilon > 0$*

$$\mathrm{PR}\left\{ \sup_{J \in \mathcal{J}} \left| \mathrm{E}(J(\boldsymbol{\Delta})) - \widehat{\mathbf{E}}_N(J(\boldsymbol{\Delta})) \right| > \epsilon \right\} \leq 8 \left(\frac{16e}{\epsilon} \log\left(\frac{16e}{\epsilon} \right) \right)^d e^{-N\epsilon^2/32}$$

holds irrespective of the underlying distribution of the data. Furthermore, for any $\delta \in (0,1)$

$$\mathrm{PR}\left\{ \sup_{J \in \mathcal{J}} \left| \mathrm{E}(J(\boldsymbol{\Delta})) - \widehat{\mathbf{E}}_N(J(\boldsymbol{\Delta})) \right| > \epsilon \right\} \leq \delta$$

holds provided that

$$N \geq \frac{32}{\epsilon^2} \left[\log \frac{8}{\delta} + d \left(\log \frac{16e}{\epsilon} + \log \log \frac{16e}{\epsilon} \right) \right].$$

Remark 10.1 (Relation between VC and P dimensions). If $\mathcal{J}$ is a family of measurable functions mapping $\mathcal{B}$ into $[0,1]$, and every function in $\mathcal{J}$ is actually binary valued, then it can be easily verified that VC $(\mathcal{J}) = $ P-DIM $(\mathcal{J})$. More generally, the VC and P dimensions are related as follows: given a family $\mathcal{J}$ of measurable functions mapping $\mathcal{B}$ into $[0,1]$, define the associated family of binary functions $\bar{\mathcal{J}}$ whose elements are the functions $\bar{J}(\Delta, c) = \mathbb{I}_{J(\Delta) \geq c}(\Delta)$, for $c \in [0,1]$, $J \in \mathcal{J}$. Then

$$\mathrm{VC}\left(\bar{\mathcal{J}} \right) = \text{P-DIM}\left(\mathcal{J} \right).$$

This relationship is explicitly proved in [178], and it is also reported in [284].

◁

10.2.1 Computing the VC dimension

From the previous results, it appears that it is of paramount importance to assess whether a given family $\mathcal{J}$ has finite VC dimension and, in this case, to determine an upper bound d on it. We next report without proof some known results on the computation of the VC dimension for special classes of sets. First, we state a simple result regarding the VC dimension of hyperrectangles, see e.g. [177].

Lemma 10.4. *If $\mathcal{S}$ is the class of all rectangles in $\mathbb{R}^d$, then $\mathrm{VC}\,(\mathcal{S}) = 2d$.*

The following result on linear functional spaces is stated in [94].

Lemma 10.5. *Let $\mathcal{G}$ be an m-dimensional vector space of real-valued functions $g : \mathbb{R}^d \to \mathbb{R}$. Define the class of sets*

$$\mathcal{S} = \left\{ \{ x \in \mathbb{R}^d : g(x) \geq 0 \} : g \in \mathcal{G} \right\}.$$

Then, $\mathrm{VC}\,(\mathcal{S}) \leq m$.

From this lemma we deduce the following corollary.

Corollary 10.1.

1. *If $\mathcal{S}$ is the class of all linear half-spaces $\{ x : a^T x \geq b \}$, $a \in \mathbb{R}^m$, $b \in \mathbb{R}$, then $\mathrm{VC}\,(\mathcal{S}) \leq m + 1$;*
2. *If $\mathcal{S}$ is the class of all closed balls $\{ x : \| x - a \|_2 \leq r \}$, $a \in \mathbb{R}^m$, $r \in \mathbb{R}_+$, then $\mathrm{VC}\,(\mathcal{S}) \leq m + 2$;*
3. *If $\mathcal{S}$ is the class of all ellipsoids in $\mathbb{R}^m$ centered at the origin $\mathcal{E}\,(0, W)$, $W \succ 0$, then $\mathrm{VC}\,(\mathcal{S}) \leq m(m + 1)/2 + 1$.*

Using a result in [73], one can show that for linear half-spaces the VC dimension actually equals $m + 1$. Similarly, in [95] it is proved that for the class of closed balls the VC dimension equals $m + 1$.

Next, we report a general result that is useful to establish upper bounds on the VC dimension of families of functions that arise as unions, intersections or other Boolean functions operating on functions belonging to families with known VC dimension.

Lemma 10.6. *Let $\mathcal{J}_1, \ldots, \mathcal{J}_k$ be k families composed of measurable functions mapping $\mathcal{B}$ into $\{0, 1\}$ and let ϕ be a given Boolean function $\phi : \{0, 1\}^k \to \{0, 1\}$. Consider the class $\mathcal{J}_\phi$*

$$\mathcal{J}_\phi \doteq \{ \phi(J_1, \ldots, J_k) : J_i \in \mathcal{J}_i, i = 1, \ldots, k \}.$$

Then, we have

$$\mathrm{VC}\,(\mathcal{J}_\phi) \leq 2k \log(e\,k) \max_{i=1,\ldots,k} \{ \mathrm{VC}\,(\mathcal{J}_i) \}.$$

This lemma is proved for instance in [250]. Results along these lines may also be found in [96, 218, 284, 287]. In particular, the following lemma deals with the case of Boolean closures of polynomial classes.

Lemma 10.7. *Let $\mathcal{J}_1, \ldots, \mathcal{J}_k$ be k families composed of measurable functions mapping $\mathcal{B}$ into $\{0, 1\}$, and let ϕ be a given Boolean function $\phi : \{0, 1\}^k \to \{0, 1\}$. Suppose further that each class $\mathcal{J}_i$ is composed of functions $J_i(\Delta, \theta)$ that are polynomials in the parameter $\theta \in \mathbb{R}^\ell$, of maximum degree β in the variables θ_i, $i = 1, \ldots, \ell$. Consider the class $\mathcal{J}_\phi$*

$$\mathcal{J}_\phi \doteq \{\phi(J_1,\ldots,J_k) : J_i \in \mathcal{J}_i, i = 1,\ldots,k\}.$$

Then

$$\mathrm{VC}\,(\mathcal{J}_\phi) \le 2\ell \log_2(4\mathrm{e}\,\beta\,k).$$

This result is proven in a more general form in [150], and it is given in the form stated above in [284].

Finally, we state a result on the pseudo dimension of the composition of functions $(h \circ J)(\Delta) \doteq h(J(\Delta))$, due to [128].

Lemma 10.8. *Let $\mathcal{J}$ be a family of measurable functions from $\mathcal{B}$ to $[0,1]$, and let h be a fixed nondecreasing measurable function $h : \mathbb{R} \to \mathbb{R}$. Then, we have*

$$\mathrm{P\text{-}DIM}\,(\{(h \circ J) : J \in \mathcal{J}\}) \le \mathrm{P\text{-}DIM}\,(\mathcal{J}).$$

10.3 A learning theory approach for control design

In this section we show how the learning theory results previously presented can be used to determine "probabilistic" solutions to average-performance control synthesis problems. This approach has been introduced in [282, 283] and leads to the formulation of the RA for average performance synthesis presented in Section 8.3.1.

Suppose that a family of uncertain plants $\{G(s,\Delta) : \Delta \in \mathcal{B}_\mathbb{D}(\rho)\}$ and a family of candidate controllers $\{K(s,\theta) : \theta \in \Theta\}$ are given, where θ represents the vector of controller parameters which is assumed to belong to a set $\Theta \subseteq \mathbb{R}^{n_\theta}$. Consider a performance function $J(\Delta,\theta) : \mathcal{B}_\mathbb{D}(\rho) \times \Theta \to [0,1]$, which measures the performance of the controlled plant, for given Δ and θ. Clearly, as θ varies over Θ, $J(\cdot,\theta)$ spans a family $\mathcal{J}$ of performance functions. Assuming now that $\Delta \in \mathcal{B}_\mathbb{D}(\rho)$ is random with given pdf, the goal is to determine a controller, i.e. $\theta \in \Theta$, such that the *average* performance (with respect to Δ) of the controlled plant is minimized. In other words, denoting the average performance by

$$\phi(\theta) \doteq \mathrm{E}_\Delta(J(\Delta,\theta))$$

the objective is to find a parameter θ^* attaining the optimal performance

$$\phi^* = \phi(\theta^*) = \inf_{\theta \in \Theta} \mathrm{E}_\Delta(J(\Delta,\theta)). \tag{10.5}$$

Notice that this "average performance" strategy differs from the usual robust design methods where the *worst-case* performance is minimized. On the other hand, a common criticism on the worst-case approach, see e.g. Chapter 5, is that worst-case synthesis leads to an overly pessimistic design, and it is in general computationally intractable.

We now formally state an assumption used in the rest of this chapter.

Assumption 10.1. *The performance function $J(\Delta, \theta)$ is a measurable function mapping $\mathcal{B}_{\mathbb{D}}(\rho) \times \Theta$ into the interval $[0,1]$.*

In the approach presented in this section, the design follows two steps: in the first step, the expected value $\phi(\theta) = \mathrm{E}_{\Delta}(J(\Delta, \theta))$ is estimated, and in the second step a minimization is performed on the estimate of $\phi(\theta)$ to obtain the "optimal" controller.

In the first step of the procedure, since the exact computation of the expected value is, in general, computationally hard, we approximate the expectation $\phi(\theta)$ by its empirical version. That is, N samples $\Delta^{(1)}, \dots, \Delta^{(N)}$ are collected and the empirical mean is obtained as

$$\widehat{\phi}_N(\theta) = \widehat{\mathrm{E}}_N(J(\Delta, \theta)) = \frac{1}{N} \sum_{i=1}^{N} J(\Delta^{(i)}, \theta).$$

Now, if $\mathcal{J}$ has finite P dimension, P-DIM $(\mathcal{J}) \leq d$, then Theorem 10.2 guarantees that for sufficiently large N the empirical mean is "close" to the actual mean, uniformly with respect to θ. Specifically, given $\epsilon_1, \delta_1 \in (0,1)$, choose N such that

$$N \geq \frac{32}{\epsilon_1^2} \left[\log \frac{8}{\delta_1} + d \left(\log \frac{16e}{\epsilon_1} + \log \log \frac{16e}{\epsilon_1} \right) \right].$$

Then, by Theorem 10.2 it can be asserted with confidence $1 - \delta_1$ that

$$|\phi(\theta) - \widehat{\phi}_N(\theta)| = |\mathrm{E}_{\Delta}(J(\Delta, \theta)) - \widehat{\mathrm{E}}_N(J(\Delta, \theta))| \leq \epsilon_1 \tag{10.6}$$

for all $\theta \in \Theta$. Notice that the uniformity of the bound with respect to θ is crucial, since the minimization with respect to θ can now be performed on the empirical mean $\widehat{\phi}_N(\theta)$ rather than on the actual (and unknown) $\phi(\theta)$. To elaborate further, let

$$\widehat{\theta}_N = \arg \inf_{\theta \in \Theta} \widehat{\phi}_N(\theta).$$

Then, if we could find (using *any* method) an exact minimizer $\widehat{\theta}_N$, we would have a so-called *approximate* near minimizer of $\phi(\theta)$ to accuracy ϵ_1. That is, we would be able to guarantee up to confidence level $1 - \delta_1$ that

$$\left| \phi(\theta^*) - \widehat{\phi}_N(\widehat{\theta}_N) \right| = \left| \mathrm{E}_{\Delta}(J(\Delta, \theta^*)) - \widehat{\mathrm{E}}_N\left(J(\Delta, \widehat{\theta}_N) \right) \right| \leq \epsilon_1. \tag{10.7}$$

The next corollary follows from the above reasoning and provides the sample size bound used in Algorithm 8.3.

Corollary 10.2. *Let Assumption 10.1 be satisfied. Given $\epsilon, \delta \in (0,1)$, let*

$$N \geq \frac{128}{\epsilon^2} \left[\log \frac{8}{\delta} + d \left(\log \frac{32e}{\epsilon} + \log \log \frac{32e}{\epsilon} \right) \right]. \tag{10.8}$$

Then, with confidence $1 - \delta$, it holds that

$$\phi(\widehat{\boldsymbol{\theta}}_N) - \phi(\theta^*) \leq \epsilon \tag{10.9}$$

where $\phi(\theta^*)$ is defined in (10.5).

Proof. The corollary is immediately proved noticing that (10.8) implies that

$$\left| \phi(\theta) - \widehat{\boldsymbol{\phi}}_N(\theta) \right| \leq \frac{\epsilon}{2}. \tag{10.10}$$

From this equation, evaluated for $\theta = \theta^*$, it follows that

$$\phi(\theta^*) \geq \widehat{\boldsymbol{\phi}}_N(\theta^*) - \frac{\epsilon}{2} \geq \widehat{\boldsymbol{\phi}}_N(\widehat{\boldsymbol{\theta}}_N) - \frac{\epsilon}{2} \tag{10.11}$$

where the last inequality holds since $\widehat{\boldsymbol{\theta}}_N$ is a minimizer of $\widehat{\boldsymbol{\phi}}_N$. From (10.10), evaluated for $\theta = \widehat{\boldsymbol{\theta}}_N$, it follows that

$$\widehat{\boldsymbol{\phi}}_N(\widehat{\boldsymbol{\theta}}_N) \geq \phi(\widehat{\boldsymbol{\theta}}_N) - \frac{\epsilon}{2}$$

which substituted in (10.11), gives

$$\phi(\theta^*) \geq \phi(\widehat{\boldsymbol{\theta}}_N) - \epsilon.$$

$\square$

As mentioned previously, in principle the minimization of $\widehat{\boldsymbol{\phi}}_N(\theta)$ over $\theta \in \Theta$ can be performed by any numerical optimization method. In particular, when the performance function $J(\Delta, \theta)$ is convex in θ for any fixed Δ, then the empirical mean $\widehat{\boldsymbol{\phi}}_N(\theta)$ is also convex, and the global minimum can (in principle) be determined efficiently. On the other hand, if $J(\Delta, \theta)$ is nonconvex, then there are obvious difficulties in finding a global minimum of $\widehat{\boldsymbol{\phi}}_N(\theta)$. Thus, a viable approach is to use a randomized algorithm also for the determination of a probable minimum of $\widehat{\boldsymbol{\phi}}_N(\theta)$. To this end, one should introduce a probability density over the set Θ of controller parameters.[1] This randomized approach is summarized in the following.

For fixed $\epsilon_2, \delta_2 \in (0, 1)$, generate M independent controller parameter samples $\theta^{(1)}, \ldots, \theta^{(M)}$, with

$$M \geq \frac{\log(1/\delta_2)}{\log(1/(1 - \epsilon_2))}.$$

Correspondingly, the function $\widehat{\boldsymbol{\phi}}_N(\theta^{(i)})$ is evaluated and the empirical minimum is obtained as

---

[1] The introduction of a probability density on controller parameters may give rise to some perplexities, since there is in principle no "physical" reason for doing so. This density is therefore viewed as a technical artifice, needed by the randomized minimization algorithm.

$$\widehat{\boldsymbol{\theta}}_{NM} = \arg \min_{i=1,\dots,M} \widehat{\phi}_N(\boldsymbol{\theta}^{(i)}).$$

By Theorem 9.1, we guarantee up to confidence $1 - \delta_2$ that

$$\mathrm{PR}\left\{\widehat{\phi}_N(\boldsymbol{\theta}) < \widehat{\phi}_N(\widehat{\boldsymbol{\theta}}_{NM})\right\} \le \epsilon_2. \tag{10.12}$$

Finally, combining the bounds (10.7) and (10.12), we obtain the next corollary.

Corollary 10.3. *Let Assumption 10.1 be satisfied. Given $\epsilon_1, \epsilon_2, \delta \in (0,1)$, let*

$$M \ge \frac{\log(2/\delta)}{\log(1/(1 - \epsilon_2))}$$

and

$$N \ge \frac{32}{\epsilon_1^2}\left[\log\frac{16}{\delta} + d\left(\log\frac{16e}{\epsilon_1} + \log\log\frac{16e}{\epsilon_1}\right)\right].$$

Then, with confidence $1 - \delta$, it holds that

$$\mathrm{PR}\left\{\phi(\boldsymbol{\theta}) < \widehat{\phi}_N(\widehat{\boldsymbol{\theta}}_{NM}) - \epsilon_1\right\} \le \epsilon_2.$$

Proof. The choice of $\delta_1 = \delta_2 = \delta/2$ in the bounds for M and N guarantees that the inequalities (10.6) and (10.12) each hold with probability at least $1 - \delta/2$. Therefore, they jointly hold with probability $(1 - \delta/2)^2 > (1 - \delta)$. From (10.6) we have that, for all $\theta \in \Theta$

$$\widehat{\phi}_N(\theta) \le \phi(\theta) + \epsilon_1.$$

Therefore, we obtain

$$\mathrm{PR}\left\{\phi(\boldsymbol{\theta}) + \epsilon_1 < \widehat{\phi}_N(\widehat{\boldsymbol{\theta}}_{NM})\right\} \le \mathrm{PR}\left\{\widehat{\phi}_N(\boldsymbol{\theta}) < \widehat{\phi}_N(\widehat{\boldsymbol{\theta}}_{NM})\right\}.$$

Combining this inequality with (10.12), it follows that

$$\mathrm{PR}\left\{\phi(\boldsymbol{\theta}) < \widehat{\phi}_N(\widehat{\boldsymbol{\theta}}_{NM}) - \epsilon_1\right\} \le \epsilon_2$$

holds with confidence at least $1 - \delta$. $\qquad\square$

The quantity $\widehat{\boldsymbol{\theta}}_{NM}$ in (10.12) is called in [282] a probably approximate near minimizer of $\phi(\theta)$. In the same paper, it is observed that the bounds derived from VC theory are not necessarily required in this context. Indeed, if we use a randomized approach to minimize the function $\phi(\theta)$ over the controller parameters, then it is not strictly necessary to guarantee uniformity of the bound (10.6) for *all* values of $\theta \in \Theta$. Actually, it is sufficient to require that the inequality

$$|\phi(\theta) - \widehat{\phi}_N(\theta)| \leq \epsilon_1, \quad \text{for } \theta \in \{\theta^{(1)}, \dots, \theta^{(M)}\}$$

holds with confidence $1 - \delta_1$. This amounts to guaranteeing the convergence of the empirical estimate uniformly with respect to a family of performance functions having *finite* cardinality M. From (10.2) it follows that the above bound is guaranteed if we draw at least

$$N \geq \frac{\log \frac{2M}{\delta_1}}{2\epsilon_1^2} \tag{10.13}$$

samples of Δ. This leads to the following theorem, also stated in [282].

Theorem 10.3. *Let Assumption 10.1 be satisfied. Given $\epsilon_1, \epsilon_2, \delta \in (0,1)$, let*

$$M \geq \frac{\log(2/\delta)}{\log(1/(1 - \epsilon_2))} \tag{10.14}$$

and

$$N \geq \frac{\log \frac{4M}{\delta}}{2\epsilon_1^2}. \tag{10.15}$$

Then, with confidence $1 - \delta$, it holds that

$$\mathrm{PR}\left\{\phi(\theta) < \widehat{\phi}_N(\widehat{\theta}_{NM}) - \epsilon_1\right\} \leq \epsilon_2. \tag{10.16}$$

Proof. The result can be easily proved using the same reasoning of the proof of Corollary 10.3, choosing $\delta_1 = \frac{\delta}{2}$ in (10.13). $\qquad\square$

It is easy to verify that the bounds required by this latter result are substantially better than those required by Corollary 10.3, which is based on the Vapnik–Chervonenkis theory. For example, taking $\delta = \epsilon_1 = \epsilon_2 = 0.01$, Theorem 10.3 leads to $M \geq 528$ and $N \geq 61,295$, whereas Corollary 10.3, with $d = 1$, would require $N \geq 4,248,297$.

We next report a randomized algorithm for controller design based on Theorem 10.3.

Algorithm 10.1 (Average performance synthesis).
Let Assumption 10.1 be satisfied. Given $\epsilon_1, \epsilon_2, \delta \in (0,1)$, *this RA returns*
with probability at least $1 - \delta$ *a design vector* $\widehat{\boldsymbol{\theta}}_{NM}$ *such that (10.16) holds.*

1. Determine $M \geq \overline{M}(\epsilon_2, \delta)$ and $N \geq \overline{N}(\epsilon_1, \delta, M)$ according to (10.14) and (10.15);
2. Draw M iid samples $\boldsymbol{\theta}^{(1)}, \ldots, \boldsymbol{\theta}^{(M)}$;
3. Draw N iid samples $\boldsymbol{\Delta}^{(1)}, \ldots, \boldsymbol{\Delta}^{(N)}$;
4. Return the empirical controller

$$
\widehat{\boldsymbol{\theta}}_{NM} = \arg\min_{i=1,\ldots,M} \frac{1}{N} \sum_{k=1}^{N} J(\boldsymbol{\Delta}^{(k)}, \boldsymbol{\theta}^{(i)}).
$$

Remark 10.2 (Conservatism of the bounds for controller design). One of the main critical issues of Corollary 10.3 is the fact that the bounds obtained are very conservative and of difficult practical application. Therefore, a challenging topic of great interest is to reduce this conservatism. Along this direction, an attempt is given in [160], where a bootstrap technique is used to determine a stopping rule for the number of required samples. However, we notice that the bounds given in Theorem 10.3, based on the Hoeffding inequality, still result in being computationally more efficient, unless extremely small values of ϵ_2 are selected. ◁

Example 10.3 (Probabilistic static output feedback).
    In this example, following the approach discussed in [285], we formulate a probabilistic version of a robust static output feedback problem, which is now briefly described. Consider a strictly proper plant of the form

$$\dot{x} = A(q)x + B_2(q)u;$$
$$y = C(q)u$$

where $x \in \mathbb{R}^{n_s}$ is the state, $u \in \mathbb{R}^{n_i}$ is the control input, $y \in \mathbb{R}^{n_o}$ is the measurement output and q represents real parametric uncertainty. The objective is to find, if it exists, a static output feedback law

$$u = Ky$$

which stabilizes the system for all q varying in the hyperrectangle $\mathcal{B}_q$. Equivalently, we seek for a matrix $K \in \mathbb{R}^{n_i, n_o}$ such that the closed-loop matrix

$$A_{cl}(q, K) = A(q) + B_2(q)KC(q)$$

has all its eigenvalues in a specified region of the complex plane (typically the open left-half plane) for all $q \in \mathcal{B}_q$. As discussed in Chapter 5, the static output feedback problem has been shown to be NP-hard, even if no uncertainty affects the plant, when bounds on the gains are given. Hence, a probabilistic solution to this problem is of interest. In particular, to illustrate the average performance minimization approach introduced in this chapter, we study a numerical example taken from [39]. Consider

$$A(q) = \begin{bmatrix} -0.0366 & 0.0271 & 0.0188 & -0.4555 \\ 0.0482 & -1.01 & 0.0024 & -4.0208 \\ 0.1002 & q_1 & -0.707 & q_2 \\ 0 & 0 & 1 & 0 \end{bmatrix}, \ B_2(q) = \begin{bmatrix} 0.4422 & 0.1761 \\ q_3 & -7.5922 \\ -5.52 & 4.49 \\ 0 & 0 \end{bmatrix};$$

$$C(q) = \begin{bmatrix} 0 & 1 & 0 & 0 \end{bmatrix}$$

and

$$\mathcal{B}_q = \left\{ q \in \mathbb{R}^3 : |q_1 - 0.3681| \leq 0.05, |q_2 - 1.4200| \leq 0.01, |q_3 - 3.5446| \leq 0.04 \right\}.$$

Suppose it is desired to determine a static controller K guaranteeing that the closed-loop system is stable with a decay rate of at least $\alpha = 0.1$, for all values of the uncertain parameter $q \in \mathcal{B}_q$. Even if determining an exact solution to this problem is NP-hard, convex relaxations leading to sufficient solvability conditions for the existence of a stabilizing gain matrix K can be found in the literature. For example, in [104] a solution was determined using an LMI approach, obtaining the feedback gain

$$K = \begin{bmatrix} -0.4357 \\ 9.5652 \end{bmatrix}.$$

To use the approach based on statistical learning theory, one needs to introduce probability distributions over the uncertain parameter vector q and the controller vector

$$K = K(\theta) = \begin{bmatrix} \theta_1 \\ \theta_2 \end{bmatrix}.$$

In particular, in this example, we assume that $\mathbf{q}$ is uniformly distributed within $\mathcal{B}_q$, and that the controller gains θ_1, θ_2 are uniform random variables in the interval $[-100, 100]$. Then, we consider the family $\mathcal{J}$ of binary-valued performance functions of the form

$$J(\mathbf{q}, \boldsymbol{\theta}) = \begin{cases} 0 \text{ if } \mathrm{Re}(\lambda_{\max}(A_{cl}(\mathbf{q}, \boldsymbol{\theta}))) \leq -\alpha; \\ 1 \text{ otherwise} \end{cases}$$

where $\alpha \geq 0.1$ is the desired decay rate, and $\lambda_{\max}$ denotes the eigenvalue of $A_{cl}(\mathbf{q}, \boldsymbol{\theta})$ of largest real part.

In [285], a bound on the VC dimension of the class $\mathcal{J}$ is computed[2]

---

[2] In [285] the VC dimension for Hurwitz stability is computed for the special case $n_o = n_i = n_s = n$. The bound used here is a straightforward extension to an α-shifted version of A_{cl}.

$$\text{VC}\,(\mathcal{J}) \leq 2n_i n_o \log_2[2en_s^2(n_s + 1)].$$

Then, substituting $n_s = 4$, $n_i = 2$ and $n_o = 1$, we immediately obtain

$$\text{VC}\,(\mathcal{J}) \leq 35.$$

Next, setting $\epsilon_1 = \epsilon_2 = \delta = 0.01$, the design of a static probabilistic controller according to Corollary 10.3 requires

$$N \geq 1.23{\times}10^8$$

samples of the uncertainty $\mathbf{q}$, and

$$M \geq 528$$

samples of the controller parameter $\boldsymbol{\theta}$. From these numerical values, we notice that this approach is hardly viable in practice. Instead, using the bounds given in Theorem 10.3, we obtain

$$N \geq 61,295 \text{ and } M \geq 528.$$

Algorithm 10.1 then yields the static output feedback controller $K(\widehat{\boldsymbol{\theta}}_{NM})$, which satisfies with confidence $1 - \delta$

$$\text{PR}\left\{\text{E}_{\mathbf{q}}(J(\mathbf{q}, \boldsymbol{\theta})) < \widehat{\text{E}}_N\left(J(\mathbf{q}, \widehat{\boldsymbol{\theta}}_{NM})\right) - \epsilon_1\right\} \leq \epsilon_2.$$

In this example, the RA led to the numerical value

$$K(\widehat{\boldsymbol{\theta}}_{NM}) = \begin{bmatrix} 12.6426 \\ 90.7250 \end{bmatrix}.$$

★

Example 10.4 (Average sensitivity minimization).
   We next present an example of probabilistic control design for the longitudinal dynamics of an aircraft. This example was originally presented in [282] and involves the minimization of the average sensitivity of the control loop, subject to a bound on the nominal complementary sensitivity.
   The linearized model for the longitudinal axis of an aircraft is given by

$$\dot{x} = \begin{bmatrix} q_1 & 1 - q_2 \\ q_3 & q_4 \end{bmatrix} x + \begin{bmatrix} q_5 \\ q_6 \end{bmatrix} u;$$

$$y = \begin{bmatrix} 1 & 0 \\ 0 & 1 \end{bmatrix} x$$

where the uncertain parameters $\mathbf{q} = [\mathbf{q}_1 \cdots \mathbf{q}_6]^T$ are experimentally available in the form of Gaussian random variables, with mean and standard deviation specified in Table 10.1.

Table 10.1. Mean and standard deviation for the uncertain parameters of the plant in Example 10.4.

	mean	standard deviation
q_1	-0.9381	0.0736
q_2	0.0424	0.0035
q_3	1.6630	0.1385
q_4	-0.8120	0.0676
q_5	-0.3765	0.0314
q_6	-10.879	3.4695

To account for other structural elements, such as actuators and sensors, the input to the above system is pre-filtered through the transfer function

$$H(s) = \frac{0.000697s^2 - 0.0397s + 1}{0.000867s^2 + 0.0591s + 1}.$$

The previous system, together with the pre-filter $H(s)$, constitutes the plant family $G(s, \mathbf{q})$.

The design objective is to synthesize a first-order controller of the form

$$K(s, \theta) = \left[\theta_1 \; \frac{1 + \theta_3 s}{1 + \theta_4 s} \theta_2 \right]$$

where $\theta = [\theta_1 \; \cdots \; \theta_4]^T$ are the design parameters. The design objectives are closed-loop stability, minimization of the average sensitivity, and the satisfaction of a bound on a modified complementary sensitivity. That is

$$\min_{\theta} \; E_{\mathbf{q}}\big(\|W(s)(I + G(s, \mathbf{q})K(s, \theta))^{-1}\|_\infty\big), \text{ subject to}$$

$$\left\| \frac{0.75K(s, \theta)G_0(s)}{1 + 1.25K(s, \theta)G_0(s)} \right\|_\infty \leq 1$$

where $G_0(s)$ denotes the nominal plant (i.e. $G(s, \mathbf{q})$ evaluated for the mean value of $\mathbf{q}$), and $W(s)$ is a given filter

$$W(s) = \begin{bmatrix} \frac{552.1376}{(s+6.28)(s+31.4)} & 0 \\ 0 & \frac{552.1376}{(s+6.28)(s+31.4)} \end{bmatrix}.$$

From practical considerations, the controller parameters are a priori assumed to be bounded in the intervals

$$\theta_1 \in [0, 2], \; \theta_2 \in [0, 1], \; \theta_3 \in [0.01, 0.1], \; \theta_4 \in [0.01, 0.1].$$

In order to apply the randomized design method to this problem, we construct the following performance function

$$J(\mathbf{q}, \theta) = \max(\psi_1(\theta), \psi_2(\mathbf{q}, \theta))$$

where

$$\psi_1(\theta) = \begin{cases} 1 & \text{if } \left\| \frac{0.75K(s,\theta)G_0(s)}{1+1.25K(s,\theta)G_0(s)} \right\|_\infty > 1; \\ 0 & \text{otherwise} \end{cases}$$

and

$$\psi_2(\mathbf{q}, \theta) = \begin{cases} 1 & \text{if closed loop is unstable;} \\ \frac{\|W(s)(I+G(s,\mathbf{q})K(s,\theta))^{-1}\|_\infty}{1+\|W(s)(I+G(s,\mathbf{q})K(s,\theta))^{-1}\|_\infty} & \text{otherwise.} \end{cases}$$

Now, assuming uniform distribution for the controller parameters, and setting probability levels $\epsilon_1 = \epsilon_2 = 0.1$, $\delta = 0.01$, from Theorem 10.3 we need $M \geq 51$ controller parameter samples, and $N \geq 496$ samples of the uncertainty. Finally, Algorithm 10.1 yields the stabilizing controller

$$K(s, \widehat{\theta}_{NM}) = \left[0.2421 \ 0.4508 \frac{1+0.0744s}{1+0.0904s} \right]$$

for which

$$\|W(s)(I + G_0(s)K(s, \widehat{\theta}_{NM}))^{-1}\|_\infty = 2.91;$$

$$\left\| \frac{0.75K(s, \widehat{\theta}_{NM})G_0(s)}{1 + 1.25K(s, \widehat{\theta}_{NM})G_0(s)} \right\|_\infty = 0.55$$

and $\widehat{\mathbb{E}}_N \left(J(\mathbf{q}, \widehat{\theta}_{NM}) \right) = 0.744$.   ★

11. Sequential Algorithms for Probabilistic Robust Design

In this chapter we move on to robust controller design of uncertain systems with probabilistic techniques, and develop stochastic approximation algorithms which represent specific instances of the meta-algorithms given in Chapter 8. The main difference with respect to the techniques discussed in the previous chapter is that, here, we pursue probabilistic robust design, as opposed to the "average case" design treated in Chapter 10.

11.1 A paradigm for probabilistic robust design

When dealing with design of uncertain systems, the classical M–Δ configuration studied in Chapter 4 is modified in order to highlight the feedback connection of plant $G(s)$ and controller $K(s)$. This design configuration is shown in Figure 11.1.

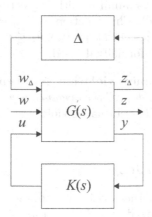

Fig. 11.1. Interconnection for robust design.

The uncertainty Δ has a structure $\mathbb{D}$ and is bounded in the ball $\mathcal{B}_{\mathbb{D}}(\rho)$, according to (3.28). Furthermore, the controller is assumed to depend on a vector of "design" parameters $\theta \in \Theta \subseteq \mathbb{R}^{n_\theta}$. The objective of standard robust design is then to select the parameters θ so that the closed-loop configuration is robustly stable, and a given controller-dependent performance specification

$$J(\Delta, \theta) \leq \gamma$$

is guaranteed for all values of $\Delta \in \mathcal{B}_{\mathbb{D}}(\rho)$, as detailed in Chapter 6. On the other hand, in a probabilistic robust setting, Δ is a random matrix with support $\mathcal{B}_{\mathbb{D}}(\rho)$, see Assumption 6.1, and we aim to construct a controller so that the performance specification is violated only for uncertainties belonging to a bad set $\mathcal{B}_B$ of "small" measure.

One of the earliest rigorous attempts to address design problems in a probabilistic setting is outlined in Section 10.3. This approach is based upon randomization of both uncertainty and controller parameters. For example, in SISO design problems the controller numerator and denominator coefficients are assumed to be random variables with an assigned probability distribution. Then, plant samples are generated in order to construct an empirical version $\widehat{\mathrm{E}}_N(J(\Delta, \theta))$ of the average performance $\mathrm{E}_\Delta(J(\Delta, \theta))$. Subsequently, a minimizer of $\widehat{\mathrm{E}}_N(J(\Delta, \theta))$ is computed by pure random search over the controller parameters. The theoretical underpinning of this method is provided by statistical learning theory, see Chapter 10.

In this chapter we present an alternative paradigm for probabilistic design that follows the robust performance synthesis philosophy of Section 8.3.2. The main idea behind this design paradigm consists of randomization to handle uncertainty, and convex optimization to select the design parameters. More precisely, we concentrate on design algorithms based on sequential random update techniques that aim to minimize, with respect to θ, a particular function $v(\Delta, \theta)$ related to $J(\Delta, \theta)$. This function, which measures the extent of violation of the performance specification $J(\Delta, \theta) \leq \gamma$, is called *performance violation function* and is now formally defined.

Definition 11.1 (Performance violation function). *Given a performance function* $J(\Delta, \theta)$ *and a desired performance level* γ, *a function* $v(\Delta, \theta)$: $\mathcal{B}_{\mathbb{D}}(\rho) \times \Theta \to \mathbb{R}$ *is said to be a* performance violation function *if the following properties hold:*

1. *$v(\Delta, \theta) \geq 0$ for all $\Delta \in \mathcal{B}_{\mathbb{D}}(\rho)$ and $\theta \in \Theta$;*
2. *$v(\Delta, \theta) = 0$ if and only if $J(\Delta, \theta) \leq \gamma$.*

We shall be mainly concerned with performance specifications $J(\Delta, \theta)$ that are convex in θ, for all fixed $\Delta \in \mathcal{B}_{\mathbb{D}}(\rho)$. More precisely, we now state a formal assumption.

Assumption 11.1 (Convexity of the performance function). *The performance function* $J(\Delta, \theta) : \mathcal{B}_{\mathbb{D}}(\rho) \times \Theta \to \mathbb{R}$ *is convex in* θ, *for all* $\Delta \in \mathcal{B}_{\mathbb{D}}(\rho)$, *and* Θ *is a convex and closed subset of* $\mathbb{R}^{n_\theta}$.

Example 11.1 (Performance violation function for $J(\Delta, \theta)$ convex).
Consider a performance functions $J(\Delta, \theta)$ which is convex in θ for any allowable value of $\Delta \in \mathcal{B}_{\mathbb{D}}(\rho)$. Then, a simple example of violation function is given by

$$v(\Delta, \theta) = \begin{cases} J(\Delta, \theta) - \gamma & \text{if } J(\Delta, \theta) > \gamma; \\ 0 & \text{otherwise.} \end{cases}$$

Different violation functions, specifically tailored to design sequential algorithms for liner quadratic regulators and LMI problems, are introduced in Sections 11.2.1 and 11.3.   ⋆

Notice that if a design parameter $\theta^* \in \Theta$ is found such that $v(\Delta, \theta^*) = 0$, for all $\Delta \in \mathcal{B}_{\mathbb{D}}(\rho)$, then the corresponding controller robustly satisfies (in the classical deterministic sense) the performance specification $J(\Delta, \theta^*) \leq \gamma$, for all values of Δ in $\mathcal{B}_{\mathbb{D}}(\rho)$.

In the following sections, we develop iterative randomized algorithms whose finite termination time can be assessed whenever the violation function $v(\Delta, \theta)$ is convex in θ for any fixed $\Delta \in \mathcal{B}_{\mathbb{D}}(\rho)$. In several cases of interest, we show that by using random samples of Δ one can construct a random sequence of candidate design parameters $\theta^{(1)}, \theta^{(2)}, \ldots$ that converges with probability one in a finite number of iterations to a deterministically robust solution. On the other hand, the number of iterations is itself a random variable, and one does not know in advance how many uncertainty samples are required to achieve such a robust solution. However, if the design goals are posed in a probabilistic setting, then we are satisfied with a solution that meets the performance objective for most of the uncertainty instances. In this case, for $v(\Delta, \theta)$ convex in θ, we can assess a priori the number of steps needed for the sequence $\theta^{(1)}, \theta^{(2)}, \ldots$ to converge to a probabilistic robust solution.

In particular, in Section 11.2 we first describe sequential algorithms for the case of a guaranteed-cost linear quadratic regulator, which has been studied in [221] and subsequent refinements in [113]. In Section 11.3 we present a more general formulation of these techniques for the solution of problems that can be cast in the general form of feasibility problems involving a finite or infinite number of linear matrix inequalities, as developed in [61]. Prior to these developments, the next section presents some ancillary technical results.

11.1.1 Technical preliminaries

The projection $[A]_S$ of a real matrix $A \in \mathbb{R}^{m,n}$ onto a closed set S is defined as

$$[A]_S \doteq \arg \min_{X \in S} \|A - X\| \tag{11.1}$$

where S is a subset of $\mathbb{R}^{m,n}$ and $\|\cdot\|$ is any matrix norm.

The projection $[\cdot]_S$ can be easily computed in many cases of interest, as shown in the next example.

Example 11.2 (Projection on $\mathbb{R}_+^n$).
If the set S is the nonnegative orthant, defined as

$$\mathbb{R}_+^n = \{x \in \mathbb{R}^n : x_i \geq 0, i = 1, \ldots, n\}$$

and $\| \cdot \|$ is the ℓ_2 norm, then the projection of a vector x onto $\mathbb{R}_+^n$ is

$$[x]_{\mathbb{R}_+^n} = \arg \min_{y \in \mathbb{R}_+^n} \|x - y\|_2.$$

This projection can be simply computed taking the positive part of each component of x

$$[x]_{\mathbb{R}_+^n} = [\max\{0, x_1\} \cdots \max\{0, x_n\}]^T.$$

⋆

The projection $[\cdot]_S$ can also be easily computed in other situations, which include the case when S is a half-space or an ellipsoid. Another case of interest is when S is the cone $\mathbb{S}_+^n$ of symmetric positive semidefinite matrices, equipped with the inner product $\langle A, B \rangle = \operatorname{Tr} AB$ and the corresponding Frobenius norm. In this case, we denote by $[\cdot]_+$ the projection $[\cdot]_{\mathbb{S}_+^n}$. That is, given a real symmetric matrix $A \in \mathbb{S}^n$, we write

$$[A]_+ \doteq \arg \min_{X \in \mathbb{S}_+^n} \|A - X\|. \tag{11.2}$$

The projection $[A]_+$ can be found explicitly by means of the eigenvalue decomposition, see e.g. [61, 219]. For a real symmetric matrix $A \in \mathbb{S}^n$, we write

$$A = T \Lambda T^T$$

where T is an orthogonal matrix and

$$\Lambda \doteq \operatorname{diag}([\lambda_1 \cdots \lambda_n])$$

and where λ_i, $i = 1, \ldots, n$ are the eigenvalues of A. Then

$$[A]_+ = T[\Lambda]_{\mathbb{R}_+^n} T^T$$

where

$$[\Lambda]_{\mathbb{R}_+^n} = \operatorname{diag}([\max\{0, \lambda_1\} \cdots \max\{0, \lambda_n\}]).$$

Remark 11.1 (Projection on $\mathbb{S}_+^n$). Another method to compute the projection $[\cdot]_+$ makes use of the real Schur decomposition, see [244, 296]. That is, we write a real symmetric matrix $A \in \mathbb{S}^n$ as

$$A = [U_p \ U_n] \begin{bmatrix} \Lambda_p & 0 \\ 0 & \Lambda_n \end{bmatrix} \begin{bmatrix} U_p^T \\ U_n^T \end{bmatrix}$$

where Λ_p and Λ_n are diagonal matrices containing, respectively, the nonnegative and the strictly negative eigenvalues of A, and $\begin{bmatrix} U_p & U_n \end{bmatrix}$ is an orthogonal matrix. Then, the projection is given by

$$[A]_+ = \begin{cases} 0 & \text{if } A \preceq 0; \\ U_p \Lambda_p U_p^T & \text{otherwise.} \end{cases}$$

$\triangleleft$

11.2 Sequential algorithms for LQR design

In this section, we study linear quadratic regulator design with guaranteed cost, discussed in a deterministic context in Sections 4.2.1 and 4.2.2. In particular, we consider a state space description with structured uncertainty $\Delta \in \mathcal{B}_{\mathbb{D}}(\rho)$ affecting the state matrix of the system

$$\dot{x} = A(\Delta)x + B_2 u, \quad x(0) = x_0 \in \mathbb{R}^{n_s}.$$

We then take the standard quadratic cost for linear quadratic regulators

$$\mathcal{C}(\Delta) = \int_0^{\infty} \left(x^T(t) Q_{xx} x(t) + u^T(t) Q_{uu} u(t) \right) dt$$

with $Q_{xx} \succeq 0$ and $Q_{uu} \succ 0$ and consider a state feedback law of the form

$$u = -Q_{uu}^{-1} B_2^T P^{-1} x.$$

Then, from Lemma 4.5 we have that, for given level γ, the system is quadratically stable and the design objective

$$\mathcal{C}(\Delta) \leq \gamma^{-1} x_0^T P^{-1} x_0$$

is guaranteed for all $\Delta \in \mathcal{B}_{\mathbb{D}}(\rho)$, if the quadratic matrix inequality (QMI) in the matrix variable $P \succ 0$

$$A(\Delta)P + PA^T(\Delta) - 2B_2 Q_{uu}^{-1} B_2^T + \gamma(B_2 Q_{uu}^{-1} B_2^T + PQ_{xx}P) \prec 0 \quad (11.3)$$

is satisfied for all $\Delta \in \mathcal{B}_{\mathbb{D}}(\rho)$.

To determine a solution to the above synthesis QMI, we propose a sequential randomized algorithm that yields a quadratic regulator with probabilistically guaranteed cost. The algorithm is a specific implementation of Algorithm 8.4 for robust synthesis, where the matrix variable $P \succ 0$ plays the role of design parameter θ. In particular, in Section 11.2.1 we introduce some technical assumptions, in Section 11.2.2 we present the specific update rule ψ_{upd} for the randomized algorithm, and in Section 11.2.3 we present probabilistic convergence results.

11.2.1 Violation function for LQR

For $\Delta \in \mathcal{B}_{\mathbb{D}}(\rho)$ and $P \succ 0$, we define the matrix-valued function

$$V(\Delta, P) \doteq A(\Delta)P + PA^T(\Delta) - 2B_2 Q_{uu}^{-1} B_2^T + \gamma(B_2 Q_{uu}^{-1} B_2^T + PQ_{xx}P).$$

Then, we introduce a specific violation function for the LQR problem

$$v(\Delta, P) = \|[V(\Delta, P)]_+\|$$

where $[\cdot]_+$ is the projection defined in (11.2) and $\|\cdot\|$ denotes Frobenius norm. It is easy to check that this violation function satisfies the requirements of Definition 11.1, and is convex in P for fixed Δ. In particular, we notice that $v(\Delta, P) \geq 0$ holds for all $P \succ 0$ and $\Delta \in \mathcal{B}_{\mathbb{D}}(\rho)$. Thus, the problem of finding a solution $P \succ 0$ of the matrix inequality (11.3) is recast as the problem of minimizing, with respect to $P \succ 0$, the scalar function $\sup_{\Delta \in \mathcal{B}_{\mathbb{D}}(\rho)} v(\Delta, P)$ and checking if the resulting minimum is equal to zero. This minimization is performed by means of the randomized algorithms presented in this chapter.

Following a probabilistic approach, we now consider Δ as a random matrix with pdf $f_{\Delta}(\Delta)$ and support $\mathcal{B}_{\mathbb{D}}(\rho)$, and seek for a randomized solution of the LQR design problem. To this aim, we first formally state a feasibility assumption for the set of solutions. This assumption is subsequently relaxed in Section 11.5, which is focused on possibly unfeasible problems.

Assumption 11.2 (Strict feasibility). *The solution set*

$$S_P \doteq \{P \succ 0 : V(\Delta, P) \preceq 0 \ \text{for all} \ \Delta \in \mathcal{B}_{\mathbb{D}}(\rho)\}$$

has a nonempty interior.

In other words, we require that the solution set S_P contains a ball

$$\{P \succ 0 : \|P - P^*\| \leq r\} \tag{11.4}$$

with center $P^* \succ 0$ and radius $r > 0$. Next, we introduce a probabilistic assumption on the set S_P.

Assumption 11.3 (Nonzero probability of detecting unfeasibility). *We assume that $f_{\Delta}(\Delta)$ is such that*

$$\mathrm{PR}_{\Delta} \{v(\mathbf{\Delta}, P) > 0\} > 0$$

for any $P \notin S_P$.

This is a very mild assumption that can be intuitively explained as follows: For any $P \notin S_P$, we assume that there is a nonzero probability of generating a random sample $\mathbf{\Delta}^{(i)}$ according to $f_{\Delta}(\Delta)$ so that P is unfeasible, i.e. $v(\mathbf{\Delta}^{(i)}, P) > 0$.

Remark 11.2 (Relaxing Assumption 11.3). If $A(\Delta)$ is a continuous function of Δ, then $v(\Delta, P)$ is a continuous function of Δ for any P and, in turn, S_P is a compact set. If, in addition, the probability density function $f_{\Delta}(\Delta)$ is nonzero for all $\Delta \in \mathcal{B}_{\mathbb{D}}(\rho)$, then the set of $\mathbf{\Delta}$ satisfying $v(\mathbf{\Delta}, P) > 0$ has nonzero measure. This implies that Assumption 11.3 is satisfied. $\lhd$

11.2.2 Update rule and subgradient computation

Following the philosophy of Algorithm 8.4, the randomized algorithm studied here iteratively updates the solution $\mathbf{P}^{(k)}$. In particular, the algorithm presents an "outer" and an "inner" loop. For the current solution $\mathbf{P}^{(k)}$, the inner iterations consist of randomly sampling the uncertainty $\mathbf{\Delta}^{(i)}$ and checking whether $v(\mathbf{\Delta}^{(i)}, \mathbf{P}^{(k)}) \leq 0$. If this inequality is satisfied, then another $\mathbf{\Delta}^{(i)}$ is extracted, and the check is repeated until a random sample is found such that $v(\mathbf{\Delta}^{(i)}, \mathbf{P}^{(k)}) > 0$ (i.e. such that the performance specification is violated). At this point, an outer iteration step is performed, updating the current solution according to the update rule

$$\mathbf{P}^{(k+1)} = \psi_{\text{upd}}(\mathbf{\Delta}^{(i)}, \mathbf{P}^{(k)}).$$

For the linear quadratic regulator, we consider the specific update rule

$$\mathbf{P}^{(k+1)} = \psi_{\text{upd}}(\mathbf{\Delta}^{(i)}, \mathbf{P}^{(k)}) = \left[\mathbf{P}^{(k)} - \eta^{(k)}\partial_P\{v(\mathbf{\Delta}^{(i)}, \mathbf{P}^{(k)})\}\right]_+ \qquad (11.5)$$

where ∂_P denotes the subgradient of v with respect to P, $[\cdot]_+$ is the projection (11.2) onto the cone $\mathbb{S}^n_+$ of symmetric positive semidefinite matrices, and $\eta^{(k)}$ is a suitably selected stepsize. In the next lemma, we compute the subgradient with respect to P of $v(\Delta, P)$, which is needed in the iterative algorithm.

Lemma 11.1. *The function $v(\Delta, P)$ is convex in P and its subgradient $\partial_P\{v(\Delta, P)\}$ is given by*

$$\partial_P\{v(\Delta, P)\} = \qquad (11.6)$$

$$\frac{[V(\Delta, P)]_+(A(\Delta) + \gamma PQ_{xx}) + (A(\Delta) + \gamma PQ_{xx})^T[V(\Delta, P)]_+}{v(\Delta, P)}$$

if $v(\Delta, P) > 0$, or

$$\partial_P\{v(\Delta, P)\} = 0$$

otherwise.

Proof. The convexity of $v(\Delta, P)$ can be checked by direct calculation, observing that the quadratic term in $V(\Delta, P)$, which is $\gamma PQ_{xx}P$, is positive semidefinite. To show that the subgradient is given by (11.6), we observe that $v(\Delta, P)$ is differentiable if $v(\Delta, P) > 0$. Indeed, for $v(\Delta, P) > 0$ and $P + P_\Delta \succ 0$, we write the first-order approximation

$$v(\Delta, (P + P_\Delta)) = \|[A(\Delta)P + PA^T(\Delta) + (\gamma - 2)B_2Q_{uu}^{-1}B_2^T + \gamma PQ_{xx}P$$
$$+ A(\Delta)P_\Delta + P_\Delta A^T(\Delta) + \gamma P_\Delta Q_{xx}P + \gamma PQ_{xx}P_\Delta$$
$$+ o(\|P_\Delta\|)]_+\|$$

where $o(\cdot)$ represents higher-order terms. Then, we obtain

$$v(\Delta, (P + P_\Delta)) = v(\Delta, P)$$
$$+ \left\langle \frac{[V(\Delta, P)]_+}{v(\Delta, P)}, \left(A(\Delta)P_\Delta + P_\Delta A^T(\Delta) + \gamma P_\Delta Q_{xx} P + \gamma P Q_{xx} P_\Delta \right) \right\rangle$$
$$+ o(\|P_\Delta\|).$$

Using the fact that $\langle A, B \rangle = \langle B, A \rangle$, we finally get

$$v(\Delta, (P + P_\Delta)) = v(\Delta, P) + \mathrm{Tr}\,[R(\Delta, P)P_\Delta] + o(\|P_\Delta\|)$$

where

$$R(\Delta, P) \doteq \frac{[V(\Delta, P)]_+}{v(\Delta, P)}(A(\Delta) + \gamma P Q_{xx})^T + (A(\Delta) + \gamma P Q_{xx})\frac{[V(\Delta, P)]_+}{v(\Delta, P)}.$$

The proof of the lemma is completed by observing that the relation above coincides with the definition of gradient. Finally, the proof for the case $v(\Delta, P) = 0$ is immediate.   □

11.2.3 Iterative algorithms and convergence results

We now state an iterative algorithm which is based on the update rule (11.5).

Algorithm 11.1 (Iterative algorithm for LQR design).

1. Initialization.
   ▷ Set $i = 0$, $k = 0$, and choose $P^{(0)} \succ 0$;
2. Feasibility loop.
   ▷ Set $i = i + 1$
      – Draw $\mathbf{\Delta}^{(i)}$ according to f_Δ;
      – If $v(\mathbf{\Delta}^{(i)}, \mathbf{P}^{(k)}) > 0$ goto 3; Else goto 2;
3. Update.
   ▷ Compute

$$\mathbf{P}^{(k+1)} = \psi_{\mathrm{upd}}(\mathbf{\Delta}^{(i)}, \mathbf{P}^{(k)}) = \left[\mathbf{P}^{(k)} - \boldsymbol{\eta}^{(k)} \partial_P \{v(\mathbf{\Delta}^{(i)}, \mathbf{P}^{(k)})\} \right]_+$$

   where the stepsize $\boldsymbol{\eta}^{(k)}$ is given in (11.7);
   ▷ Set $k = k + 1$ and goto 2.

The stepsize in this algorithm is given by

$$\boldsymbol{\eta}^{(k)} = \frac{v(\mathbf{\Delta}^{(i)}, \mathbf{P}^{(k)}) + r\|\partial_P \{v(\mathbf{\Delta}^{(i)}, \mathbf{P}^{(k)})\}\|}{\|\partial_P \{v(\mathbf{\Delta}^{(i)}, \mathbf{P}^{(k)})\}\|^2} \tag{11.7}$$

where $r > 0$ is the radius of a ball (11.4) with center P^* contained in the feasibility set S_P. Notice that in the algorithm we make a distinction between "inner" iterations (those related to the feasibility check loop in step 2) and "outer" iterations, which provide the actual parameter updates.

We now state a result which guarantees that Algorithm 11.1 converges to a feasible solution in a finite number of outer iterations.

Theorem 11.1 (Finite-time convergence). *Let Assumptions 11.2 and 11.3 be satisfied. Then, with probability one, Algorithm 11.1 executes a finite number of updates to determine a robustly feasible solution within S_P. Moreover, if an upper bound D is known on the initial distance*

$$\|P^{(0)} - P^*\| \leq D$$

then the total number of updates is bounded by $\overline{k} \doteq \lfloor (D/r)^2 \rfloor$, with $r > 0$ and $P^* \succ 0$ the radius and the center of a ball (11.4) contained in S_P.

Proof. First, we define

$$\bar{\mathbf{P}} = P^* + r \frac{\partial_P \{ v(\mathbf{\Delta}^{(i)}, \mathbf{P}^{(k)}) \}}{\|\partial_P \{ v(\mathbf{\Delta}^{(i)}, \mathbf{P}^{(k)}) \}\|}.$$

Then, owing to Assumption 11.2, $\bar{\mathbf{P}} \in S_P$, i.e. it is a feasible solution and $v(\mathbf{\Delta}^{(i)}, \bar{\mathbf{P}}) = 0$. Now, if $v(\mathbf{\Delta}^{(i)}, \mathbf{P}^{(k)}) > 0$, then it follows from the properties of a projection that

$$\begin{aligned}
\|\mathbf{P}^{(k+1)} - P^*\|^2 &\leq \|\mathbf{P}^{(k)} - P^* - \eta^{(k)} \partial_P \{ v(\mathbf{\Delta}^{(i)}, \mathbf{P}^{(k)}) \}\|^2 \\
&= \|\mathbf{P}^{(k)} - P^*\|^2 + (\eta^{(k)})^2 \|\partial_P \{ v(\mathbf{\Delta}^{(i)}, \mathbf{P}^{(k)}) \}\|^2 \\
&\quad -2\eta^{(k)} \langle \partial_P \{ v(\mathbf{\Delta}^{(i)}, \mathbf{P}^{(k)}) \}, (\mathbf{P}^{(k)} - \bar{\mathbf{P}}) \rangle \\
&\quad -2\eta^{(k)} \langle \partial_P \{ v(\mathbf{\Delta}^{(i)}, \mathbf{P}^{(k)}) \}, (\bar{\mathbf{P}} - P^*) \rangle.
\end{aligned}$$

We now consider the last two terms in this inequality. Owing to convexity of $v(\Delta, P)$ and to the feasibility of $\bar{\mathbf{P}}$, we obtain

$$\langle \partial_P \{ v(\mathbf{\Delta}^{(i)}, \mathbf{P}^{(k)}) \}, (\mathbf{P}^{(k)} - \bar{\mathbf{P}}) \rangle \geq v(\mathbf{\Delta}^{(i)}, \mathbf{P}^{(k)})$$

and, owing to the definition of $\bar{\mathbf{P}}$, we have

$$\langle \partial_P \{ v(\mathbf{\Delta}^{(i)}, \mathbf{P}^{(k)}) \}, (\bar{\mathbf{P}} - P^*) \rangle = r \|\partial_P \{ v(\mathbf{\Delta}^{(i)}, \mathbf{P}^{(k)}) \}\|.$$

Thus, we write

$$\begin{aligned}
\|\mathbf{P}^{(k+1)} - P^*\|^2 &\leq \|\mathbf{P}^{(k)} - P^*\|^2 + (\eta^{(k)})^2 \|\partial_P \{ v(\mathbf{\Delta}^{(i)}, \mathbf{P}^{(k)}) \}\|^2 \\
&\quad -2\eta^{(k)} (v(\mathbf{\Delta}^{(i)}, \mathbf{P}^{(k)}) + r \|\partial_P \{ v(\mathbf{\Delta}^{(i)}, \mathbf{P}^{(k)}) \}\|).
\end{aligned}$$

Now, substituting the value of $\eta^{(k)}$ given in (11.7), we get

$$\begin{aligned}
\|\mathbf{P}^{(k+1)} - P^*\|^2 &\leq \|\mathbf{P}^{(k)} - P^*\|^2 - \frac{(v(\mathbf{\Delta}^{(i)}, \mathbf{P}^{(k)}) + r \|\partial_P \{ v(\mathbf{\Delta}^{(i)}, \mathbf{P}^{(k)}) \}\|)^2}{\|\partial_P \{ v(\mathbf{\Delta}^{(i)}, \mathbf{P}^{(k)}) \}\|^2} \\
&\leq \|\mathbf{P}^{(k)} - P^*\|^2 - r^2.
\end{aligned}$$

Therefore, if $v(\mathbf{\Delta}^{(i)}, \mathbf{P}^{(k)}) > 0$, we obtain

$$\|\mathbf{P}^{(k+1)} - P^*\|^2 \leq \|\mathbf{P}^{(k)} - P^*\|^2 - r^2. \tag{11.8}$$

From this formula we conclude that the total number of updates is bounded by $\lfloor \|P^{(0)} - P^*\|^2 / r^2 \rfloor$.

On the other hand, if $\mathbf{P}^{(k)}$ is unfeasible, then, owing to Assumption 11.3, there is a nonzero probability of performing a correction step. Thus, with probability one, the method cannot terminate at an unfeasible point. Therefore, we conclude that the algorithm must terminate after a finite number of iterations at a feasible solution.   □

Remark 11.3 (Related literature). We observe that sequential algorithms similar to those proposed in this chapter have been used in a deterministic setting for adaptive control and unknown but bounded identification, see e.g. [18, 46]. In particular, these iterative algorithms can be traced back to the Kaczmarz projection method [146] for solving linear equations, which was studied in 1937, and to the methods given in [3, 193] for studying linear inequalities, which were presented in 1954. A finite-convergent version of the latter method was proposed later, see [46] and references therein. A similar approach for solving convex inequalities can also be found in [110, 219]. Finally, if compared with the method of alternating projections, see [244], we observe that the update rule in the iterative algorithm for LQR design does not require the computation of projections on sets defined by quadratic matrix inequalities. From the numerical point of view, this is a crucial advantage, since this operation is generally difficult to perform. More generally, stochastic approximation algorithms have been widely studied in the stochastic optimization literature, see e.g. [163] for detailed discussions on this topic.   ◁

Theorem 11.1 guarantees convergence of Algorithm 11.1 to a robustly feasible solution (if one exists) in a finite number of iterations, but does not provide an explicit a priori bound on the total number of samples (sample complexity) needed in the inner iterations of the algorithm. In words, what the theorem says is that if in an actual run of the algorithm we observe $\lfloor (D/r)^2 \rfloor$ updates, then we can stop the algorithm and be sure to have a robustly feasible solution. However, it can happen that the algorithm gets seemingly stuck in loop 2, for some $k < \lfloor (D/r)^2 \rfloor$. In this case, we do not know how long to wait before stopping the algorithm anyways and claiming to have determined a robustly feasible solution. Indeed, to make this claim we should run loop 2 forever, which is clearly unacceptable.

Nevertheless, if we are satisfied with a probabilistic robust solution, it is possible to bound a priori the number of inner iterations and, therefore, obtain a modified version of Algorithm 11.1. In this case, we need to relax the request of seeking a deterministically feasible solution and be satisfied with a probabilistically robust one. The modified version of the algorithm, which conforms to the requirements for randomized algorithms for robust performance synthesis given in Definition 8.4, has the same structure as Algorithm 8.4 and is presented next.

Algorithm 11.2 (RA for LQR design).

1. Initialization.
    ▷ Let $N(k) = \overline{N}_{ss}(p^*, \delta, k)$ given in (11.9);
    ▷ Set $k = 0$, $i = 0$, and choose $P^{(0)} \succ 0$;
2. Feasibility loop.
    ▷ Set $\ell = 0$ and feas=true;
    ▷ While $\ell < N(k)$ and feas=true
        − Set $i = i + 1$, $\ell = \ell + 1$;
        − Draw $\mathbf{\Delta}^{(i)}$ according to $f_{\mathbf{\Delta}}$;
        − If $v(\mathbf{\Delta}^{(i)}, \mathbf{P}^{(k)}) > 0$ set feas=false;
    ▷ End While;
3. Exit condition.
    ▷ If feas=true
        − Set $N = i$;
        − Return $\widehat{\mathbf{P}}_N = \mathbf{P}^{(k)}$ and Exit;
    ▷ End If;
4. Update.
    ▷ Compute

$$\mathbf{P}^{(k+1)} = \psi_{\text{upd}}(\mathbf{\Delta}^{(i)}, \mathbf{P}^{(k)}) = \left[\mathbf{P}^{(k)} - \boldsymbol{\eta}^{(k)} \partial_P \{ v(\mathbf{\Delta}^{(i)}, \mathbf{P}^{(k)}) \} \right]_+$$

where the stepsize $\boldsymbol{\eta}^{(k)}$ is given in (11.7);
    ▷ Set $k = k + 1$ and goto 2.

We notice that, in this algorithm, for any candidate $\mathbf{P}^{(k)}$ the inner loop ("While" loop in step 2) performs a randomized check of robust feasibility of the current solution. If the loop performs all the $\ell = N(k)$ checks without detecting unfeasibility, then the algorithm exits at step 3 with the current solution. Otherwise, if a random sample $\mathbf{\Delta}^{(i)}$ is found such that $v(\mathbf{\Delta}^{(i)}, \mathbf{P}^{(k)}) > 0$, i.e. the performance is violated, then the current solution $\mathbf{P}^{(k)}$ is updated and the process is repeated. We next present two results related to Algorithm 11.2. First, Theorem 11.2 below shows that if a suitable bound is imposed on the number of samples to be drawn in the feasibility loop, then with high probability, if the algorithm stops (i.e. it exits at step 3), it returns a probabilistically robust solution. This first result holds under very general conditions, and in particular it does not require the existence of a robustly feasible solution. Subsequently, Corollary 11.1 combines Theorem 11.2 and the results of Theorem 11.1 on the convergence of the outer loop in a finite and explicitly bounded number of iterations. These results are stated in [206] for the ellipsoid algorithm. We present here minor modifications that apply to the iterative gradient methods discussed in this section.

Theorem 11.2 (Sample size function). *Let the sample size function* $N(k)$ *be defined as*

$$N(k) = \overline{N}_{\mathrm{ss}}(p^*, \delta, k) \doteq \left\lceil \frac{\log \frac{\pi^2 (k+1)^2}{6\delta}}{\log \frac{1}{p^*}} \right\rceil . \tag{11.9}$$

Then, with probability greater than $1 - \delta$, *if Algorithm 11.2 exits, the returned solution* $\widehat{\mathbf{P}}_N \succ 0$ *satisfies*

$$\mathrm{PR}\left\{ v(\boldsymbol{\Delta}, \widehat{\mathbf{P}}_N) \le 0 \right\} \ge p^*. \tag{11.10}$$

Proof. This proof mimics the one given in [206]. Let $\mathbf{P}^{(k)}$ be the solution just after the kth update. Notice that $\mathbf{P}^{(k)}$ remains unchanged until the next update occurs. Define the following two events for $k = 0, 1, 2, \dots$

$$\mathcal{M}_k \doteq \left\{ \text{the } k^{\mathrm{th}} \text{ update has been performed and no update} \right.$$
$$\left. \text{is made for the successive } N(k) \text{ iterations} \right\};$$

$$\mathcal{B}_k \doteq \left\{ \text{the } k^{\mathrm{th}} \text{ update has been performed and} \right.$$
$$\left. \mathrm{PR}_{\boldsymbol{\Delta}}\left\{ v(\boldsymbol{\Delta}, \mathbf{P}^{(k)}) \le 0 \right\} < p^* \right\}.$$

Notice that the event $\mathcal{M}_k$ can occur at one k at most, since occurrence of $\mathcal{M}_k$ causes the algorithm to exit with the current solution $\mathbf{P}^{(k)}$. Consider now the probability of the joint event $\mathcal{M}_k \cap \mathcal{B}_k$

$$\mathrm{PR}_{\boldsymbol{\Delta}^{(1,\dots)}}\left\{ \mathcal{M}_k \cap \mathcal{B}_k \right\} = \mathrm{PR}_{\boldsymbol{\Delta}^{(1,\dots)}}\left\{ \mathcal{M}_k | \mathcal{B}_k \right\} \mathrm{PR}_{\boldsymbol{\Delta}^{(1,\dots)}}\left\{ \mathcal{B}_k \right\}$$
$$\le \mathrm{PR}_{\boldsymbol{\Delta}^{(1,\dots)}}\left\{ \mathcal{M}_k | \mathcal{B}_k \right\} < (p^*)^{N(k)}.$$

We next evaluate the probability that, at some outer iteration, the algorithm exits with a "bad" solution, i.e. a $\mathbf{P}^{(k)}$ such that $\mathrm{PR}_{\boldsymbol{\Delta}}\left\{ v(\boldsymbol{\Delta}, \mathbf{P}^{(k)}) \le 0 \right\} < p^*$

$$\mathrm{PR}_{\boldsymbol{\Delta}^{(1,\dots)}}\left\{ (\mathcal{M}_0 \cap \mathcal{B}_0) \cup (\mathcal{M}_1 \cap \mathcal{B}_1) \cup \cdots \right\} < (p^*)^{N(0)} + (p^*)^{N(1)} + \cdots$$
$$\le (p^*)^{\widetilde{N}(0)} + (p^*)^{\widetilde{N}(1)} + \cdots$$

where the last inequality follows from

$$\widetilde{N}(k) \doteq \frac{\log \frac{\pi^2 (k+1)^2}{6\delta}}{\log \frac{1}{p^*}} \le \lceil \widetilde{N}(k) \rceil = N(k).$$

Since

$$(p^*)^{\widetilde{N}(k)} = \frac{6\delta}{\pi^2 (k+1)^2}$$

we have

$$(p^*)^{\widetilde{N}(0)} + (p^*)^{\widetilde{N}(1)} + (p^*)^{\widetilde{N}(2)} + \cdots = \frac{6\delta}{\pi^2} \left(1 + \frac{1}{2^2} + \frac{1}{3^2} + \cdots \right) = \delta.$$

Hence

$$\mathrm{PR}_{\mathbf{\Delta}^{(1,\ldots)}} \left\{ (\mathcal{M}_0 \cap \mathcal{B}_0) \cup (\mathcal{M}_1 \cap \mathcal{B}_1) \cup \cdots \right\} < \delta$$

and considering the complementary event, we have

$$\mathrm{PR}_{\mathbf{\Delta}^{(1,\ldots)}} \left\{ (\overline{\mathcal{M}}_0 \cup \overline{\mathcal{B}}_0) \cap (\overline{\mathcal{M}}_1 \cup \overline{\mathcal{B}}_1) \cap \cdots \right\} > 1 - \delta$$

where $\overline{\mathcal{M}}_k$ and $\overline{\mathcal{B}}_k$ stand for the complements of $\mathcal{M}_k$ and $\mathcal{B}_k$ respectively. In words, this last statement means that, with probability greater than $1 - \delta$, at every outer iteration either we update the current solution $\mathbf{P}^{(k)}$, or this solution is "good", i.e. $\mathrm{PR}_{\mathbf{\Delta}} \left\{ v(\mathbf{\Delta}, \mathbf{P}^{(k)}) \leq 0 \right\} \geq p^*$. Therefore, we conclude that, with probability greater than $1 - \delta$, if the algorithm exits after the kth update (and hence the solution $\mathbf{P}^{(k)}$ is not updated) then $\mathbf{P}^{(k)}$ satisfies (11.9).

$\square$

Corollary 11.1 (Convergence of Algorithm 11.2). *Let Assumptions 11.2 and 11.3 be satisfied. Then, Algorithm 11.2 terminates in a finite number N of total iterations.*

Moreover, if an upper bound D is known on the initial distance $\| P^{(0)} - P^* \| \leq D$, then Algorithm 11.2 executes at most $\overline{k} \doteq \lfloor (D/r)^2 \rfloor$ updates, and at most

$$N \leq 1 + \sum_{k=0,\ldots,\overline{k}} \overline{N}_{\mathrm{ss}}(p^*, \delta, k) \leq 1 + (\overline{k} + 1)\overline{N}_{\mathrm{ss}}(p^*, \delta, \overline{k})$$

total iterations. Upon termination, we have two cases:

1. *If Algorithm 11.2 executes $\overline{k}$ updates, then $\widehat{\mathbf{P}}_N \in S_P$, i.e. it is a robustly feasible solution;*
2. *With probability greater than $1 - \delta$, if Algorithm 11.2 terminates before $\overline{k}$ updates, then $\widehat{\mathbf{P}}_N$ satisfies*

$$\mathrm{PR} \left\{ v(\mathbf{\Delta}, \widehat{\mathbf{P}}_N) \leq 0 \right\} \geq p^* \tag{11.11}$$

i.e. it is a probabilistically feasible solution.

The proof of this result can be easily obtained by combining Theorems 11.1 and 11.2.

Remark 11.4 (Probabilistic measure). Notice that the probabilistic behavior of Algorithm 11.2 is determined by the sequence of random samples $\mathbf{\Delta}^{(1,\ldots)}$. In particular, the confidence on statement (11.11) is evaluated according to the probabilistic measure of the sequence $\mathbf{\Delta}^{(1,\ldots)}$, see also [274] for further discussions on this topic. To this extent, the result of Corollary 11.1 can be written more precisely as

$$\text{PR}_{\Delta^{(1,\ldots)}} \left\{ \text{PR}_\Delta \left\{ v(\Delta, \widehat{\mathbf{P}}_N) \leq 0 \right\} \geq p^* \right\} > 1 - \delta.$$

$\lhd$

Remark 11.5 (Feasibility radius). Notice that, for running Algorithms 11.1 and 11.2, the radius r of a ball (11.4) contained in the feasibility set S_P needs to be known in advance.[1] Unfortunately, in general the value of r may not be known a priori. However, this should not prevent the use of these algorithms. In fact, following standard techniques in stochastic gradient optimization, see e.g. [163], we can replace the fixed r in the stepsize (11.7) with a sequence of $r^{(k)}$, such that $r^{(k)} > 0$, $r^{(k)} \to 0$ as $k \to \infty$, and $\sum_{k=0}^{\infty} r^{(k)} = \infty$. Then, inequality (11.8) in the proof of Theorem 11.1 becomes

$$\|\mathbf{P}^{(k+1)} - P^*\|^2 \leq \|\mathbf{P}^{(k)} - P^*\|^2 - \sum_{\ell=0}^{k} (r^{(\ell)})^2.$$

From this inequality we can again conclude that the modified algorithm terminates in a finite number of steps.

More generally, when facing a specific problem, we may not know in advance whether the strict feasibility assumption holds. In this case, we may simply treat r as an adjustable parameter of the algorithm, and fix it to some small value. Then, if the algorithm converges to a solution, Theorem 11.2 guarantees that this solution has the desired properties of probabilistic robustness, regardless of whether the original deterministic problem is strictly feasible or not. Strict feasibility and knowledge of r are indeed only used in Corollary 11.1 to prove a priori (i.e. without running any experiment) that under these hypotheses the algorithm will always converge. On the other hand, for unfeasible problems, one may fail to determine a suitable r and/or a suitable initial condition that makes the algorithm converge. In this case, an algorithm such as the one presented in Section 11.5, which is specifically tailored to deal with unfeasible problems, may be used. $\lhd$

We now present an example, taken from [221], showing an application of Algorithm 11.2.

Example 11.3 (LQR design of a lateral motion of an aircraft).
We consider a multivariable example given in [11] (see also the original paper [273] for a slightly different model and set of data) which studies the design of a controller for the lateral motion of an aircraft. The state space equation consists of four states and two inputs and is given by

---

[1] Clearly, the center P^* need not be known in advance, otherwise there would be no point in solving the problem.

$$\dot{x} = \begin{bmatrix} 0 & 1 & 0 & 0 \\ 0 & L_p & L_\beta & L_r \\ g/V & 0 & Y_\beta & -1 \\ N_{\dot\beta}(g/V) & N_p & N_\beta + N_{\dot\beta}Y_\beta & N_r - N_{\dot\beta} \end{bmatrix} x + $$

$$\begin{bmatrix} 0 & 0 \\ 0 & -3.91 \\ 0.035 & 0 \\ -2.53 & 0.31 \end{bmatrix} u$$

where x_1 is the bank angle, x_2 its derivative, x_3 is the sideslip angle, x_4 the yaw rate, u_1 the rudder deflection and u_2 the aileron deflection.

The following nominal values for the nine aircraft parameters entering into the state matrix are taken: $L_p = -2.93$, $L_\beta = -4.75$, $L_r = 0.78$, $g/V = 0.086$, $Y_\beta = -0.11$, $N_{\dot\beta} = 0.1$, $N_p = -0.042$, $N_\beta = 2.601$ and $N_r = -0.29$. Then, each nominal parameter is perturbed by a relative uncertainty which is taken to be equal to 15%; e.g. the parameter L_p is bounded in the interval $[-3.3695, -2.4905]$. That is, we consider the uncertain system

$$\dot{x}(t) = A(q)x(t) + B_2u(t)$$

where

$$q = [q_1 \cdots q_\ell]^T = \left[L_p\ L_\beta\ L_r\ g/V\ Y_\beta\ N_{\dot\beta}\ N_p\ N_\beta \right]^T.$$

The set $\mathcal{B}_q$ is defined accordingly and we assume that q is a random vector with uniform pdf within $\mathcal{B}_q$. Finally, we study quadratic stabilization, setting $\gamma = 0$ and taking weights $Q_{uu} = I$ and $Q_{xx} = 0.01I$.

The results of the simulations are now described. The initial condition $P^{(0)} \succ 0$ is (randomly) chosen as follows

$$P^{(0)} = \begin{bmatrix} 0.6995 & 0.7106 & 0.6822 & 0.5428 \\ 0.7106 & 0.8338 & 0.9575 & 0.6448 \\ 0.6822 & 0.9575 & 1.3380 & 0.8838 \\ 0.5428 & 0.6448 & 0.8838 & 0.8799 \end{bmatrix}.$$

Then, Algorithm 11.2 is applied with the choice of $\delta = 0.01$ and $p^* = 0.99$, obtaining the sample size function

$$N(k) = \lceil 507.7313 + 198.9983 \log(k+1) \rceil.$$

This choice guarantees that the algorithm terminates, with probability greater than $1 - \delta = 0.99$, to a solution $\widehat{\mathbf{P}}_N$ which is feasible with probability at least $p^* = 0.99$. We also considered $r = 0.001$ in the stepsize (11.7). This yielded a realization of the solution $\widehat{\mathbf{P}}_N$ and corresponding controller $\widehat{\mathbf{K}}_N = B_2^T \widehat{\mathbf{P}}_N^{-1}$

$$\widehat{P}_N = \begin{bmatrix} 0.7560 & -0.0843 & 0.1645 & 0.7338 \\ -0.0843 & 1.0927 & 0.7020 & 0.4452 \\ 0.1645 & 0.7020 & 0.7798 & 0.7382 \\ 0.7338 & 0.4452 & 0.7382 & 1.2162 \end{bmatrix}$$

and

$$\widehat{K}_N = \begin{bmatrix} 38.6191 & -4.3731 & 43.1284 & -49.9587 \\ -2.8814 & -10.1758 & 10.2370 & -0.4954 \end{bmatrix}.$$

Moreover, we verified a posteriori that the solution $\widehat{P}_N$ satisfies all the QMIs

$$A(q^{(\nu)})\widehat{P}_N + \widehat{P}_N A^T(q^{(\nu)}) - 2B_2 B_2^T \prec 0$$

obtained considering the $2^9 = 512$ vertices $q^{(\nu)}$ of the hyperrectangle $\mathcal{B}_q$. Since the uncertain parameters enter into $A(q)$ multiaffinely, we conclude that $\widehat{K}_N$ is also a quadratic stabilizing controller in a classical worst-case sense. The performance of the standard optimal controller has also been verified. That is, we take $K_0 = B_2^T X_0$, where X_0 is given by the solution of the Riccati equation corresponding to the nominal system

$$A_0^T X_0 + X_0 A_0 - X_0 B_2 B_2^T X_0 + 0.01I = 0$$

where A_0 is the state matrix obtained by setting the aircraft parameters to their nominal values. In this case, the optimal controller K_0 quadratically stabilizes a subset of 240 vertex systems.   ⋆

Remark 11.6 (Uncertain input matrix). The previous convergence results can be easily extended to the case when the uncertainty also enters into $B_2 = B_2(\Delta)$. In this case, we study the system

$$\dot{x} = A(\Delta)x + B_2(\Delta)u, \quad x(0) = x_0 \in \mathbb{R}^{n_s}$$

where $\Delta \in \mathcal{B}_\mathbb{D}(\rho)$, and consider the feedback law

$$u = -WP^{-1}x$$

where both $P = P^T$ and W are design variables. In this case, quadratic stabilization is equivalent to the solution of the LMI

$$V(\Delta, P, W) \doteq A(\Delta)P + PA^T(\Delta) + B_2(\Delta)W + W^T B_2^T(\Delta) \prec 0$$

for all $\Delta \in \mathcal{B}_\mathbb{D}(\rho)$. Then, we set the violation function to

$$v(\Delta, P, W) = \|[V(\Delta, P, W)]_+\|$$

where $[\cdot]_+$ is the projection defined in (11.2) and $\|\cdot\|$ denotes Frobenius norm. The algorithms given in Section 11.2.2 can be applied to the pair of matrix variables P, W with suitable modifications in the subgradient computations.

◁

11.2.4 Further convergence properties

In this section we study the properties of the norm of the solution $\mathbf{P}^{(k)}$. We recall that Theorem 11.1 guarantees that Algorithm 11.1 terminates in a finite number of iterations. This implies that the norm of the final solution never becomes infinity even if the feasibility set S_P includes an element whose norm is arbitrarily large. In the next corollary, we show that the norm of the solution $\mathbf{P}^{(k)}$ is bounded independently of the number of iterations, see [112].

Corollary 11.2 (Convergence properties). *Let Assumptions 11.2 and 11.3 be satisfied. Then, the norm of the solution $\mathbf{P}^{(k)}$ given by Algorithm 11.1 is bounded independently of k. That is, the inequalities*

$$\|\mathbf{P}^{(k)}\| \le \|P^{(0)}\| + 2 \inf_{\widetilde{P} \in S_P} \|\widetilde{P}\|; \tag{11.12}$$

$$\|\mathbf{P}^{(k+1)}\| \le \|\mathbf{P}^{(k)}\| + 2 \inf_{\widetilde{P} \in S_P} \|\widetilde{P}\| \tag{11.13}$$

hold for all $k \ge 0$.

Proof. From the proof of Theorem 11.1, we notice that if $v(\mathbf{P}^{(k)}, \mathbf{\Delta}^{(i)}) > 0$ then the inequality (11.8) holds for any $r > 0$. On the other hand, if $v(\mathbf{P}^{(k)}, \mathbf{\Delta}^{(i)}) = 0$ then $\mathbf{P}^{(k+1)} = \mathbf{P}^{(k)}$. Thus, it follows that

$$\|\mathbf{P}^{(k+1)} - \widetilde{P}\| \le \|\mathbf{P}^{(k)} - \widetilde{P}\|$$

holds for all $k \ge 0$ and for all $\widetilde{P} \in S_P$. This immediately implies that

$$\|\mathbf{P}^{(k)} - \widetilde{P}\| \le \|P^{(0)} - \widetilde{P}\|$$

holds for all $k \ge 0$ and for all $\widetilde{P} \in S_P$. Using this fact and standard norm properties, we easily compute the bounds

$$\|\mathbf{P}^{(k)}\| = \|\mathbf{P}^{(k)} - \widetilde{P} + \widetilde{P}\| \le \|\mathbf{P}^{(k)} - \widetilde{P}\| + \|\widetilde{P}\|$$
$$\le \|P^{(0)} - \widetilde{P}\| + \|\widetilde{P}\| \le \|P^{(0)}\| + 2\|\widetilde{P}\|$$

for all $k \ge 0$ and for all $\widetilde{P} \in S_P$. Therefore, we conclude that inequality (11.12) holds. The derivation of (11.13) proceeds along similar lines.   □

To summarize, Corollary 11.2 says that the upper bound of the norm of $\mathbf{P}^{(k)}$ is determined by a feasible solution having the smallest norm and is independent of k. In addition, from this result we also conclude that, at each iteration, the norm of $\mathbf{P}^{(k)}$ does not increase beyond a constant, which is determined independently of k. This fact implies a monotone-like convergence.

11.3 Sequential algorithm for uncertain LMIs

The approach presented in the previous section for design of probabilistic linear quadratic regulators is now extended to the general uncertain LMI setting, following the developments in [61]. In particular, we discuss iterative randomized algorithms for computing probabilistically feasible solutions of general uncertain LMIs of the form

$$F(\theta, \Delta) \preceq 0, \quad \Delta \in \mathcal{B}_{\mathbb{D}}(\rho) \tag{11.14}$$

where

$$F(\theta, \Delta) \doteq F_0(\Delta) + \sum_{i=1}^{n_\theta} \theta_i F_i(\Delta)$$

and $F_i(\Delta) \in \mathbb{S}^n$, $i = 0, 1, \ldots, n_\theta$, $\Delta \in \mathcal{B}_{\mathbb{D}}(\rho)$ and $\theta \in \Theta$.

Remark 11.7 (Robust optimization). In the optimization area, mathematical programming problems subject to an infinite number of constraints, such as (11.14), are usually referred to as *semi-infinite programs*, see e.g. [226]. Recently, a new paradigm in this area termed *robust optimization*, has gained increasing interest among researchers. Pioneering studies in this field are [34, 105]. In particular, [105] deals with optimization problems subject to constraints of type (11.14), which are termed robust semidefinite programs. Determining an exact solution for these problems is computationally intractable in the general case, and this motivates the study of randomized techniques to determine approximate probabilistic solutions.   ◁

In this section, in parallel with the developments of Section 11.2, we discuss the issue of determining a robustly feasible solution for the convex constraints (11.14), assuming that such a feasible solution exists. Then, in Section 11.5, we present a different class of algorithms whose purpose is to determine *approximately feasible* solutions for the system of LMIs, in case no robustly feasible solution actually exists.

Assume Θ is some a priori admissible convex membership set for the decision variables θ. For instance, the set Θ can represent bounds on the maximum allowed values of the elements of θ, or positivity constraint on θ, as well as any other convex constraints on the design variables θ that do not depend on the uncertainty Δ. The restrictions on θ that depend on the uncertainties are instead described by the uncertain LMI constraints (11.14).

First, we consider a performance violation function defined as

$$v(\Delta, \theta) = \|[F(\theta, \Delta)]_+\|$$

where $[\cdot]_+$ is the projection defined in (11.2) and $\|\cdot\|$ denotes Frobenius norm. We now state a strict feasibility assumption regarding the set of solutions similar to Assumption 11.2.

Assumption 11.4 (Strict feasibility). *The solution set*

$$S_\theta = \{\theta \in \Theta : F(\theta, \Delta) \preceq 0 \text{ for all } \Delta \in \mathcal{B}_\mathbb{D}(\rho)\}$$

has a nonempty interior.

This assumption implies that there exist a radius $r > 0$ and center θ^*, such that the ball

$$\{\theta \in \Theta : \|\theta - \theta^*\| \leq r\} \tag{11.15}$$

is contained in S_θ.

Next, we introduce a probabilistic assumption on the set S_θ that plays the same role of Assumption 11.3.

Assumption 11.5 (Nonzero probability of detecting unfeasibility). *We assume that $f_\Delta(\Delta)$ is such that*

$$\mathrm{PR}\{v(\Delta, \theta) > 0\} > 0$$

for any $\theta \notin S_\theta$.

Given initial conditions $\theta^{(0)}$, consider the update rule

$$\theta^{(k+1)} = \psi_{\text{upd}}(\Delta^{(i)}, \theta^{(k)}) = \left[\theta^{(k)} - \eta^{(k)}\partial_\theta\{v(\Delta^{(i)}, \theta^{(k)})\}\right]_\Theta$$

where ∂_θ denotes the subgradient of $v(\Delta, \theta)$ with respect to θ, and $[\cdot]_\Theta$ is the projection on the set Θ, defined in (11.1).

The next lemma, stated in [61], shows an explicit expression for the subgradient of $v(\Delta, \theta)$.

Lemma 11.2. *The function $v(\Delta, \theta)$ is convex in θ and its subgradient $\partial_\theta\{v(\Delta, \theta)\}$ is given by*

$$\partial_\theta\{v(\Delta, \theta)\} = \begin{cases} \nabla v(\Delta, \theta) & \text{if } v(\Delta, \theta) > 0; \\ 0 & \text{otherwise} \end{cases}$$

where

$$\nabla v(\Delta, \theta) \doteq \frac{1}{v(\Delta, \theta)} \begin{bmatrix} \mathrm{Tr}\,[F_1(\Delta)[F(\Delta, \theta)]_+] \\ \vdots \\ \mathrm{Tr}\,[F_{n_\theta}(\Delta)[F(\Delta, \theta)]_+] \end{bmatrix}.$$

As an immediate consequence of this lemma, we observe that the subgradient computation basically requires the solution of a symmetric eigenvalue problem and, therefore, it can be efficiently performed for various medium/moderate-scale problems. We introduce the following algorithm, which has the same structure of Algorithm 8.4.

Algorithm 11.3 (RA for uncertain LMI).

1. Initialization.
   ▷ Let $N(k) = \overline{N}_{ss}(p^*, \delta, k)$ given in (11.9);
   ▷ Set $k = 0$, $i = 0$, and choose $\theta^{(0)} \in \Theta$;
2. Feasibility loop.
   ▷ Set $\ell = 0$ and feas=true;
   ▷ While $\ell < N(k)$ and feas=true
     – Set $i = i + 1$, $\ell = \ell + 1$;
     – Draw $\Delta^{(i)}$ according to f_Δ;
     – If $v(\Delta^{(i)}, \theta^{(k)}) > 0$ set feas=false;
   ▷ End While;
3. Exit condition.
   ▷ If feas=true
     – Set $N = i$;
     – Return $\widehat{\theta}_N = \theta^{(k)}$ and Exit;
   ▷ End If;
4. Update.
   ▷ Compute

$$\theta^{(k+1)} = \psi_{upd}(\Delta^{(i)}, \theta^{(k)}) = \left[\theta^{(k)} - \eta^{(k)}\partial_\theta\{v(\Delta^{(i)}, \theta^{(k)})\}\right]_\Theta$$

   where the stepsize $\eta^{(k)}$ is given in (11.16);
   ▷ Set $k = k + 1$ and goto 2.

The stepsize $\eta^{(k)}$ in this algorithm is given by

$$\eta^{(k)} = \frac{v(\Delta^{(i)}, \theta^{(k)}) + r\|\partial_\theta\{v(\Delta^{(i)}, \theta^{(k)})\}\|}{\|\partial_\theta\{v(\Delta^{(i)}, \theta^{(k)})\}\|^2} \tag{11.16}$$

where $r > 0$ is the radius of a ball contained in S_θ with center θ^*.

We now state a finite-time convergence result, whose proof is similar to that of Theorem 11.1 and, therefore, is not given here. The interested reader may refer to [61]. Notice that the theorem below considers a conceptual variation of the algorithm, in which $N(k)$ is left unspecified, i.e. the feasibility loop in step 2 runs indefinitely, until an unfeasible point is found. We recall that if $\theta^{(k)}$ is not robustly feasible, then by Assumption 11.5 loop 2 will detect unfeasibility in a finite number of iterations. Otherwise, if $\theta^{(k)}$ is robustly feasible, then loop 2 will run forever. In this case, it means that a robustly feasible solution has been determined in k outer iterations.

Theorem 11.3 (Finite-time convergence). *Let Assumptions 11.4 and 11.5 be satisfied. Then, with probability one, Algorithm 11.3 with unspecified $N(k)$ executes a finite number of updates to determine a robustly feasible solution within S_θ. Moreover, if an upper bound D is known on the initial distance*

$$\|\theta^{(0)} - \theta^*\| \leq D$$

then the total number of updates is bounded by $\overline{k} \doteq \lfloor (D/r)^2 \rfloor$, with $r > 0$ and θ^* the radius and the center of a ball (11.15) contained in S_θ.

Remark 11.8 (Achievement time). **As discussed previously, without a stopping criterion for the inner loop, Algorithm 11.3 could run indefinitely. In [208], the authors analyze this phenomenon in closer detail. In particular, they show that, when the algorithm is started with a randomly drawn initial point, the expected value of the number of *total iterations* to reach a robustly feasible solution is indeed infinite. The total number of iterations coincides with the number of uncertainty samples to be drawn. Clearly, this problem is avoided if we accept a probabilistically feasible solution in place of a deterministically robust one, by imposing the exit condition via the sample size bound (11.9).** ◁

The following corollary, stated without proof, is analogous to Corollary 11.1 and combines Theorem 11.3 with the bound (11.9) on the sample complexity needed for convergence to a probabilistic solution.

Corollary 11.3 (Convergence of Algorithm 11.3). *Let Assumptions 11.4 and 11.5 be satisfied. Then, Algorithm 11.3 terminates in a finite number N of total iterations.*

Moreover, if an upper bound D is known on the initial distance $\|\theta^{(0)} - \theta^*\| \leq D$ then Algorithm 11.3 executes at most $\overline{k} \doteq \lfloor (D/r)^2 \rfloor$ updates, and at most

$$N \leq 1 + \sum_{k=0,\ldots,\overline{k}} \overline{N}_{\mathrm{ss}}(p^*, \delta, k) \leq 1 + (\overline{k} + 1)\overline{N}_{\mathrm{ss}}(p^*, \delta, \overline{k})$$

total iterations. Upon termination, we have two cases:

1. *If Algorithm 11.3 executes $\overline{k}$ updates, then $\widehat{\theta}_N \in S_\theta$, i.e. it is a robustly feasible solution;*
2. *With probability greater than $1 - \delta$, if Algorithm 11.3 terminates before $\overline{k}$ updates, then $\widehat{\theta}_N$ satisfies*

$$\mathrm{PR}\left\{ v(\Delta, \widehat{\theta}_N) \leq 0 \right\} \geq p^*$$

i.e. it is a probabilistically feasible solution.

Example 11.4 (Gain scheduling control).

We present an example of application of Algorithm 11.3 to a control problem that is expressed in the form of a robust LMI feasibility problem. Consider a linear system depending on a time-varying parameter $q(t) = [q_1(t) \cdots q_\ell(t)]^T$

$$\dot{x} = A(q(t))x + B_2 u.$$

For fixed t, the parameter vector q is constrained in the box $\mathcal{B}_q = \{q \in \mathbb{R}^\ell : |q_i| \leq \rho_i, i = 1, \ldots, \ell\}$, where ρ_i denote the allowed limits of variations of the parameters. We assume that the parameter $q(t)$ can be measured on-line, and hence this information is available to the controller. This setup is indeed a special case of a more general class of control problems usually known as linear parameter varying (LPV) problems, that are treated in detail Chapter 12.

Here, we specifically consider a state-feedback controller of the form

$$u = K(q(t))x$$

where we set

$$K(q(t)) = K_0 + \sum_{i=1}^{\ell} K_i q_i(t).$$

The control objective is to determine the controller gains K_i, $i = 0, 1, \ldots, \ell$, such that the controlled system has a guaranteed exponential decay rate $\alpha > 0$. Defining the closed-loop matrix

$$A_{cl}(q(t)) = A(q(t)) + B_2 K(q(t))$$

the control objective is met if there exists a symmetric matrix $P \succ 0$ such that the matrix inequality

$$A_{cl}(q)P + PA_{cl}^T(q) + 2\alpha P \prec 0 \tag{11.17}$$

holds for all $q \in \mathcal{B}_q$. We notice that while there is no actual "uncertainty" in the plant (since the parameter $q(t)$ is measured on-line), the resulting design condition (11.17) is a robust matrix inequality, i.e. a condition that should hold for all possible values of a "formal" uncertain parameter $q \in \mathcal{B}_q$. In particular, introducing the new variables

$$Y_i \doteq K_i P$$

for $i = 0, 1, \ldots, \ell$, the control design problem is expressed as a robust LMI feasibility problem. That is, determine $P = P^T$, $Y_i = K_i P$, for $i = 0, 1, \ldots, \ell$, such that

$$P \succ 0 \tag{11.18}$$

$$0 \succ A(q)P + PA^T(q) + 2\alpha P$$

$$+ B_2 Y_0 + Y_0^T B_2^T + \sum_{i=1}^{\ell} q_i B_2 Y_i + \sum_{i=1}^{\ell} q_i Y_i^T B_2^T \tag{11.19}$$

for all $q \in \mathcal{B}_q$. Conditions (11.18) and (11.19) represent a robust LMI condition of the form $F(\theta, \Delta) \preceq 0$ considered in (11.14), where the design variable θ contains the free entries of P and Y_i, $i = 0, 1, \ldots, \ell$, and the uncertainty Δ is here represented by the parameter q.

For a numerical test, we considered the data in Example 11.3, assuming that the values of the $\ell = 9$ parameters of the plant could now be measured on-line, and that the bounds ρ_i are equal to 15% of the nominal (central) value. Setting the desired decay rate to $\alpha = 0.5$, we applied Algorithm 11.3 in order to determine a probabilistically feasible solution for the system of LMIs (11.18) and (11.19). The probability levels were set to $p^* = 0.9$ and $\delta = 0.01$, and we selected feasibility radius $r = 0.05$. The probability distribution of the parameter was set to the uniform distribution over $\mathcal{B}_q$. Notice also that since (11.18) and (11.19) are homogeneous in P, we added a condition number constraint on P of the form $I \preceq P \preceq \kappa P$, with $\kappa = 1,000$. Algorithm 11.3 converged in 48 outer iterations to the solution

$$
P = \begin{bmatrix} 1.2675 & -1.1846 & -0.0142 & 0.1013 \\ -1.1846 & 8.3174 & 0.2765 & -0.5450 \\ -0.0142 & 0.2765 & 1.2810 & 0.5769 \\ 0.1013 & -0.5450 & 0.5769 & 2.4329 \end{bmatrix};
$$

$$
K_0 = \begin{bmatrix} 0.0023 & 0.0768 & -1.0892 & 0.5657 \\ 1.3496 & -0.1164 & -0.9659 & 0.1213 \end{bmatrix};
$$

$$
K_1 = \begin{bmatrix} 0.0826 & 0.0353 & -0.1685 & 0.2000 \\ 0.3875 & 0.0808 & -0.0610 & 0.0226 \end{bmatrix};
$$

$$
K_2 = \begin{bmatrix} -0.0127 & -0.0090 & 0.0359 & -0.0975 \\ 0.1151 & 0.0074 & 0.2536 & -0.0489 \end{bmatrix};
$$

$$
K_3 = \begin{bmatrix} 0.0207 & 0.0039 & 0.0023 & 0.0039 \\ 0.0291 & 0.0075 & -0.0301 & 0.0107 \end{bmatrix};
$$

$$
K_4 = \begin{bmatrix} 0.0011 & 0.0005 & -0.0026 & 0.0026 \\ 0.0042 & 0.0009 & -0.0016 & 0.0004 \end{bmatrix};
$$

$$
K_5 = \begin{bmatrix} -0.0016 & -0.0006 & 0.0039 & -0.0006 \\ -0.0014 & -0.0004 & 0.0018 & -0.0018 \end{bmatrix};
$$

$$
K_6 = \begin{bmatrix} -0.1134 & -0.0029 & -0.1616 & 0.2304 \\ 0.0544 & 0.0162 & -0.0916 & -0.0388 \end{bmatrix};
$$

$$
K_7 = \begin{bmatrix} 0.0023 & 0.0002 & 0.0025 & -0.0021 \\ 0.0004 & 0.0001 & -0.0004 & 0.0007 \end{bmatrix};
$$

$$
K_8 = \begin{bmatrix} -0.1134 & -0.0029 & -0.1616 & 0.2304 \\ 0.0544 & 0.0162 & -0.0916 & -0.0388 \end{bmatrix};
$$

$$
K_9 = \begin{bmatrix} -0.0024 & -0.0001 & -0.0026 & 0.0083 \\ 0.0012 & -0.0001 & 0.0021 & -0.0035 \end{bmatrix}.
$$

This solution was a posteriori tested using the Monte Carlo method, showing that it actually satisfies the design LMIs with estimated probability higher than 0.998.   ★

11.4 Ellipsoid algorithm for uncertain LMIs

A modification of Algorithm 11.3 has been proposed in [148], that uses the ellipsoid algorithm (see [176]) instead of a subgradient step in the update rule. In this approach, an initial ellipsoid that contains the robust feasible set S_θ is determined, and then a sequence of ellipsoids with geometrically decreasing volumes is constructed based on random uncertainty samples. Each ellipsoid in the sequence is guaranteed to contain S_θ, and since the ellipsoid volume is decreasing, at a certain finite iteration the center of the ellipsoid is guaranteed to belong to S_θ. We now state more formally the ellipsoid algorithm for feasibility of uncertain LMIs.

This algorithm is still based on Assumptions 11.4 and 11.5, but one distinct advantage is that it does not make use of the actual value of the feasibility radius r. Moreover, in [148] the authors prove that the algorithm also converges when the volume of S_θ is zero, although convergence in this case is only asymptotic.

We assume that an initial ellipsoid $\mathcal{E}^{(0)} = \mathcal{E}\left(\theta^{(0)}, W^{(0)}\right)$ is given, such that $S_\theta \subseteq \mathcal{E}^{(0)}$. For simplicity, and without loss of generality, we assume that $\Theta \equiv \mathbb{R}^{n_\theta}$. We next explicitly state the ellipsoid algorithm for feasibility of robust LMIs. The update rule ψ_{upd} for the ellipsoid parameters is given by

$$\boldsymbol{\theta}^{(k+1)} = \boldsymbol{\theta}^{(k)} - \frac{1}{n_\theta + 1} \frac{\mathbf{W}^{(k)} \partial_\theta^{(i,k)}}{\sqrt{\partial_\theta^{(i,k)T} \mathbf{W}^{(k)} \partial_\theta^{(i,k)}}}; \tag{11.20}$$

$$\mathbf{W}^{(k+1)} = \frac{n_\theta{}^2}{n_\theta{}^2 - 1} \left(\mathbf{W}^{(k)} - \frac{2}{n_\theta + 1} \frac{\mathbf{W}^{(k)} \partial_\theta^{(i,k)} \partial_\theta^{(i,k)T} \mathbf{W}^{(k)}}{\partial_\theta^{(i,k)T} \mathbf{W}^{(k)} \partial_\theta^{(i,k)}} \right) \tag{11.21}$$

where for notational convenience we set $\partial_\theta^{(i,k)} \doteq \partial_\theta \{v(\boldsymbol{\Delta}^{(i)}, \boldsymbol{\theta}^{(k)})\}$.

Algorithm 11.4 (Ellipsoid algorithm for uncertain LMI).

1. Initialization.
   ▷ Let $N(k) = \overline{N}_{\mathrm{ss}}(p^*, \delta, k)$ given in (11.9);
   ▷ Set $k = 0$, $i = 0$;
   ▷ Choose initial ellipsoid $\mathcal{E}^{(0)} = \mathcal{E}\left(\theta^{(0)}, W^{(0)}\right) \supseteq S_\theta$;
2. Feasibility loop.
   ▷ Set $\ell = 0$ and feas=true;
   ▷ While $\ell < N(k)$ and feas=true
       − Set $i = i + 1$, $\ell = \ell + 1$;
       − Draw $\boldsymbol{\Delta}^{(i)}$ according to f_Δ;
       − If $v(\boldsymbol{\Delta}^{(i)}, \boldsymbol{\theta}^{(k)}) > 0$ set feas=false;
   ▷ End While;
3. Exit condition.
   ▷ If feas=true
       − Set $N = i$;
       − Return $\widehat{\boldsymbol{\theta}}_N = \boldsymbol{\theta}^{(k)}$ and Exit;
   ▷ End If;
4. Update.
   ▷ Compute

$$(\boldsymbol{\theta}^{(k+1)}, \mathbf{W}^{(k+1)}) = \psi_{\mathrm{upd}}(\boldsymbol{\Delta}^{(i)}, \boldsymbol{\theta}^{(k)}, \mathbf{W}^{(k)})$$

   where the update rule ψ_{upd} is specified in (11.20) and (11.21);
   ▷ Set $k = k + 1$ and goto 2.

Remark 11.9 (Initial ellipsoid). Notice that if an ellipsoid contains the set $\{\theta : F(\theta, \Delta) \preceq 0\}$ for *any fixed* $\Delta \in \mathcal{B}_{\mathbb{D}}(\rho)$, then it also contains S_θ. Therefore, determining the initial ellipsoid $\mathcal{E}^{(0)}$ simply amounts to solving a standard convex optimization problem where we minimize the volume of $\mathcal{E}^{(0)}$, subject to the constraint that $\{\theta : F(\theta, \Delta) \preceq 0\} \subseteq \mathcal{E}^{(0)}$, for fixed $\Delta \in \mathcal{B}_{\mathbb{D}}(\rho)$.   ◁

The following result adapted from [148] establishes the convergence properties of the above algorithm.

Theorem 11.4 (Finite-time convergence). *Let Assumptions 11.4 and 11.5 be satisfied. Then, with probability one, Algorithm 11.4 with unspecified $N(k)$ executes a finite number of updates to determine a robustly feasible solution within S_θ. Moreover, if a lower bound V is known on the volume of the solution set*

$$\mathrm{Vol}(S_\theta) \geq V > 0$$

then the total number of updates is bounded by

$$2n_\theta \left\lfloor \log \frac{\mathrm{Vol}(\mathcal{E}^{(0)})}{V} \right\rfloor .$$

In order to assure that the algorithm actually stops returning a solution, however, we again impose an exit criterion on the feasibility check loop. That is, in Algorithm 11.4, the sample size function is chosen according to the bound (11.9), in order to guarantee the satisfaction of the probabilistic requirements, see [206]. We have the following corollary analogous to Corollaries 11.1 and 11.3.

Corollary 11.4 (Convergence of Algorithm 11.4). *Let Assumptions 11.4 and 11.5 be satisfied. Then, Algorithm 11.4 terminates in a finite number N of total iterations.*

Moreover, if a lower bound V is known on the volume of the solution set $\mathrm{Vol}(S_\theta) \geq V > 0$, *then Algorithm 11.4 executes at most* $\overline{k} = 2n_\theta \lfloor \log(\mathrm{Vol}(\mathcal{E}^{(0)})/V) \rfloor$ *updates, and at most*

$$N \leq 1 + \sum_{k=0,\dots,\overline{k}} \overline{N}_{ss}(p^*, \delta, k) \leq 1 + (\overline{k}+1)\overline{N}_{ss}(p^*, \delta, \overline{k})$$

total iterations. Upon termination, we have two cases:

1. *If Algorithm 11.4 executes* $\overline{k}$ *updates, then* $\widehat{\boldsymbol{\theta}}_N \in S_\theta$, *i.e. it is a robustly feasible solution;*
2. *With probability greater than* $1 - \delta$, *if Algorithm 11.4 terminates before* $\overline{k}$ *updates, then* $\widehat{\boldsymbol{\theta}}_N$ *satisfies*

$$\mathrm{PR}\left\{v(\boldsymbol{\Delta}, \widehat{\boldsymbol{\theta}}_N) \leq 0\right\} \geq p^*$$

i.e. it is a probabilistically feasible solution.

11.5 Sequential algorithm for (possibly) unfeasible LMIs

To prove convergence of the algorithms discussed in the previous section, we require the existence of a solution that is strictly feasible *for all* the matrix inequalities brought about by the uncertainty. In practice, however, one may not know in advance whether or not a robustly feasible solution exists, or such a solution may actually fail to exist. In this latter case, Algorithm 11.3 may fail to converge to an acceptable probabilistic solution, see Remark 11.5. Unfeasibility could be due to a set of uncertainty instances having "small" probability measure, and hence, in a probabilistic approach, we would like to neglect this set and determine anyways a solution that is at least "approximately feasible." The concept of approximate feasibility in the context of uncertain semidefinite programs has been proposed in [28], while the derivation presented here is based on [61]. The algorithm in this section is actually an application of a general technique in stochastic optimization proposed in [198], and also more recently discussed in [200].

The intuition behind the method now presented is that if one is not able to find a solution θ such that $v(\Delta, \theta) = 0$, for all $\Delta \in \mathcal{B}_{\mathbb{D}}(\rho)$ (because no such solution exists), then one could relax its goals and seek a solution that minimizes the expectation of $v(\Delta, \theta)$. Notice that if a robustly feasible solution actually exists, then it must be among the minimizers of $\mathrm{E}_{\Delta}(v(\Delta, \theta))$. In this case the attained minimum is zero and, therefore, the approach described in this section could be used for both robustly feasible and unfeasible problems. To summarize, the method presented in this section does not require as an hypothesis the existence of a (deterministically) robustly feasible solution to the family of LMIs (11.14) in order to converge asymptotically, and it is based on the minimization of the expected value of the violation function $v(\Delta, \theta) = \|[F(\theta, \Delta)]_+\|$. Hence we face a stochastic optimization problem where we seek to determine θ such that

$$\theta_{\min} = \arg \min_{\theta \in \Theta} f(\theta)$$

where

$$f(\theta) \doteq \mathrm{E}_{\Delta}(v(\Delta, \theta))$$

being $\mathrm{E}_{\Delta}(v(\Delta, \theta))$ the expected value of $v(\Delta, \theta)$.

We show that a sequence of random variables $\bar{\theta}^{(1)}, \bar{\theta}^{(2)}, \dots$ can be constructed such that the related sequence $f(\bar{\theta}^{(1)}), f(\bar{\theta}^{(2)}), \dots$ converges asymptotically in expectation to $f(\theta_{\min})$. This sequence is obtained by a modification of the sequential gradient algorithms discussed in the previous sections, which includes an averaging smoothing over the sequence of candidate solutions. Given an initial condition $\theta^{(0)}$, consider the recursion

$$\theta^{(k+1)} = \left[\theta^{(k)} - \eta^{(k)} \partial_\theta \{v(\Delta^{(k)}, \theta^{(k)})\} \right]_{\Theta}$$

where $\partial_\theta \{v(\Delta^{(k)}, \theta^{(k)})\}$ denotes the subgradient with respect to θ computed in Lemma 11.2, $[\cdot]_\Theta$ is the projection defined in (11.1), and $\eta^{(k)}$ is a suitable stepsize. We notice that $\eta^{(k)}$ is not random, as opposed to the cases discussed in the previous sections, see further details in Remark 11.10.

The subgradient $\partial_\theta \{v(\Delta, \theta)\}$ and the set Θ should satisfy the following assumption.

Assumption 11.6 (Boundedness of the subgradient). *The set Θ is closed and convex and the subgradient is bounded as*

$$\|\partial_\theta \{v(\Delta, \theta)\}\| \leq \nu$$

for all $\theta \in \Theta$ and $\Delta \in \mathcal{B}_{\mathbb{D}}(\rho)$.

This is a very mild assumption. In fact, we notice that the subgradient is bounded when θ is "nearly" feasible, i.e. when $v(\Delta, \theta) \to 0$. It can also be shown that if the matrices $F_i(\theta, \Delta)$ are bounded for $\Delta \in \mathcal{B}_\mathbb{D}(\rho)$, then the subgradient is bounded.

Next, given initial conditions $\bar{\theta}^{(0)} = 0$ and $m^{(0)} = 0$, define the averaged sequence

$$\bar{\theta}^{(k)} \doteq \frac{m^{(k-1)}}{m^{(k)}} \bar{\theta}^{(k-1)} + \frac{\eta^{(k)}}{m^{(k)}} \theta^{(k)} \tag{11.22}$$

where $m^{(k)} = m^{(k-1)} + \eta^{(k)}$ and $\eta^{(k)} > 0$.

Observe that $\theta^{(k)}$ and $\bar{\theta}^{(k)}$ depend on the realizations $\Delta^{(1)}, \dots, \Delta^{(k-1)}$ of the random matrix Δ. Therefore, for any function $g(\cdot)$ we denote by $E_{\Delta^{(1 \dots k-1)}} \left(g(\bar{\theta}^{(k)}) \right)$ and $E_{\Delta^{(1 \dots k-1)}} \left(g(\theta^{(k)}) \right)$ the expectations with respect to the multisample $\Delta^{(1 \dots k-1)}$ of $g(\bar{\theta}^{(k)})$ and $g(\theta^{(k)})$ respectively.

We are now ready to state a theorem which shows "how good" is the approximation $E_{\Delta^{(1 \dots k-1)}} \left(f(\bar{\theta}^{(k)}) \right)$ of $f(\theta_{\min}) = E_\Delta (v(\Delta, \theta_{\min}))$.

Theorem 11.5. *Let Δ be a random matrix with given pdf $f_\Delta(\Delta)$ and support $\mathcal{B}_\mathbb{D}(\rho)$ and let Assumption 11.6 be satisfied. Then, for any sequence of stepsizes $\eta^{(k)} > 0$ we have*

$$E_{\Delta^{(1 \dots k-1)}} \left(f(\bar{\theta}^{(k)}) \right) - f(\theta_{\min}) \le \frac{\|\theta^{(0)} - \theta_{\min}\|^2 + \nu^2 \sum_{i=0}^k (\eta^{(i)})^2}{2 \sum_{i=0}^k \eta^{(i)}}. \tag{11.23}$$

Proof. By definition of projection, we have that

$$\|[\theta]_\Theta - \theta^*\| \le \|\theta - \theta^*\|$$

for any θ and for any $\theta^* \in \Theta$. Therefore

$$\|\theta^{(k+1)} - \theta_{\min}\|^2 \le \|\theta^{(k)} - \theta_{\min}\|^2 - 2\eta^{(k)}(\theta^{(k)} - \theta_{\min})^T \partial_\theta \{v(\Delta^{(k)}, \theta^{(k)})\}$$
$$+ (\eta^{(k)})^2 \|\partial_\theta \{v(\Delta^{(k)}, \theta^{(k)})\}\|^2. \tag{11.24}$$

Now, for a convex function $h(\theta)$ and for any θ and θ^* in Θ, it holds that

$$(\theta - \theta^*)^T \partial_\theta \{h(\theta)\} \ge h(\theta) - h(\theta^*).$$

Hence, we have

$$(\theta^{(k)} - \theta_{\min})^T \partial_\theta \{v(\Delta^{(k)}, \theta^{(k)})\} \ge v(\Delta^{(k)}, \theta^{(k)}) - v(\Delta^{(k)}, \theta_{\min}).$$

On the other hand, from the boundedness condition of the subgradient stated in Assumption 11.6, we have that $\|\partial_\theta \{v(\Delta^{(k)}, \theta^{(k)})\}\|^2 \le \nu^2$. Therefore, from (11.24), we obtain

$$\|\boldsymbol{\theta}^{(k+1)} - \theta_{\min}\|^2 \leq \|\boldsymbol{\theta}^{(k)} - \theta_{\min}\|^2 - 2\eta^{(k)}\left(v(\boldsymbol{\Delta}^{(k)}, \boldsymbol{\theta}^{(k)}) - v(\boldsymbol{\Delta}^{(k)}, \theta_{\min})\right)$$
$$+(\eta^{(k)})^2\nu^2. \tag{11.25}$$

Next, letting

$$\mathbf{u}^{(k)} \doteq \mathrm{E}_{\boldsymbol{\Delta}^{(1\ldots k-1)}}\left(\|\boldsymbol{\theta}^{(k)} - \theta_{\min}\|^2\right)$$

and taking the expectation of both sides of (11.25), we get

$$\mathbf{u}^{(k+1)} \leq \mathbf{u}^{(k)} - 2\eta^{(k)}\left(\mathrm{E}_{\boldsymbol{\Delta}^{(1\ldots k-1)}}\left(f(\boldsymbol{\theta}^{(k)})\right) - f(\theta_{\min})\right) + (\eta^{(k)})^2\nu^2$$

and

$$\mathbf{u}^{(k+1)} \leq u^{(0)} - 2\sum_{i=0}^{k}\eta^{(i)}\left(\mathrm{E}_{\boldsymbol{\Delta}^{(1\ldots i-1)}}\left(f(\boldsymbol{\theta}^{(i)})\right) - f(\theta_{\min})\right) + \nu^2\sum_{i=0}^{k}(\eta^{(i)})^2.$$

Subsequently, we obtain

$$\sum_{i=0}^{k}\eta^{(i)}\left(\mathrm{E}_{\boldsymbol{\Delta}^{(1\ldots i-1)}}\left(f(\boldsymbol{\theta}^{(i)})\right) - f(\theta_{\min})\right) \leq \frac{1}{2}\left(u^{(0)} + \nu^2\sum_{i=0}^{k}(\eta^{(i)})^2\right).$$
$$\tag{11.26}$$

Clearly, the iteration (11.22) is a recursive version of a Cesàro mean. Therefore, $\bar{\boldsymbol{\theta}}^{(k)}$ is an averaged version of $\boldsymbol{\theta}^{(0)}, \boldsymbol{\theta}^{(1)}, \ldots, \boldsymbol{\theta}^{(k)}$

$$\bar{\boldsymbol{\theta}}^{(k)} = \frac{\sum_{i=0}^{k}\eta^{(i)}\boldsymbol{\theta}^{(i)}}{\sum_{i=0}^{k}\eta^{(i)}}.$$

From the Jensen inequality for convex functions, see e.g. [114], we have

$$f(\bar{\boldsymbol{\theta}}^{(k)}) = f\left(\frac{\sum_{i=0}^{k}\eta^{(i)}\boldsymbol{\theta}^{(i)}}{\sum_{i=0}^{k}\eta^{(i)}}\right) \leq \frac{\sum_{i=0}^{k}\eta^{(i)}f(\boldsymbol{\theta}^{(i)})}{\sum_{i=0}^{k}\eta^{(i)}}.$$

Hence, we have

$$\mathrm{E}_{\boldsymbol{\Delta}^{(1\ldots k-1)}}\left(f(\bar{\boldsymbol{\theta}}^{(k)})\right) - f(\theta_{\min}) \leq \frac{\sum_{i=0}^{k}\eta^{(i)}\left(\mathrm{E}_{\boldsymbol{\Delta}^{(1\ldots i-1)}}\left(f(\boldsymbol{\theta}^{(i)})\right) - f(\theta_{\min})\right)}{\sum_{i=0}^{k}\eta^{(i)}}$$

and, using (11.26), we obtain

$$\mathrm{E}_{\boldsymbol{\Delta}^{(1\ldots k-1)}}\left(f(\bar{\boldsymbol{\theta}}^{(k)})\right) - f(\theta_{\min}) \leq \frac{u^{(0)} + \nu^2\sum_{i=0}^{k}(\eta^{(i)})^2}{2\sum_{i=0}^{k}\eta^{(i)}}$$

which proves (11.23). $\qquad\square$

Remark 11.10 (Choice of the stepsize). Theorem 11.5 gives an estimate of the accuracy of the solution for a finite sample size. In particular, we notice that if k is fixed in advance, then the best choice for the sequence $\eta^{(i)}$ is to take a constant stepsize

$$\eta = \frac{\|\theta^{(0)} - \theta_{\min}\|}{\nu\sqrt{k+1}}$$

which yields

$$\mathrm{E}_{\Delta^{(1\ldots k-1)}}\left(f(\bar{\theta}^{(k)})\right) - f(\theta_{\min}) \leq \frac{\nu\|\theta^{(0)} - \theta_{\min}\|}{\sqrt{k+1}}.$$

On the other hand, if the number of iterations is not fixed a priori and $c > 0$ is a constant, a good choice for the stepsize is

$$\eta^{(i)} = \frac{c}{\sqrt{i+1}}$$

which provides asymptotically the estimate

$$\mathrm{E}_{\Delta^{(1\ldots k-1)}}\left(f(\bar{\theta}^{(k)})\right) - f(\theta_{\min}) = O\left(\frac{1}{\sqrt{k+1}}\right).$$

$\lhd$

An asymptotic result for $k \to \infty$ is established in the next corollary.

Corollary 11.5. *Let Δ be a random matrix with given pdf $f_\Delta(\Delta)$ and support $\mathcal{B}_{\mathbb{D}}(\rho)$ and let Assumption 11.6 be satisfied. If the sequence of stepsizes is such that*

$$\eta^{(k)} \to 0 \text{ as } k \to \infty \text{ and } \sum_{k=0}^{\infty} \eta^{(k)} = \infty \qquad (11.27)$$

then

$$\lim_{k\to\infty} \mathrm{E}_{\Delta^{(1\ldots k-1)}}\left(f(\bar{\theta}^{(k)})\right) = f(\theta_{\min}).$$

Proof. From Theorem 11.5, with the assumptions on the stepsize (11.27), it follows immediately that

$$\mathrm{E}_{\Delta^{(1\ldots k-1)}}\left(f(\bar{\theta}^{(k)})\right) \to f(\theta_{\min})$$

for $k \to \infty$. $\square$

12. Sequential Algorithms for LPV Systems

In this chapter we develop iterative methods similar to those presented in Chapter 11, with the specific aim of solving output feedback design problems in the context of linear parameter varying (LPV) systems.

12.1 Classical and probabilistic LPV settings

According to standard literature on the subject, see e.g. the survey papers [173, 229], we consider LTI systems that depend on time-varying (scheduling) parameters ξ. In the LPV setting these parameters are not known in advance, but they can be measured on-line and their time value is made available to the controller. In contrast to the main focus of this book, for LPV systems there is no uncertainty Δ but only scheduling parameters ξ, which are restricted within a set Ξ. The general LPV design setting is shown in Figure 12.1.

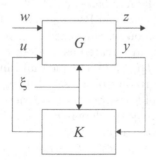

Fig. 12.1. Interconnection for LPV design.

LPV systems provide a good starting point for analysis and design of more general gain scheduling problems. One of the main motivations for introducing gain scheduling is to deal with various plant nonlinearities. Hence, a critical issue is that the resulting LPV model generally depends nonlinearly on the scheduling parameters.

The design of an LPV controller that guarantees performance in the face of variation of ξ is often referred to as "robust performance" problem. This problem has been shown to be equivalent to the solution of some parameter-dependent linear matrix inequalities. Clearly, the LMI decision variables are nonlinear functions of the scheduling parameters, and it is hence impossible to solve the robust performance problem exactly. Therefore, in the LPV literature, two different approaches are generally followed.

The first approach, also denoted as "approximation," is based on the restriction to a specific class of functions of the scheduling parameters. For example, one can assume that the matrices of the LPV model are multiaffine functions or linear fractional transformations of the underlying parameters, see for instance [15, 31, 161]. The robust performance problem is then reduced to more tractable formulae which involve a finite number of LMIs. Unfortunately, in this case, some conservatism is introduced by the approximation.

The second approach, often denoted as "gridding," is to grid the set Ξ of parameters, see e.g. [14, 31, 289]. In this case, robust performance is shown to be equivalent to the solution of a finite number of LMIs. However, this number depends on the grid points and, therefore, it may increase dramatically with the number of scheduling parameters. Moreover, the satisfaction of these LMIs at the grid points does not give any guarantee of the achievement of robust performance for the entire set Ξ. Obviously, depending on the specific problem under attention, other methods which combine both approaches may also be used. In any case, the key problem is to solve simultaneously a fairly large number of LMIs. Therefore, computational complexity of LPV systems becomes a critical issue, as discussed in [267].

In this chapter, which is largely based on [112], we study an alternative probabilistic-based method to handle LPV systems with state space matrices depending in a nonlinear fashion on the scheduling parameters. In particular, the scheduling parameters are treated as random variables $\boldsymbol{\xi}$ with a given probability density function $f_{\boldsymbol{\xi}}(\xi)$ and a sequential randomized algorithm is proposed. We observe that a gridding of the scheduling parameter set Ξ or approximations, such as LFTs of the state space matrices, are not required and only mild assumptions regarding the dependence on the scheduling parameters are needed. The randomized algorithm consists of two phases. In the first phase, we randomly generate samples $\boldsymbol{\xi}^{(i)}$ within the set Ξ according to the pdf $f_{\boldsymbol{\xi}}(\xi)$. Second, we introduce a sequential gradient-based algorithm which computes approximate solutions of LMIs at each step. Hence, each iteration does not require heavy computational effort, such as solving simultaneously a large number of LMIs. Obviously, in contrast to classical approaches for handling LPV systems, a probabilistic solution is provided and, therefore, a certain level of risk should be accepted by the user. One of the main results stated in this chapter shows that the randomized algorithm provides a candidate solution with probability one in a finite number of steps. This result is similar in spirit to Theorem 11.1, but the randomized algorithm has a different structure than Algorithms 11.1 and 11.2, so that two matrix recursions need to be performed simultaneously.

12.2 Quadratic LPV $\mathcal{L}_2$ control problem

In this section we present some standard material on the so-called quadratic LPV $\mathcal{L}_2$ control problem. Additional details may be found for example in [17, 31, 229, 239]. First, we introduce the general setting, then we focus on the problem formulation and solvability conditions for the special case where some regularity assumptions hold.

We consider an LPV plant G with the state space representation

$$\dot{x} = A(\xi(t))x + B_1(\xi(t))w + B_2(\xi(t))u;$$
$$z = C_1(\xi(t))x + D_{12}(\xi(t))u;$$
$$y = C_2(\xi(t))x + D_{21}(\xi(t))w$$

where $A \in \mathbb{R}^{n_s,n_s}, B_1 \in \mathbb{R}^{n_s,q}, B_2 \in \mathbb{R}^{n_s,n_i}, C_1 \in \mathbb{R}^{p,n_s}, D_{12} \in \mathbb{R}^{p,n_i}, C_2 \in \mathbb{R}^{n_o,n_s}, D_{21} \in \mathbb{R}^{n_o,q}$. We remark that the LPV system matrices $A(\xi)$, $B_1(\xi)$, $B_2(\xi)$, $C_1(\xi)$, $C_2(\xi)$ and $D_{12}(\xi)$ and $D_{21}(\xi)$ may be nonlinear functions of $\xi \in \mathbb{R}^\ell$. We assume that the scheduling parameters ξ entering in the system matrices can be measured on-line and are bounded within a given set for all $t \geq 0$. That is, we assume

$$\xi(t) \in \Xi$$

where Ξ is a subset of $\mathbb{R}^\ell$. We remark that we do not use any information on the derivative $\dot{\xi}(t)$. A consequence of this fact is that the closed-loop matrices depend only on ξ and not on its derivative. In the general case, discussed in [112], the derivative of ξ is also used and assumed to lie in a given subset of $\mathbb{R}^\ell$.

Next, we consider a full-order strictly proper controller K of the form

$$\dot{x}_c = A_c(\xi(t))x_c + B_c(\xi(t))y;$$
$$u = C_c(\xi(t))x_c \tag{12.1}$$

where $A_c \in \mathbb{R}^{n_s,n_s}$, $B_c \in \mathbb{R}^{n_s,n_o}$ and $C_c \in \mathbb{R}^{n_i,n_s}$ are the controller matrices. Then, we write the closed-loop system

$$\dot{x}_{cl} = A_{cl}(\xi(t))x_{cl} + B_{cl}(\xi(t))w;$$
$$z = C_{cl}(\xi(t))x_{cl}$$

where the augmented state $x_{cl} \in \mathbb{R}^{2n_s}$ is

$$x_{cl} = \begin{bmatrix} x \\ x_c \end{bmatrix}$$

and the closed-loop matrices are given by

$$A_{cl}(\xi(t)) = \begin{bmatrix} A(\xi(t)) & B_2(\xi(t))C_c(\xi(t)) \\ B_c(\xi(t))C_2(\xi(t)) & A_c(\xi(t)) \end{bmatrix};$$

$$B_{cl}(\xi(t)) = \begin{bmatrix} B_1(\xi(t)) \\ B_c(\xi(t))D_{21}(\xi(t)) \end{bmatrix}; \tag{12.2}$$

$$C_{cl}(\xi(t)) = \begin{bmatrix} C_1(\xi(t)) & D_{12}(\xi(t))C_c(\xi(t)) \end{bmatrix}.$$

We now introduce some mild regularity assumptions, used in the remainder of the chapter, which are needed to simplify the resulting equations. The reader interested in the general problem formulation, where the regularity conditions are relaxed, may refer to [112] for additional details.

Assumption 12.1 (LPV regularity conditions). *We assume that the following conditions hold:*

1. *The orthogonality conditions*

$$D_{12}^T(\xi)\begin{bmatrix} C_1(\xi) & D_{12}(\xi) \end{bmatrix} = \begin{bmatrix} 0 & I \end{bmatrix};$$

$$\begin{bmatrix} B_1(\xi) \\ D_{21}(\xi) \end{bmatrix} D_{21}^T(\xi) = \begin{bmatrix} 0 \\ I \end{bmatrix}$$

   are satisfied for all $\xi \in \Xi$;
2. *The matrices $A(\xi)$, $B_1(\xi)$, $B_2(\xi)$, $C_1(\xi)$, $C_2(\xi)$, $D_{12}(\xi)$, $D_{21}(\xi)$, and the controller matrices $A_c(\xi)$, $B_c(\xi)$, $C_c(\xi)$ are continuous functions of ξ;*
3. *The vector $\xi(t)$ is a piecewise continuous function of t with a finite number of discontinuities.*

The objective is to design a controller of the form (12.1) such that the closed-loop system (12.2) is quadratically stable and the performance

$$\sup_{0 \neq w \in \mathcal{L}_2^+} \frac{(\int_0^\infty z^T(t)z(t)\,dt)^{1/2}}{(\int_0^\infty w^T(t)w(t)\,dt)^{1/2}} < \gamma \tag{12.3}$$

is satisfied for all $\xi(t) \in \Xi$ and $x(0)$. This objective is now formally stated, and this leads to the so-called regularized quadratic LPV $\mathcal{L}_2$ problem.

Problem 12.1 (Regularized quadratic LPV $\mathcal{L}_2$). *Let Assumption 12.1 be satisfied. Then, given $\gamma > 0$, find matrices $A_c(\xi)$, $B_c(\xi)$ and $C_c(\xi)$ such that there exist $X_{cl} \in \mathbb{R}^{2n_s, 2n_s}$, $X_{cl} \succeq 0$, and $\alpha > 0$ satisfying*

$$A_{cl}(\xi)X_{cl} + X_{cl}A_{cl}^T(\xi) + X_{cl}C_{cl}^T(\xi)C_{cl}(\xi)X_{cl} + \gamma^{-2}B_{cl}(\xi)B_{cl}^T(\xi) + \alpha I \preceq 0 \tag{12.4}$$

for all $\xi \in \Xi$.

We observe that inequality (12.4) implies that the closed-loop system is quadratically stable and its $\mathcal{L}_2$ norm satisfies (12.3) for all $\xi(t) \in \Xi$.

We now state a lemma which shows that the solvability condition of the regularized quadratic LPV $\mathcal{L}_2$ control problem is equivalent to the solution of two coupled matrix inequalities. This lemma, stated without proof, is a minor variation of a result given in [31]. In particular, in the statement given here there are two QMIs and one LMI with two symmetric matrix variables, whereas in the original statement three LMIs with two symmetric positive definite matrix variables are considered.

Lemma 12.1. *The regularized quadratic LPV $\mathcal{L}_2$ control problem is solvable if and only if there exist $X \in \mathbb{R}^{n_s,n_s}$, $Y \in \mathbb{R}^{n_s,n_s}$ and $\alpha > 0$ satisfying*

$$Q(\xi, X) = A(\xi)X + XA^T(\xi) + XC_1^T(\xi)C_1(\xi)X + \gamma^{-2}B_1(\xi)B_1^T(\xi)$$
$$-B_2(\xi)B_2^T(\xi) + \alpha I \preceq 0; \tag{12.5}$$

$$R(\xi, Y) = A^T(\xi)Y + YA(\xi) + YB_1(\xi)B_1^T(\xi)Y + \gamma^{-2}C_1^T(\xi)C_1(\xi)$$
$$-C_2^T(\xi)C_2(\xi) + \alpha I \preceq 0; \tag{12.6}$$

$$T(X,Y) = -\begin{bmatrix} X & \gamma^{-1}I \\ \gamma^{-1}I & Y \end{bmatrix} \preceq 0 \tag{12.7}$$

for all $\xi \in \Xi$. Furthermore, if the above conditions hold, then there exist $X \succ 0$, $Y \succ 0$, and $\alpha > 0$ which satisfy (12.5), (12.6) and $T(X,Y) \prec 0$. Subsequently, the matrices of an LPV controller are given by

$$A_c(\xi) = A(\xi) - Y^{-1}C_2^T(\xi)C_2(\xi) - B_2(\xi)B_2^T(\xi)(X - \gamma^{-2}Y^{-1})^{-1}$$
$$+\gamma^{-2}Y^{-1}C_1^T(\xi)C_1(\xi) + \gamma^{-2}Y^{-1}(R(Y,\xi) - \alpha I)(XY - \gamma^{-2}I)^{-1};$$

$$B_c(\xi) = Y^{-1}C_2^T(\xi);$$

$$C_c(\xi) = -B_2^T(\xi)(X - \gamma^{-2}Y^{-1})^{-1}. \tag{12.8}$$

We observe that solving the matrix equations (12.5)-(12.7) is a quite difficult task. In fact, these matrix equations should be solved *for all* values of the scheduling parameters $\xi \in \Xi$. As discussed previously, in the existing literature this difficulty is generally worked out by introducing a conservative overbounding of the matrices involved or a gridding of the set Ξ. In this way, these equations are then reduced to a *finite* number of conditions which should be solved simultaneously. Contrary to these methods, in this chapter we study a probabilistic approach which can handle directly the conditions given in Lemma 12.1.

12.3 Randomized algorithm for LPV systems

In this section we propose a randomized algorithm for the regularized quadratic LPV $\mathcal{L}_2$ problem. In particular, we assume that ξ is a random vector with given probability density function $f_\xi(\xi)$ and support Ξ. The proposed randomized algorithm is sequential. At each step of the sequence, first a random vector sample $\xi^{(k)}$ is generated and then a gradient-based minimization defined by the convex inequalities given in Lemma 12.1 for $\xi^{(k)}$ is executed. In particular, in the next section we reformulate the original feasibility problem defined in Lemma 12.1, and in Section 12.3.3 we present a randomized algorithm and state finite-time convergence results.

12.3.1 Violation function for LPV systems

For $\xi \in \Xi$, $X \succ 0$ and $Y \succ 0$, we define the matrix-valued function

$$V(\xi, X, Y) = \begin{bmatrix} Q(\xi, X) & 0 & 0 \\ 0 & R(\xi, Y) & 0 \\ 0 & 0 & T(X, Y) \end{bmatrix}.$$

Then, we introduce the performance violation function for the LPV problem

$$v(\xi, X, Y) = \|[V(\xi, X, Y)]_+\|$$

where $\|\cdot\|$ denotes Frobenius norm and $[\cdot]_+$ is the projection defined in (11.2). Then, we immediately have

$$v(\xi, X, Y) =$$

$$\left(\|[Q(\xi, X, Y)]_+\|^2 + \|[R(\xi, X, Y)]_+\|^2 + \|[T(X, Y)]_+\|^2 \right)^{1/2}.$$

We now state two technical assumptions which play the same role of Assumptions 11.2 and 11.3 in Chapter 11. First, we assume that the problem defined by the solvability conditions (12.5)–(12.7) with fixed $\alpha > 0$ is strictly feasible.

Assumption 12.2 (Strict feasibility). *For fixed $\alpha > 0$, the solution set*

$$S_{X,Y} \doteq \{X \succ 0, Y \succ 0 : V(\xi, X, Y) \preceq 0 \text{ for all } \xi \in \Xi\}$$

has a nonempty interior.

Remark 12.1 (Choice of α in the regularized LPV problem). We notice that, for fixed $\alpha > 0$, $S_{X,Y} \subset \bar{S}_{X,Y}$, where $\bar{S}_{X,Y}$ is the feasible solution set of Lemma 12.1. This set is defined as the set of (X, Y) such that there exists $\alpha > 0$ satisfying equations (12.5)–(12.7). In the following, we assume that such an appropriate α, whose existence is guaranteed whenever $\bar{S}_{X,Y}$ is not empty, is selected and fixed. In [208] a randomized algorithm for LMIs similar to Algorithm 12.1, presented in Section 12.3.3, is studied, and the expected number of iterations for decreasing values of α is determined.   ◁

Next, we introduce a probabilistic assumption on the set $S_{X,Y}$. Comments similar to those written in Remark 11.2 can also be made here.

Assumption 12.3 (Nonzero probability of detecting unfeasibility). *We assume that $f_{\boldsymbol{\xi}}(\xi)$ is such that*

$$\text{PR}_{\boldsymbol{\xi}}\{v(\boldsymbol{\xi}, X, Y) > 0\} > 0$$

for any $X, Y \notin S_{X,Y}$.

12.3.2 Update rule and subgradient computation

We now present the specific update rule defined in Algorithm 8.4. For LPV systems, this update rule consists of two coupled recursions which should be executed simultaneously. Each of these recursions has a structure similar to that used for linear quadratic regulator, see (11.5). More precisely, for the matrix variables X and Y, we have

$$\mathbf{X}^{(k+1)} = \psi_{\text{upd}}(\boldsymbol{\xi}^{(i)}, \mathbf{X}^{(k)}, \mathbf{Y}^{(k)}) = \mathbf{X}^{(k)} - \boldsymbol{\eta}^{(k)}\partial_X\{v(\boldsymbol{\xi}^{(i)}, \mathbf{X}^{(k)}, \mathbf{Y}^{(k)})\};$$
$$\mathbf{Y}^{(k+1)} = \psi_{\text{upd}}(\boldsymbol{\xi}^{(i)}, \mathbf{X}^{(k)}, \mathbf{Y}^{(k)}) = \mathbf{Y}^{(k)} - \boldsymbol{\eta}^{(k)}\partial_Y\{v(\boldsymbol{\xi}^{(i)}, \mathbf{X}^{(k)}, \mathbf{Y}^{(k)})\}$$

$$(12.9)$$

where ∂_X and ∂_Y denote the subgradients of $v(\xi, X, Y)$ with respect to X and Y respectively, and $\eta^{(k)}$ is a stepsize. We notice that the coupling between the recursions appears in the subgradients ∂_X and ∂_Y, see Remark 12.2, and also on the stepsize.

The next lemma is a technical tool which provides the computation of the subgradients of $v(\xi, X, Y)$ with respect to X and Y. These subgradients are subsequently used in the implementation of the randomized algorithm.

Lemma 12.2. *The function $v(\xi, X, Y)$ is convex in X, Y and its subgradients are given by*

$$\partial_X\{v(\xi, X, Y)\} = \frac{[Q(\xi, X)]_+}{v(\xi, X, Y)}Z_X(\xi) + Z_X(\xi)^T\frac{[Q(\xi, X)]_+}{v(\xi, X, Y)} - [I\ 0]\frac{[T(X, Y)]_+}{v(\xi, X, Y)}\begin{bmatrix}I\\0\end{bmatrix};$$
$$\partial_Y\{v(\xi, X, Y)\} = \frac{[R(\xi, Y)]_+}{v(\xi, X, Y)}Z_Y(\xi) + Z_Y(\xi)^T\frac{[R(\xi, Y)]_+}{v(\xi, X, Y)} - [0\ I]\frac{[T(X, Y)]_+}{v(\xi, X, Y)}\begin{bmatrix}0\\I\end{bmatrix}$$

where

$$Z_X(\xi) \doteq A(\xi) + XC_1^T(\xi)C_1(\xi);$$
$$Z_Y(\xi) \doteq A^T(\xi) + YB_1(\xi)B_1^T(\xi)$$

if $v(\xi, X, Y) > 0$, or

$$\partial_X\{v(\xi, X, Y)\} = 0 \text{ and } \partial_Y\{v(\xi, X, Y)\} = 0$$

otherwise.

Remark 12.2 (Coupling between $Q(\xi, X)$ and $R(\xi, Y)$). We observe that the solvability condition in Lemma 12.1 is given by the two decoupled QMIs

$$Q(\xi, X) \preceq 0 \text{ and } R(\xi, Y) \preceq 0$$

and by the LMI

$$T(X, Y) \preceq 0$$

with coupling on X and Y. Hence, the subgradients computed in Lemma 12.2 are coupled through the term $[T(X, Y)]_+$. If $T(X, Y) \preceq 0$, then we see immediately that this coupling disappears. In this case, each subgradient has the same form as that computed in Lemma 11.1 for the LQR problem.   ◁

12.3.3 Iterative algorithm and convergence results

We now introduce a randomized algorithm that has the same structure as Algorithm 8.4 and utilizes the update rule defined in (12.9).

Algorithm 12.1 (Iterative algorithm for LPV design).

1. Initialization.
   ▷ Set $i = 0$, $k = 0$, and choose $X^{(0)} \succ 0, Y^{(0)} \succ 0$;
2. Feasibility loop.
   ▷ Set $i = i + 1$
      – Draw $\mathbf{\Delta}^{(i)}$ according to f_Δ;
      – If $v(\boldsymbol{\xi}^{(i)}, \mathbf{X}^{(k)}, \mathbf{Y}^{(k)}) > 0$ goto 3; Else goto 2;
3. Update.
   ▷ Compute
   $$\mathbf{X}^{(k+1)} = \mathbf{X}^{(k)} - \eta^{(k)} \partial_X \{v(\boldsymbol{\xi}^{(i)}, \mathbf{X}^{(k)}, \mathbf{Y}^{(k)})\};$$
   $$\mathbf{Y}^{(k+1)} = \mathbf{Y}^{(k)} - \eta^{(k)} \partial_Y \{v(\boldsymbol{\xi}^{(i)}, \mathbf{X}^{(k)}, \mathbf{Y}^{(k)})\}$$
   where the stepsize $\eta^{(k)}$ is given in (12.10);
   ▷ Set $k = k + 1$ and goto 2.

In the update rule (12.9) of the previous iterative algorithm, we take a stepsize given by

$$\eta^{(k)} = \frac{v(\boldsymbol{\xi}^{(k)}, \mathbf{X}^{(k)}, \mathbf{Y}^{(k)}) + rw(\boldsymbol{\xi}^{(k)}, \mathbf{X}^{(k)}, \mathbf{Y}^{(k)})}{w(\boldsymbol{\xi}^{(k)}, \mathbf{X}^{(k)}, \mathbf{Y}^{(k)})^2} \tag{12.10}$$

where $r > 0$ is the radius of a ball contained in the set $S_{X,Y}$ with center X^*, Y^* and $w(\boldsymbol{\xi}^{(k)}, \mathbf{X}^{(k)}, \mathbf{Y}^{(k)})$ is given by

$$w(\boldsymbol{\xi}^{(k)}, \mathbf{X}^{(k)}, \mathbf{Y}^{(k)})$$

$$\doteq \left(\|\partial_X\{v(\pmb{\xi}^{(k)}, \mathbf{X}^{(k)}, \mathbf{Y}^{(k)})\}\|^2 + \|\partial_Y\{v(\pmb{\xi}^{(k)}, \mathbf{X}^{(k)}, \mathbf{Y}^{(k)})\}\|^2 \right)^{1/2}.$$

Algorithm 12.1 provides matrices $\widehat{\mathbf{X}}_N$ and $\widehat{\mathbf{Y}}_N$. Subsequently, the matrices $A_c(\xi)$, $B_c(\xi)$ and $C_c(\xi)$ of an LPV controller can be obtained using the formulae (12.8). We now state a finite-time convergence result, see [112] for proof. In this case, $\overline{N}(k)$ is left unspecified, i.e. the feasibility loop is run until an unfeasible point is found.

Theorem 12.1 (Finite-time convergence). *Let Assumptions 12.1, 12.2 and 12.3 be satisfied. Then, with probability one, Algorithm 12.1 with unspecified $\overline{N}(k)$ executes a finite number of updates to determine a robustly feasible solution within $S_{X,Y}$. Moreover, if an upper bound D is known on the initial distance*

$$\|X^{(0)} - X^*\| + \|Y^{(0)} - Y^*\| \le D$$

then the total number of updates is bounded by $\lfloor (D/r)^2 \rfloor$, with $r > 0$ the radius and X^*, Y^* the center of a ball contained in $S_{X,Y}$.

Remark 12.3 (Further convergence properties). The convergence results of this section are extensions to LPV systems of the theorems stated in Section 11.2 for linear quadratic regulators. A key difference, however, is the fact that the update rule (12.9) does not require the projection $[\cdot]_+$ on the set of positive definite matrices. In fact, convergence of the solutions to positive semidefinite matrices is automatically guaranteed by the coupling condition (12.7). However, we remark that projection is necessary in Algorithm 11.2 presented in the previous chapter, since no coupling condition is present.

We also observe that Algorithm 12.1 can be modified so that the update rule takes the form

$$\mathbf{X}^{(k+1)} = \psi_{\mathrm{upd}}(\pmb{\xi}^{(i)}, \mathbf{X}^{(k)}, \mathbf{Y}^{(k)}) = \left[\mathbf{X}^{(k)} - \eta^{(k)}\partial_X\{v(\pmb{\xi}^{(i)}, \mathbf{X}^{(k)}, \mathbf{Y}^{(k)})\}\right]_+;$$

$$\mathbf{Y}^{(k+1)} = \psi_{\mathrm{upd}}(\pmb{\xi}^{(i)}, \mathbf{X}^{(k)}, \mathbf{Y}^{(k)}) = \left[\mathbf{Y}^{(k)} - \eta^{(k)}\partial_Y\{v(\pmb{\xi}^{(i)}, \mathbf{X}^{(k)}, \mathbf{Y}^{(k)})\}\right]_+.$$

That is, at each step, the projection operation $[\cdot]_+$ is performed. It can be shown that this modification has the property to improve the rate of convergence of the randomized algorithm.   ◁

Example 12.1 (LPV design of a lateral motion of an aircraft).
The multivariable system studied in Example 11.3 regarding the lateral motion of an aircraft is suitably modified in order to fit within the gain scheduling and LPV setting of this chapter. In particular, the uncertain parameters are treated as scheduling parameters. The state space equation coincides with that previously studied and is given by

$$\dot{x} = \begin{bmatrix} 0 & 1 & 0 & 0 \\ 0 & L_p & L_\beta & L_r \\ g/V & 0 & Y_\beta & -1 \\ N_{\dot{\beta}}(g/V) & N_p & N_\beta + N_{\dot{\beta}}Y_\beta & N_r - N_{\dot{\beta}} \end{bmatrix} x + \begin{bmatrix} 0 & 0 \\ 0 & -3.91 \\ 0.035 & 0 \\ -2.53 & 0.31 \end{bmatrix} u.$$

In addition, we select x_1, x_3, x_4 as outputs. That is, we take

$$C_2 = \begin{bmatrix} 1 & 0 & 0 & 0 \\ 0 & 0 & 1 & 0 \\ 0 & 0 & 0 & 1 \end{bmatrix}.$$

The nominal values of the parameters coincide with those given in Example 11.3. Subsequently, the matrix $A(\xi)$ is constructed so that each scheduling parameter is allowed to vary 10% around its nominal value. Then, we assume that the pdf $f_\xi(\xi)$ is uniform in these intervals. We also choose

$$B_1 = \begin{bmatrix} 0.1 B_2 & 0 \end{bmatrix}, \quad D_{21} = \begin{bmatrix} 0 & I \end{bmatrix}, \quad C_1 = \begin{bmatrix} 0.1 C_2 \\ 0 \end{bmatrix}, \quad D_{12} = \begin{bmatrix} 0 \\ I \end{bmatrix}$$

and set $\gamma = 3$.

We now describe the results obtained using Algorithm 12.1. First, we (randomly) generated initial conditions of the form

$$X^{(0)} = \begin{bmatrix} 0.458026 & -0.154166 & 0.164914 & 0.054525 \\ -0.154166 & 0.122168 & -0.141305 & 0.076440 \\ 0.164914 & -0.141305 & 0.250909 & 0.038064 \\ 0.054525 & 0.076440 & 0.038064 & 0.471667 \end{bmatrix};$$

$$Y^{(0)} = \begin{bmatrix} 0.311724 & -0.088885 & -0.102318 & 0.342056 \\ -0.088885 & 0.350925 & -0.165670 & -0.133283 \\ -0.102318 & -0.165670 & 0.337603 & -0.213017 \\ 0.342056 & -0.133283 & -0.213017 & 0.473112 \end{bmatrix}.$$

We set $r = 0.001$ in the stepsize (12.10) and $\alpha = 0.08$ in (12.4). Then, we sequentially generated random vectors $\xi^{(k)}$ obtaining a sequence of solutions $X^{(k)}$ and $Y^{(k)}$. We observed 28 updates and the final solutions were given by

$$\widehat{X}_N = \begin{bmatrix} 0.525869 & -0.088285 & 0.125412 & -0.019091 \\ -0.088285 & 0.354168 & -0.068721 & 0.015752 \\ 0.125412 & -0.068721 & 0.333923 & 0.114620 \\ -0.019091 & 0.015752 & 0.114620 & 0.514689 \end{bmatrix};$$

$$\widehat{Y}_N = \begin{bmatrix} 0.415584 & 0.049482 & -0.126227 & 0.164431 \\ 0.049482 & 0.347744 & 0.033614 & 0.013021 \\ -0.126227 & 0.033614 & 0.518057 & -0.222386 \\ 0.164431 & 0.013021 & -0.222386 & 0.449475 \end{bmatrix}.$$

These solutions satisfy all 512 equations given by the vertex set. Since the scheduling parameters ξ enter into the state space equation in a multiaffine manner, we conclude that quadratic $\mathcal{L}_2$ performance in worst-case sense is also achieved. $\star$

13. Scenario Approach for Probabilistic Robust Design

This chapter formally presents the so-called *scenario approach* for robust design, as an alternative to the sequential methods discussed in Chapters 11 and 12, and shows how probabilistic robustness can be achieved via the scenario approach.

As widely discussed in this book, the prime motivation for studying robustness problems in engineering comes from the fact that the actual system (a "plant," or in general a problem involving physical data) upon which the engineer should act, is realistically not fixed but rather it entails some level of uncertainty. For instance, the characterization of some chemical plant G depends on the value of physical parameters. If measurements of these parameters are performed, say, on different days or under different operating conditions, it is likely that we will end up not with a single plant G, but rather with a collection $\mathcal{G}_N = \{G^{(1)}, \ldots, G^{(N)}\}$ of possible plants, each corresponding to a different realization (scenario) of the uncertain parameters upon which the plant depends. If the problem solver task is to devise a once and for all fixed policy that performs well on the actual (unknown) plant, a sensible strategy would be to design this policy such that it performs well on all the collected scenarios $\mathcal{G}_N$. This is of course a well-established technique which is widely used in practical problems, and it is for instance the standard way in which uncertainty is dealt with in difficult financial planning problems, such as multi-stage stochastic portfolio optimization, see e.g. [76].

While simple and effective in practice, the scenario approach also raises interesting theoretical questions. First, it is clear that a design that is robust for given scenarios is not robust in the standard deterministic worst-case sense, unless the considered scenarios $\mathcal{G}_N$ actually contain all possible realizations of the uncertain parameters. Then, it becomes natural to ask what is the relation between robustness in the scenario sense and the probabilistic robustness discussed extensively in this book. It turns out that a design based on scenarios actually guarantees a specified level of probabilistic robustness, provided that the number N of scenarios is chosen properly. In this sense, a robust design based on a number of randomly chosen scenarios fits exactly in the definitions of randomized algorithm for probabilistic robust design given in Chapter 8. The developments of this chapter are mainly based on [56, 57].

13.1 Three robustness paradigms

With the notation set in Chapter 11, let $\mathbf{\Delta} \in \mathcal{B}_{\mathbb{D}}(\rho)$ be the random uncertainty acting on the system, and let $\theta \in \Theta \subseteq \mathbb{R}^{n_\theta}$ be the design vector, which includes controller parameters, as well as possibly other additional variables. Let further $J(\mathbf{\Delta}, \theta)$ be a performance function for the closed-loop system, and γ an assigned performance level. We again make a standing assumption on the convexity of $J(\mathbf{\Delta}, \theta)$, which was also enforced in all the algorithms presented in Chapter 11, see Assumption 11.1.

In this context, a "robust" design problem is to determine a design parameter θ such that

$$J(\mathbf{\Delta}, \theta) \leq \gamma$$

holds "robustly" with respect to the possible realizations of $\mathbf{\Delta}$. We remained voluntarily vague as to the meaning of robustness, since we shall next define three different ways in which this robustness can be intended.

The first paradigm is of course the deterministic worst-case one, in which we seek θ such that

$$J(\mathbf{\Delta}, \theta) \leq \gamma, \quad \text{for all } \mathbf{\Delta} \in \mathcal{B}_{\mathbb{D}}(\rho).$$

To add some generality to this problem, we next consider a situation where an optimization can be performed over the design parameters θ. This gives rise to the *worst-case design problem* defined below.

Problem 13.1 (Worst-case design). Let Assumption 11.1 be satisfied. The worst-case robust design problem is to determine $\theta \in \Theta$ that solves

$$\min_{\theta, \gamma} \quad c^T [\theta^T \ \gamma]^T \tag{13.1}$$
$$\text{subject to } J(\mathbf{\Delta}, \theta) \leq \gamma \quad \text{for all } \mathbf{\Delta} \in \mathcal{B}_{\mathbb{D}}(\rho).$$

Clearly, the feasibility problem, when one is only interested in determining a solution θ that satisfies the constraints, is simply recovered as a special case of the above optimization problem, introducing a slack variable and rewriting the constraint in epigraph form.

A second paradigm for robustness is the probabilistic one: if $\mathbf{\Delta}$ is a random variable with assigned probability distribution over $\mathcal{B}_{\mathbb{D}}(\rho)$, then the probabilistic design objective is to determine a parameter θ such that

$$\mathrm{PR}\{J(\mathbf{\Delta}, \theta) \leq \gamma\} \geq p^*$$

where $p^* \in (0, 1)$ is a given desired level of probabilistic guarantee of robustness. Again, considering the more general situation in which we optimize over the design parameters, we can state the following definition of *probabilistic design problem*.

Problem 13.2 (Probabilistic design). Let Assumption 11.1 be satisfied. The probabilistic robust design problem is to determine $\theta \in \Theta$ that solves

$$\min_{\theta,\gamma} \quad c^T [\theta^T \ \gamma]^T \tag{13.2}$$

$$\text{subject to} \quad \Pr\{J(\Delta,\theta) \le \gamma\} \ge p^* \tag{13.3}$$

for some assigned probability level $p^* \in (0,1)$.

Finally, we define a third approach to robustness, which is the scenario approach: let $\Delta^{(1,\dots,N)}$ denote a multisample $\Delta^{(1)},\dots,\Delta^{(N)}$ of independent samples of Δ extracted according to some probability distribution. Then, $\Delta^{(1,\dots,N)}$ represents the randomly selected scenarios from the uncertain system, and we can define the following *scenario design problem*.

Problem 13.3 (Scenario design). Let Assumption 11.1 be satisfied. For randomly extracted scenarios $\Delta^{(1,\dots,N)}$, the scenario-based robust design problem is to determine $\theta \in \Theta$ that solves

$$\min_{\theta,\gamma} \quad c^T [\theta^T \ \gamma]^T \tag{13.4}$$

$$\text{subject to} \quad J(\Delta^{(i)},\theta) \le \gamma, \quad i = 1,\dots,N.$$

13.1.1 Advantages of scenario design

The scenario design has a distinctive advantage over its two competitors: it is computationally tractable. Indeed, we already know (see e.g. Chapter 5) that worst-case design problems are, in general, computationally hard. Even under the convexity assumptions (Assumption 11.1), the convex optimization problem (13.1) resulting from the worst-case design approach entails a usually infinite number of constraints. This class of problem goes under the name of robust convex programs, which are known to be NP-hard, [34, 105].

It is also worth stressing that the probabilistic design approach does *not* alleviate the computational complexity issue. In fact, even under the convexity assumption, problem (13.2) can be extremely hard to solve exactly in general. This is due to the fact that the probability in the so-called "chance constraint" (13.3) can be hard to compute explicitly and, more fundamentally, to the fact that the restriction imposed on θ by (13.3) is in general *nonconvex*, even though $J(\Delta,\theta)$ is convex in θ, see for instance [197, 223, 275].

Contrary to the two situations above, the scenario problem (13.4) is a standard convex program, with a finite number of constraints, and as such it is usually computationally tractable. For instance, in control problems where the performance condition is frequently expressed (for each given Δ) as linear matrix inequality constraints on the θ parameter, the scenario-robust design simply amounts to solving a standard semidefinite program.

In addition to being efficiently computable, the design $\widehat{\theta}_N$ resulting from a scenario-robust optimal design has an additional interesting property: with high probability, it also satisfies the probability constraint (13.3). In other words, if the number N of scenarios is chosen properly, then the optimal solution returned by the scenario-robust design is (with high probability) also robust in the probabilistic sense required by constraint (13.3). We explore this key issue next.

13.2 Scenario-robust design

In this section we analyze in further detail the properties of the solution of the scenario-robust design problem (13.4).

Denote with $\widehat{\theta}_N$ the optimal solution of (13.4), assuming that the problem is feasible and the solution is attained. Notice that problem (13.4) is certainly feasible whenever the worst-case problem (13.1) is feasible, since the former involves a subset of the constraints of the latter. Notice further that the optimal solution $\widehat{\theta}_N$ is a random variable, since it depends on the sampled random scenarios $\Delta^{(1)}, \ldots, \Delta^{(N)}$.

The following key result from [57] establishes the connection between the scenario approach and the probabilistic approach to robust design.

Theorem 13.1 (Scenario optimization). *Let Assumption 11.1 be satisfied and let $p^*, \delta \in (0,1)$ be given probability levels. Further, let $\widehat{\theta}_N$ denote the optimal solution[1] of problem (13.4), where the number N of scenarios has been selected so that*

$$N \geq \frac{n_\theta}{(1-p^*)\delta}. \tag{13.5}$$

Then, it holds with probability at least $1 - \delta$ that

$$\mathrm{PR}\left\{ J(\Delta, \widehat{\theta}_N) \leq \gamma \right\} \geq p^*.$$

What it is claimed in the above theorem is that if the number of scenarios is selected according to the bound (13.5), then the optimal solution returned by the scenario-robust design has, with high probability $1 - \delta$, a guaranteed level p^* of probabilistic robustness. A particular case of the above result, when J is affine in θ, has also been studied in the context of system identification in [55, 207], based on the idea of leave-one-out estimation presented in [279].

The scenario design method exactly fits the definition of randomized algorithms for probabilistic robust design (Definition 8.4). We summarize the scenario-based design approach in the following algorithm.

---

[1] Here, we assume for simplicity that the optimal solution exists and it is unique. These hypotheses can be removed without harming the result, see [57].

Algorithm 13.1 (Scenario RA for probabilistic robust design).
Let the hypotheses of Theorem 13.1 be satisfied. Then, given $p^*, \delta \in (0, 1)$, the following RA returns with probability at least $1 - \delta$ a decision variable $\widehat{\boldsymbol{\theta}}_N \in \Theta$ such that

$$\mathrm{PR}\left\{J(\boldsymbol{\Delta}, \widehat{\boldsymbol{\theta}}_N) \leq \gamma\right\} \geq p^*.$$

1. Initialization.
   ▷ Set $N \geq \frac{n_\theta}{(1-p^*)\delta}$;
2. Scenario generation.
   ▷ Generate N random iid scenarios $\boldsymbol{\Delta}^{(1)}, \ldots, \boldsymbol{\Delta}^{(N)}$;
3. Solve (13.4) and (13.5) with the given scenarios;
4. Return $\widehat{\boldsymbol{\theta}}_N$.

Remark 13.1 (Sample complexity and a priori vs. a posteriori probabilities).
From Theorem 13.1, we see that the sample complexity, and hence the number of constraints in the resulting optimization problem (13.4), is inversely proportional to the probability levels $(1 - p^*), \delta$. While this number is first-order polynomial in the number of variables and (the inverse of) the probability levels, it can still be large in practice. The main limitation here is of a technological nature, and relates to the maximum number of constraints that a specific optimization solver can deal with. This limit could be currently in the order of millions, if we are dealing with linear programs, or in the order of thousands or tens of thousands, if we deal with semidefinite or conic programs. The software limitation thus imposes a limit on the probability levels that we may ask *a priori*.

It is important to stress the difference between a priori and a posteriori levels of probability. Specifically, the a priori levels in Theorem 13.1 are guaranteed *before* we run any actual optimization experiment. In contrast, once we run the optimization and hence have in our hands a *fixed* candidate design $\widehat{\boldsymbol{\theta}}_N$, we can (and should) test this solution a posteriori, using a Monte Carlo test. This latter analysis gives us a more precise estimate of the actual probability levels attached to the design, since these tests do not involve the solution of an optimization problem and can, therefore, be performed using a very large sample size.

Numerical experience on many examples shows that the a posteriori probabilistic robustness levels are far better than those imposed a priori. In practice, this suggests not being too severe with the a priori requirements, thus easing the computational issues related to design.   ◁

Example 13.1 (Fixed-order SISO robust control design).
The example reported below is taken from [56] and it is an adaptation of a SISO fixed-order robust controller design problem originally presented in

[151]. Consider a SISO plant described by the uncertain transfer function

$$G(s,q) = 2(1+q_1)\frac{s^2 + 1.5(1+q_2)s + 1}{(s-(2+q_3))(s+(1+q_4))(s+0.236)}$$

where $q = [q_1 \ q_2 \ q_3 \ q_4]^T$ collects the uncertainty terms acting respectively on the DC-gain, the numerator damping, and the pole locations of the plant. In this example, we assume

$$\mathcal{B}_q = \{q : |q_1| \le 0.05, |q_2| \le 0.05, |q_3| \le 0.1, |q_4| \le 0.05\}.$$

The above uncertain plant can be rewritten in the form .

$$G(s,q) = \frac{N_G(s,q)}{D_G(s,q)} = \frac{b_0(q) + b_1(q)s + b_2(q)s^2}{a_0(q) + a_1(q)s + a_2(q)s^2 + s^3}$$

where

$$b_0(q) = 2(1+q_1);$$
$$b_1(q) = 3(1+q_1)(1+q_2);$$
$$b_2(q) = 2(1+q_1);$$
$$a_0(q) = -0.236(2+q_3)(1+q_4);$$
$$a_1(q) = -(2+q_3)(1+q_4) + 0.236(q_4 - q_3) - 0.236;$$
$$a_2(q) = q_4 - q_3 - 0.764.$$

Now define the following target stable interval polynomial family

$$\mathcal{P} = \{p(s) : p(s) = c_0 + c_1 s + c_2 s^2 + c_3 s^3 + s^4, \quad c_i \in [c_i^-, c_i^+], i = 0, 1, \ldots, 3\}$$

with

$$c^- \doteq \begin{bmatrix} c_0^- \\ c_1^- \\ c_2^- \\ c_3^- \end{bmatrix} = \begin{bmatrix} 38.25 \\ 57 \\ 31.25 \\ 6 \end{bmatrix}, \quad c^+ \doteq \begin{bmatrix} c_0^+ \\ c_1^+ \\ c_2^+ \\ c_3^+ \end{bmatrix} = \begin{bmatrix} 54.25 \\ 77 \\ 45.25 \\ 14 \end{bmatrix}.$$

The robust synthesis problem we consider is to determine (if one exists) a first-order controller

$$K(s,\theta) = \frac{N_K(s)}{D_K(s)} = \frac{\theta_1 + \theta_2 s}{\theta_3 + s}$$

depending on the design parameter $\theta \doteq [\theta_1 \ \theta_2 \ \theta_3]^T$, such that the closed-loop polynomial of the system

$$p_{cl}(s,q) = N_G(s,q)N_K(s) + D_G(s,q)D_K(s)$$
$$= (b_0(q)\theta_1 + a_0(q)\theta_3) + (b_1(q)\theta_1 + b_0(q)\theta_2 + a_1(q)\theta_3 + a_0(q))s$$
$$+ (b_2(q)\theta_1 + b_1(q)\theta_2 + a_2(q)\theta_3 + a_1(q))s^2$$
$$+ (b_2(q)\theta_2 + \theta_3 + a_2(q))s^3 + s^4$$

belongs to $\mathcal{P}$, for all $q \in \mathcal{B}_q$. Then, defining

$$A(q) \doteq \begin{bmatrix} b_0(q) & 0 & a_0(q) \\ b_1(q) & b_0(q) & a_1(q) \\ b_2(q) & b_1(q) & a_2(q) \\ 0 & b_2(q) & 1 \end{bmatrix}, \quad d(q) \doteq \begin{bmatrix} 0 \\ a_0(q) \\ a_1(q) \\ a_2(q) \end{bmatrix}$$

the robust synthesis conditions are satisfied if and only if

$$c^- \le A(q)\theta + d(q) \le c^+ \quad \text{for all } q \in \mathcal{B}_q. \tag{13.6}$$

To the above robust linear constraints, we also associate a linear objective vector $c^T \doteq [0\ 1\ 0]$, which amounts to seeking the robustly stabilizing controller having the smallest high-frequency gain. We thus obtain the robust linear program

$$\min_\theta c^T \theta, \quad \text{subject to (13.6)}.$$

Solving this robust linear program corresponds to determining a worst-case design for the system, see Problem 13.1. Notice, however, that the numerical solution of this problem is not "easy," since the coefficients $a_i(q), b_i(q)$ do not lie in independent intervals and depend in a nonlinear way on q. Therefore, the approach of [151] cannot be directly applied in this case.

We hence proceed via the scenario-based solution approach: assuming a uniform density over $\mathcal{B}_q$ and fixing the desired probabilistic robustness level parameter to $p^* = 0.99$, and the confidence parameter to $\delta = 0.01$, we determine the sample bound according to Theorem 13.1

$$N \ge \frac{3}{0.01 \times 0.01} = 30,000.$$

Then, $N = 30,000$ iid scenarios $\mathbf{q}^{(1)}, \ldots, \mathbf{q}^{(N)}$ are generated, and the scenario problem

$$\min_\theta c^T \theta, \text{ subject to}$$

$$c^- \le A(\mathbf{q}^{(i)})\theta + d(\mathbf{q}^{(i)}) \le c^+, \quad i = 1, \ldots, N.$$

is formed. The numerical solution of one instance of the above scenario linear program yielded the solution

$$\widehat{\theta}_N = [9.0993\ 19.1832\ 11.7309]^T$$

and hence the controller

$$K(s, \widehat{\theta}_N) = \frac{9.0993 + 19.1832s}{11.7309 + s}.$$

Once we have solved the synthesis problem, we can proceed to an a posteriori Monte Carlo test in order to obtain a more refined estimate of the probability of constraint violation for the computed solution. As discussed in Remark 13.1, we can use a much larger sample size for this a posteriori analysis, since no numerical optimization is involved in the process. Setting for instance $\epsilon = 0.001$, and $\delta = 0.00001$, from the Chernoff bound we obtain that the test should be run using at least $N = 6.103 \times 10^6$ samples. This test yielded an estimated probability of feasibility larger than 0.999.

14. Random Number and Variate Generation

In this chapter we discuss various methods for the generation of random samples distributed according to given probability distributions, in both the univariate and multivariate cases. These methods can be traced back to the issue of generating uniform random numbers in the interval $[0, 1]$. This problem is analyzed in Section 14.1, where a summary of the main existing techniques as well as more recent algorithms are reported. Subsequently, we study the more general problem of univariate random generation. In particular, we present some standard results regarding transformations between random variables and show specific examples for various classical distributions. The second part of the chapter describes techniques for multivariate distributions, focusing in particular on rejection methods, on the recursive conditional densities method, and on asymptotic methods based on Markov chains.

14.1 Random number generators

The importance of random numbers in Monte Carlo methods has been discussed in Chapter 7. Good random number generators (RNGs) should provide uniform and independent samples, and should be reproducible and fast. Computer methods for random generation, such as the classical one based on the method of von Neumann [286] for simulating neutron transport, produce only *pseudo-random* sequences, which show cyclicities and correlations. Indeed, RNGs are deterministic algorithms that provide numbers with certain statistical properties. Roughly speaking, these numbers should behave similar to realizations of independent, identically distributed uniform random variables. However, every RNG has its deficiencies, so that no RNG is appropriate for all purposes. For example, several "good" RNGs for stochastic simulation are unsuitable for cryptographic applications, because they produce predictable output streams.

RNGs mainly consist of linear and nonlinear generators. The linear algorithms are well known and widely available. However, linear RNGs may sometimes be inadequate, since these algorithms may produce lattice structures in every dimension, as shown in Section 14.1.1, and this fact may interfere with the simulation problem at hand. For this reason, nonlinear generators have been introduced. In general, these latter methods are computationally slower than linear generators of comparable size, but they allow for the use of larger strings of samples, see for instance [129].

Random number generation constitutes a whole field of study in its own. As a starting point, the reader interested in further understanding these topics may consult the classical reference [158], as well as the special issue [72], the survey paper [170] and the edited volume [130]. However, even though this is a well-established topic, current research is performed with the objective to produce extremely fast and reliable algorithms for various applications. An example of a recent and extremely efficient RNG is the so-called Mersenne twister (MT) algorithm [184].

We now describe linear congruential generators (LCGs), which are among the earliest methods for random number generation.

14.1.1 Linear congruential generators

Linear congruential generators for uniform distribution in the interval $[0, 1)$, have been proposed by Lehmer [172], and are based on a recursion of the form

$$x^{(i+1)} = ax^{(i)} + c - mk^{(i)}, \ i = 0, 1, \ldots$$

where the *multiplier* a, the *increment* c and the *modulus* m are nonnegative integers, and $k^{(i)}$ is given by

$$k^{(i)} = \left\lfloor \frac{ax^{(i)} + c}{m} \right\rfloor.$$

This recursion is often written using the notation

$$x^{(i+1)} = (ax^{(i)} + c) \bmod m, \ i = 0, 1, \ldots \tag{14.1}$$

This linear congruential generator is denoted as $\text{LCG}(a, c, m, x^{(0)})$. The modulus m is chosen as a "large" positive integer. In particular, this value is generally set to the word length of the machine, for example $m = 2^{32}$. The multiplier $a \in \{1, \ldots, m\}$ is selected so that the greatest common divisor of (a, m) is one and $c \in \{0, 1, \ldots, m - 1\}$. If the increment c is set to zero, the RNG is called "multiplicative congruential generator," otherwise it is called "mixed congruential generator."

Given an initial value $x^{(0)} \in \{0, 1, \ldots, m - 1\}$, called the *seed*, one generates a sequence $x^{(0)}, x^{(1)}, \ldots, x^{(m-1)}$ according to the recursion (14.1); this sequence is generally called a Lehmer sequence. Notice that $x^{(i)} \in \{0, 1, \ldots, m - 1\}$ for all i; therefore, numbers $y^{(i)}$ in the interval $[0, 1)$ can be subsequently obtained by taking

$$y^{(i)} = \frac{x^{(i)}}{m} \in [0, 1), \ i = 0, 1, \ldots$$

We remark that the sequences $x^{(0)}, x^{(1)}, \ldots$ and $y^{(0)}, y^{(1)}, \ldots$ are both periodic with the same period, which is no greater than m. For example, taking $m = 31, a = 3, c = 0$ and $x^{(0)} = 5$, using (14.1), we obtain the sequence

5 15 14 11 2 6 18 23 7 21 1 3 9 27 19 26 16 17 20 29 25 13 8 24 10 30 28 22 4 12
5 15 14 11 2 6 18 23 7 21 1 3 $\cdots$

$$(14.2)$$

It can be easily observed that the period in this case is 30.

A critical issue regarding LCGs is that multidimensional samples built using successive outcomes of the generator lie on a lattice. That is, if we consider the d-dimensional vectors $w^{(k)} = [x^{(k)} \, x^{(k+1)} \cdots x^{(k+d-1)}]^T$ for different values of k, and study the distribution of the points $w^{(k)}$ in $[0, 1)^d$, we observe that the generated points are of the form

$$w^{(k)} = \sum_{i=1}^{d} z_i^{(k)} v_i$$

where $\{z_1^{(k)}, \ldots, z_d^{(k)}\}$ are integers, and $\{v_1, \ldots, v_d\}$ is a set of linearly independent vectors $v_i \in \mathbb{R}^d$, $i = 1, \ldots, d$, which constitutes a basis of the lattice. For instance, in the LCG$(3, 0, 31, 5)$ sequence (14.2), the couples obtained considering the nonoverlapping vectors

$$w^{(1)} = \begin{bmatrix} 5 \\ 15 \end{bmatrix}, \quad w^{(3)} = \begin{bmatrix} 14 \\ 11 \end{bmatrix}, \quad w^{(5)} = \begin{bmatrix} 2 \\ 6 \end{bmatrix}, \quad w^{(7)} = \begin{bmatrix} 18 \\ 23 \end{bmatrix}, \ldots \quad (14.3)$$

lie on three lines, see Figure 14.1. It is known that the IBM random generator RANDU, which was used for a number of years, is a striking example of a generator providing a lattice structure in the distribution. RANDU is LCG$(2^{16}+3, 0, 2^{31}, x^{(0)})$, see further discussions in [119, 129]. This discussion pinpoints some of the limits of LCGs and motivates the study of other more sophisticated generators.

14.1.2 Linear and nonlinear random number generators

We now briefly describe some standard linear and nonlinear generators. This description is at introductory level, and the interested reader may consult more specific references such as [119, 204].

Multiple recursive generators. A simple generalization of the multiplicative congruential generator is given by

$$x^{(i+1)} = \left(a_1 x^{(i)} + a_2 x^{(i-1)} + \cdots + a_n x^{(i-n-1)} \right) \bmod m, \quad i = 0, 1, \ldots$$

where $a_1, a_2, \ldots, a_n$ are given multipliers. One of the advantages of this generator is to exhibit a longer period than the multiplicative congruential generator. Further statistical properties and specific recommendations regarding the selection of multipliers are given in [171].

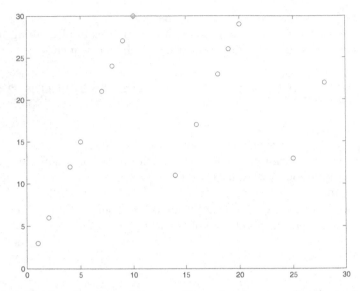

Fig. 14.1. Lattice structure of the two-dimensional vectors (14.3) obtained from points generated by LCG(3, 0, 31, 5).

Lagged Fibonacci generators. It is well known that a Fibonacci sequence is given by

$$x^{(i+1)} = x^{(i)} + x^{(i-1)}, \ i = 0, 1, \ldots$$

In principle, we can use a Fibonacci sequence as an RNG, with the simple modification

$$x^{(i+1)} = \left(x^{(i)} + x^{(i-1)} \right) \bmod m, \ i = 0, 1, \ldots$$

However, this generator does not have nice statistical properties. A simple way to modify and improve this algorithm is to introduce the so-called lagged Fibonacci congruential generator. That is, we introduce the terms $x^{(i-\ell)}$ and $x^{(i-k)}$, where $k > \ell$, obtaining

$$x^{(i+1)} = \left(x^{(i-\ell)} + x^{(i-k)} \right) \bmod m, \ i = 0, 1, \ldots$$

Clearly, in this sequence a set of seeds, rather than a single seed, needs to be specified. In [8] it is shown that if the initial sequence, ℓ, k and m are carefully selected, then this RNG "performs well." In particular, if m is a prime number and $k > \ell$, in the same paper it is shown that the period of the sequence is at most $m^k - 1$.

Nonlinear congruential generators. Knuth [158] suggested a simple generalization of the linear congruential generator proposing the nonlinear congruential generator

$$x^{(i+1)} = \left(d(x^{(i)})^2 + ax^{(i)} + c \right) \bmod m, \ i = 0, 1, \ldots$$

In general, higher order polynomials could also be used, but the advantages of this further extension and the rules for selecting the order of the polynomial are unclear. A special case of this nonlinear congruential generator has been studied in [43, 44]

$$x^{(i+1)} = \left(d(x^{(i)})^2 \right) \bmod m, \ i = 0, 1, \ldots$$

where m is the product of two "large" distinct prime numbers, and the output is the least significant bit of $x^{(i+1)}$, or its k least significant bits. This generator, known as the Blum Blum Shub (BBS) generator, has interesting theoretical properties that make it suitable for applications in cryptography.

Remark 14.1 (RNG for cryptography). It should be noted that not just any generator is appropriate for cryptographic applications, as shown in [32, 111]. In particular, linear congruential generators are not suitable, since it is possible to recover their parameters in polynomial time, given a sufficiently long observation of the output stream. Unlike LCGs, the BBS generator has very strong cryptographic properties, which relate the quality of the generator to the difficulty of the "integer factorization problem;" that is, computing the prime factors of a very large integer. When the prime factors of m are chosen appropriately, and $O(\log \log m)$ bits of each $x^{(i)}$ are considered as the output of the BBS recursion, then, for large values of m, distinguishing the output bits from random numbers becomes at least as difficult as factoring m. Since integer factorization is largely believed not to be polynomial-time solvable, then BBS with large m has an output free from any nonrandom patterns that could be discovered with a reasonable amount of calculations.   ◁

On the other hand, BBS appears not to be the preferred choice for stochastic simulations, since the required nonlinear operations cannot be performed with high computational efficiency. However, LCGs do have high computational efficiency, as do the shift register generators discussed in the following paragraph.

Feedback shift register generators. A generator that returns binary numbers $x^{(i)} \in \{0, 1\}$ is studied in [262]. This *binary generator* is of the form

$$x^{(i+1)} = \left(c_p x^{(i-p)} + c_{p-1} x^{(i-p+1)} + \cdots + c_1 x^{(i-1)} \right) \bmod 2, \ i = 0, 1, \ldots$$

where all variables take binary values $\{0, 1\}$ and p is the order of the recursion. The name of this generator follows from the fact that recursive operations of this form can be performed in a feedback shift register. Further properties of the feedback shift register generators are discussed in [158].

The Mersenne twister generator [184] is itself a twisted generalized shift feedback register generator. The "twist" is a transformation which assures equidistribution of the generated numbers in 623 dimensions, while LCGs can at best manage reasonable distribution in five dimensions. MT was proved to have a period as large as $2^{19937} - 1$, which, incidentally, explains the origin of the name: the number $2^{19937} - 1$ is a Mersenne prime. Unlike BBS, the MT algorithm in its native form is not suitable for cryptography. For many other applications, such as stochastic simulation, however, it is becoming the random number generator of preferred choice.

14.2 Non uniform random variables

In the previous section we introduced several standard methods for generating pseudo-random numbers, which can be considered uniformly distributed over the interval $[0, 1)$. Starting from these basic uniform generators, many different distributions can be obtained by means of suitable functional transformations or other operations.

In this section, we study general operations on random variables and analyze some well-known univariate random generation methods. These results can be used for constructing sample generators according to various distributions, as shown in the examples presented in this section. First, we present a fundamental result that indicates how the probability density is changed by a functional operation on a random variable, see e.g. [227] for proof.

Theorem 14.1 (Functions of scalar random variables). *Let* $\mathbf{x} \in \mathbb{R}$ *be a random variable with distribution function* $F_{\mathbf{x}}(x)$ *and pdf* $f_{\mathbf{x}}(x)$, *and let* $\mathbf{y}$ *be related to* $\mathbf{x}$ *by a strictly monotone and absolutely continuous transformation* $\mathbf{y} = g(\mathbf{x})$. *Let* $h(\cdot) \doteq g^{-1}(\cdot)$. *Then, the random variable* $\mathbf{y}$ *has distribution function*

$$F_{\mathbf{y}}(y) = \begin{cases} F_{\mathbf{x}}(h(y)) & \text{if } g(x) \text{ is increasing;} \\ 1 - F_{\mathbf{x}}(h(y)) & \text{if } g(x) \text{ is decreasing} \end{cases}$$

and density

$$f_{\mathbf{y}}(y) = f_{\mathbf{x}}(h(y)) \left| \frac{\mathrm{d}h(y)}{\mathrm{d}y} \right|$$

for almost all y.

The transformation rule of Theorem 14.1 also has a multivariate extension, which is stated in Section 14.3. Some standard applications of this result are presented in the following examples.

Example 14.1 (Linear transformation).
The simpler transformation on a random variable $\mathbf{x}$ is the linear transformation $\mathbf{y} = a\mathbf{x} + b$, $a > 0$. If $\mathbf{x}$ has distribution function $F_{\mathbf{x}}(x)$ then

$$F_{\mathbf{y}}(y) = F_{\mathbf{x}}\left(\frac{y-b}{a}\right).$$

If the corresponding density $f_{\mathbf{x}}(x)$ exists, then we also have

$$f_{\mathbf{y}}(y) = \frac{1}{a} f_{\mathbf{x}}\left(\frac{y-b}{a}\right).$$

★

Example 14.2 (Linear transformation of the Gamma density).
A density widely used in statistics is the unilateral Gamma density with parameters a, b, defined in (2.10) as

$$G_{a,b}(x) = \frac{1}{\Gamma(a)b^a} x^{a-1} e^{-x/b}, \quad x \geq 0.$$

If $\mathbf{x}$ is distributed according to $G_{a,b}$, then the random variable $\mathbf{y} = c\mathbf{x}$, $c > 0$, obtained by linear transformation, is distributed according to $G_{a,cb}$. ★

Example 14.3 (Power transformation).
If a random variable $\mathbf{x} \geq 0$ has distribution function $F_{\mathbf{x}}(x)$ and density $f_{\mathbf{x}}(x)$, then the variable $\mathbf{y} = \mathbf{x}^\lambda$, $\lambda > 0$, has distribution

$$F_{\mathbf{y}}(y) = F_{\mathbf{x}}(y^{1/\lambda})$$

and density

$$f_{\mathbf{y}}(y) = \frac{1}{\lambda} y^{\frac{1}{\lambda}-1} f_{\mathbf{x}}(y^{1/\lambda}).$$

★

Example 14.4 (Generalized Gamma density via power transformation).
The (unilateral) generalized Gamma density with parameters a, c, see (2.11), is given by

$$\overline{G}_{a,c}(x) = \frac{c}{\Gamma(a)} x^{ca-1} e^{-x^c}, \quad x \geq 0.$$

If $\mathbf{x} \sim G_{a,1}$, then the random variable $\mathbf{y} = \mathbf{x}^{1/c}$ obtained by power transformation has density function $f_{\mathbf{y}}(y) = \overline{G}_{a,c}(y)$. ★

Example 14.5 (Weibull density via power transformation).
A random variable with Weibull density (2.8)

$$W_a(x) = ax^{a-1} e^{-x^a}$$

can be obtained from a random variable distributed according to an exponential density via power transformation. In fact, if $\mathbf{x} \sim e^{-x}$, $x \geq 0$, then $\mathbf{y} = \mathbf{x}^{1/a}$, $a > 0$, has density $f_{\mathbf{y}}(y) = W_a(y)$. ★

Example 14.6 (Logarithmic transformation).
   If a random variable $\mathbf{x} \geq 0$ has distribution function $F_{\mathbf{x}}(x)$ and density $f_{\mathbf{x}}(x)$, then the variable $\mathbf{y} = -\frac{1}{\lambda} \log \mathbf{x}$, $\lambda > 0$, has distribution

$$F_{\mathbf{y}}(y) = 1 - F_{\mathbf{x}}(e^{-\lambda y})$$

and density

$$f_{\mathbf{y}}(y) = \lambda e^{-\lambda y} f_{\mathbf{x}}(e^{-\lambda y}).$$

For instance, if $\mathbf{x}$ is uniform in $[0, 1]$, then $\mathbf{y} = -\frac{1}{\lambda} \log \mathbf{x}$ has the unilateral Laplace (exponential) density (2.9), i.e.

$$f_{\mathbf{y}}(y) = \lambda e^{-\lambda y}, \quad y \geq 0.$$

$\star$

   A useful consequence of Theorem 14.1 is a standard method for generating a univariate random variable with a given distribution function. This method is known as the *inversion method*, see e.g. [84, 214], and it is stated next.

Corollary 14.1 (Inversion method). *Let* $\mathbf{x} \in \mathbb{R}$ *be a random variable with uniform distribution in the interval* $[0, 1]$. *Let* F *be a continuous distribution function on* $\mathbb{R}$ *with inverse* F^{-1} *defined by*

$$F^{-1}(x) = \inf \{y : F(y) = x, \ 0 \leq x \leq 1\}.$$

Then, the random variable $\mathbf{y} = F^{-1}(\mathbf{x})$ *has distribution function* F. *Also, if a random variable* $\mathbf{y}$ *has distribution function* F, *then the random variable* $\mathbf{x} = F(\mathbf{y})$ *is uniformly distributed on* $[0, 1]$.

Proof. The statement immediately follows from Theorem 14.1, taking $y \doteq g(x) \doteq F^{-1}(x)$, i.e. $x = h(y) \doteq F(y)$, and noticing that for the uniform distribution $F(h(y)) = h(y)$. $\qquad\qquad\Box$

A plot showing the idea behind the inversion method is shown in Figure 14.2: uniform samples of $\mathbf{x}$ in the vertical axis are mapped into samples of $\mathbf{y}$ having distribution F.
   Corollary 14.1 can be used to generate samples of a univariate random variables with an arbitrary continuous distribution function, provided that its inverse is explicitly known, or readily computable. Clearly, the numerical efficiency of the method relies on how fast the inverse can be numerically computed. Implementation refinements of the above method are discussed for instance in [84]. An application of the inversion method for generation of samples according to a polynomial density is presented next.

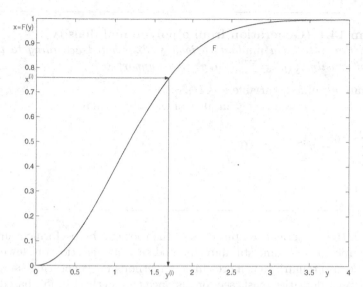

Fig. 14.2. Graphical interpretation of the inversion method.

Example 14.7 (Generation from a polynomial density).
The inversion method is useful for the generation of samples according to a generic univariate polynomial density over the interval $[0, c]$. Let

$$f_{\mathbf{y}}(y) = p(y) = \sum_{k=0}^{n} a_k y^k$$

be a polynomial density with support $[0, c]$. Notice that the distribution function of $\mathbf{y}$ can be easily computed as

$$F_{\mathbf{y}}(y) = \int_{0}^{y} p(y) dy = \sum_{k=0}^{n} \frac{a_k}{k+1} y^{k+1}.$$

The condition that the polynomial $p(y)$ is a density function requires that $p(y) \geq 0$ for all $y \in [0, c]$, and that

$$\sum_{k=0}^{n} \frac{a_k}{k+1} c^{k+1} = 1.$$

A simple algorithm can hence be used for polynomial sample generation.

Algorithm 14.1 (Generation from a polynomial density).
This algorithm returns a random variable **y** *distributed according to the polynomial density* $f_\mathbf{y}(y) = \sum_{k=0}^{n} a_k y^k$ *with support* $[0, c]$.

1. Generate a random variable $\mathbf{x} \sim \mathcal{U}_{[0,1]}$;
2. Compute the unique root **y** in $[0, c]$ of the polynomial

$$\sum_{k=0}^{n} \frac{a_k}{k+1} y^{k+1} - \mathbf{x} = 0;$$

3. Return **y**.

In step 2, the numerical computation of the root can be performed, up to a given accuracy, using some standard method such as bisection or Newton–Raphson. We also remark that more efficient methods for generating samples from polynomial densities exist, see for instance the method in [5], based on finite mixtures.   ⋆

In the next section, we briefly discuss two classical tests, the chi-square and the Kolmogorov–Smirnov (KS) test, that are used in statistics to assess whether a given batch of sample data comes from a specific distribution, see e.g. [158].

14.2.1 Statistical tests for pseudo-random numbers

Chi-square test. The chi-square goodness-of-fit test, see e.g. [247], is used to decide if a batch of sampled data comes from a specific distribution $F^0(x)$. An attractive feature of this test is that it can be applied to any univariate distribution (continuous or discrete) for which the cumulative distribution function can be calculated. The chi-square test is applied to binned data, but this is actually not a restriction, since for nonbinned data one can simply calculate a histogram or frequency table before applying the test. However, the value of the chi-square test statistic depends on how the data is binned. Another disadvantage is that this test requires sufficiently large sample size for its approximations to be valid. This method is briefly described below.

Let $\mathbf{x}^{(1)}, \dots, \mathbf{x}^{(N)}$ be a multisample of size N drawn from an unknown cdf $F_\mathbf{x}(x)$, and let $F^0(x)$ be a completely specified candidate cdf. We wish to test the null hypothesis

$$H_0 : \ F_\mathbf{x}(x) = F^0(x), \text{ for all } x$$

against the alternative

$$H_1 : \ \exists x : \ F_\mathbf{x}(x) \neq F^0(x).$$

In the chi-square test, the data is divided into n bins, i.e. into n intervals $[x_i^-, x_i^+]$. Then, we introduce the sum known as *Pearson's test statistic*

$$\widehat{\mathbf{y}} = \sum_{i=1}^{n} \frac{(\widehat{\mathbf{p}}_i - p_i)^2}{p_i} \tag{14.4}$$

where $\widehat{\mathbf{p}}_i$ is the empirical probability for the ith bin, and p_i is the probability associated with the ith bin, given by

$$p_i = F^0(x_i^+) - F^0(x_i^-).$$

Clearly, the value of $\widehat{\mathbf{y}}$ tends to be small when H_0 is true, and large when H_0 is false. For large sample size, the distribution of $\widehat{\mathbf{y}}$ is approximately chi square with $n - 1$ degrees of freedom, see (2.7). Therefore, under the H_0 hypothesis, we expect

$$\text{PR}\{\widehat{\mathbf{y}} > x_{1-\alpha}\} = \alpha$$

where $\alpha \in (0, 1)$ is the significance level, and $x_{1-\alpha}$ is the $1 - \alpha$ percentile of the chi-square distribution.

The chi-square test then goes as follows: given samples $\mathbf{x}^{(1)}, \ldots, \mathbf{x}^{(N)}$, and a candidate cdf $F^0(x)$, compute $\widehat{\mathbf{y}}$ from (14.4). Select a significance level $\alpha \in (0, 1)$ and compute the chi-square percentile[1] $x_{1-\alpha}$. If $\widehat{\mathbf{y}} \leq x_{1-\alpha}$, then the test is passed, and the cdf F^0 is a good fit for the data distribution.

Kolmogorov–Smirnov test. The Kolmogorov–Smirnov test, see e.g. [214], is an alternative to the chi-square goodness-of-fit test. Given a multisample $\mathbf{x}^{(1)}, \ldots, \mathbf{x}^{(N)}$ of size N drawn from an unknown cdf $F_{\mathbf{x}}(x)$, the empirical distribution function is defined as

$$\widehat{\mathbf{F}}_N(x) \doteq \frac{\widehat{\mathbf{k}}(x)}{N}$$

where $\widehat{\mathbf{k}}(x)$ is the number of sample points $\mathbf{x}^{(i)}$ which are smaller than x. This is a step function that increases by $1/N$ at the value of each data point. The KS test is based on the maximum distance between the empirical distribution and a candidate distribution $F^0(x)$. Formally, the random quantity

$$\widehat{\mathbf{y}} = \sup_x |\widehat{\mathbf{F}}_N(x) - F^0(x)| \tag{14.5}$$

measures how far $\widehat{\mathbf{F}}_N(x)$ deviates from $F^0(x)$, and is called the Kolmogorov–Smirnov one-sample statistic. For large sample size N, it holds that

$$\text{PR}\left\{\sqrt{N}\widehat{\mathbf{y}} \leq x\right\} \simeq H(x) \tag{14.6}$$

---

[1] Chi-square percentile tables are available in standard statistics books, e.g. [214].

where

$$H(x) = 1 - 2\sum_{k=1}^{\infty}(-1)^{k-1}e^{-2k^2x^2}.$$

The function $H(x)$ is tabulated, and the approximation (14.6) is practically good for $N > 35$.

The KS test goes as follows: given samples $\mathbf{x}^{(1)}, \dots, \mathbf{x}^{(N)}$, and a candidate cdf $F^0(x)$, compute $\widehat{\mathbf{y}}$ using equation (14.5). Select a significance level $\alpha \in (0,1)$ and compute the $1 - \alpha$ percentile of $H(x)$. If $\sqrt{N}\widehat{\mathbf{y}} \le x_{1-\alpha}$, then the test is passed, and the cdf $F^0(x)$ is a good fit for the empirical distribution. In the case when $F^0(x)$ is close to the true underlying cdf $F_{\mathbf{x}}(x)$, the probability of failing the test is smaller than α.

An attractive feature of this test is that the distribution of the KS test statistic $\widehat{\mathbf{y}}$ does not depend on the underlying cdf being tested. Another advantage is that it is an exact test. Despite these advantages, the KS test has several limitations: (1) it only applies to continuous distributions; (2) it tends to be more sensitive near the center of the distribution than at the tails; (3) the distribution should be fully specified. Owing to these limitations, many analysts prefer to use other, more sophisticated tests, such as the Anderson–Darling goodness-of-fit test [13].

14.3 Methods for multivariate random generation

In this section we discuss some standard methods for generating random samples from multivariate densities. In particular, we discuss rejection-based methods, and the method based on conditional densities.

We first present a multivariate extension of Theorem 14.1, see for instance [84]. More precisely, the following theorem can be used for obtaining a random vector $\mathbf{y} \in \mathbb{R}^n$ with desired density $f_{\mathbf{y}}(y)$, starting from a random vector $\mathbf{x} \in \mathbb{R}^n$ with pdf $f_{\mathbf{x}}(x)$. This tool is based on the functional transformation $\mathbf{y} = g(\mathbf{x})$ and it can be used whenever the inverse $\mathbf{x} = h(\mathbf{y}) \doteq g^{-1}(\mathbf{y})$ exists. The transformation rule makes use of the Jacobian of the function $x = h(y)$, defined as

$$J(x \to y) \doteq \begin{vmatrix} \frac{\partial x_1}{\partial y_1} & \frac{\partial x_2}{\partial y_1} & \dots & \frac{\partial x_n}{\partial y_1} \\ \frac{\partial x_1}{\partial y_2} & \frac{\partial x_2}{\partial y_2} & \dots & \frac{\partial x_n}{\partial y_2} \\ \vdots & \vdots & & \vdots \\ \frac{\partial x_1}{\partial y_n} & \frac{\partial x_2}{\partial y_n} & \dots & \frac{\partial x_n}{\partial y_n} \end{vmatrix} \tag{14.7}$$

where $\frac{\partial x_i}{\partial y_\ell} \doteq \frac{\partial h_i(y)}{\partial y_\ell}$ are the partial derivatives, and $|\cdot|$ denotes the absolute value of the determinant. The notation $J(x \to y)$ means that the Jacobian is to be computed taking the derivatives of x with respect to y. It also helps recalling that the Jacobian $J(x \to y)$ is to be used when determining the pdf of $\mathbf{y}$ given the pdf of $\mathbf{x}$, see Theorem 14.2 Various rules for computation of Jacobians are given in Appendix A.2.

Theorem 14.2 (Functions of random vectors). *Let $\mathbf{x} \in \mathbb{R}^n$ be a random vector with density $f_{\mathbf{x}}(x_1, \dots, x_n)$ continuous on the support $\mathcal{Y}_x \subseteq \mathbb{R}^n$, and let $\mathbf{y} = g(\mathbf{x})$, where $g : \mathcal{Y}_x \to \mathcal{Y}_y$, $\mathcal{Y}_y \subseteq \mathbb{R}^n$, is a one-to-one and onto mapping, so that the inverse $\mathbf{x} = h(\mathbf{y}) \doteq g^{-1}(\mathbf{y})$ is well-defined. Then, if the partial derivatives $\frac{\partial x_i}{\partial y_\ell} \doteq \frac{\partial h_i(y)}{\partial y_\ell}$ exist and are continuous on $\mathcal{Y}_y$, the random vector $\mathbf{y}$ has density*

$$f_{\mathbf{y}}(y) = f_{\mathbf{x}}(h(y))J(x \to y), \quad y \in \mathcal{Y}_y. \tag{14.8}$$

An extension of this theorem to transformations between random matrices is presented in Appendix A.1. A classical example of application of this result is the Box–Muller method for generating normal samples in $\mathbb{R}^2$, starting from uniform samples.

Example 14.8 (Box–Muller method for normal densities in $\mathbb{R}^2$).
Consider the transformation $y = g(x)$, defined by

$$y_1 = \sqrt{-2 \log x_1} \cos(2\pi x_2);$$

$$y_2 = \sqrt{-2 \log x_1} \sin(2\pi x_2)$$

mapping $x \in \mathcal{Y}_x = [0,1]^2$ into $y \in \mathcal{Y}_y = \mathbb{R}^2$. Then, if $\mathbf{x}$ is uniform in $\mathcal{Y}_x$, the transformed random variables $\mathbf{y}_1, \mathbf{y}_2$ have independent normal densities. This fact is easily verified using Theorem 14.2. Indeed, the inverse mapping $x = h(y)$ is

$$x_1 = e^{-\frac{R^2}{2}};$$

$$x_2 = \frac{1}{2\pi} \arcsin \frac{y_2}{R}, \quad R \doteq \sqrt{y_1^2 + y_2^2}.$$

The Jacobian of this transformation is given by

$$J(x \to y) = \begin{vmatrix} -y_1 e^{-\frac{R^2}{2}} & -y_2 e^{-\frac{R^2}{2}} \\ -\frac{1}{2\pi}\frac{y_2}{R^2} & \frac{1}{2\pi}\frac{y_1}{R^2} \end{vmatrix} = \frac{1}{2\pi} e^{-\frac{R^2}{2}}.$$

Since $f_{\mathbf{x}}(h(y)) = 1$ for all $y \in \mathbb{R}^2$, from (14.8) we have

$$f_{\mathbf{y}}(y) = f_{\mathbf{x}}(h(y))J(x \to y) = \frac{1}{2\pi} e^{-\frac{R^2}{2}} = \left(\frac{1}{\sqrt{2\pi}} e^{-\frac{y_1^2}{2}} \right) \left(\frac{1}{\sqrt{2\pi}} e^{-\frac{y_2^2}{2}} \right).$$

$\star$

14.3.1 Rejection methods

We study two classes of rejection methods: the first one is based on the concept of rejection from a "dominating density." The second is for uniform generation in given sets and it is based on rejection from a bounding set. The two methods are obviously related, and a critical issue in both cases is their numerical efficiency.

Rejection from dominating density. We now discuss the standard version of a rejection algorithm for sample generation from a multivariate density $f_{\mathbf{x}}(x)$, $\mathbf{x} \in \mathbb{R}^n$. A basic result on multivariate densities, see for instance [84], is the following.

Theorem 14.3. *Let $f_{\mathbf{x}}(x) : \mathbb{R}^n \to \mathbb{R}$ be a density function, and consider the set $S \subset \mathbb{R}^{n+1}$ defined as*

$$S = \left\{ \begin{bmatrix} x \\ u \end{bmatrix} : x \in \mathbb{R}^n, u \in \mathbb{R}, 0 \le u \le f_{\mathbf{x}}(x) \right\}.$$

If the random vector $\begin{bmatrix} \mathbf{x} \\ u \end{bmatrix}$ is uniformly distributed in S, then $\mathbf{x}$ has density function $f_{\mathbf{x}}(x)$ on $\mathbb{R}^n$. On the other hand, if $\mathbf{x} \in \mathbb{R}^n$ has pdf $g_{\mathbf{x}}(x)$ and $\mathbf{w} \in \mathbb{R}$ is uniformly distributed on the interval $[0,1]$, then the random vector $\begin{bmatrix} \mathbf{x} \\ \eta \mathbf{w} g_{\mathbf{x}}(x) \end{bmatrix}$ is uniformly distributed in the set

$$S_d = \left\{ \begin{bmatrix} x \\ u \end{bmatrix} : x \in \mathbb{R}^n, u \in \mathbb{R}, 0 \le u \le \eta g_{\mathbf{x}}(x), \eta > 0 \right\}.$$

Using this theorem, we now present the basic rejection scheme. Let $f_{\mathbf{x}}(x)$ be a density function on $\mathbb{R}^n$ and let $g_{\mathbf{x}}(x)$ be a *dominating density* for $f_{\mathbf{x}}(x)$, i.e. a density such that

$$f_{\mathbf{x}}(x) \le \eta g_{\mathbf{x}}(x) \tag{14.9}$$

for some constant $\eta \ge 1$. Random samples from $f_{\mathbf{x}}(x)$ can be obtained using the following algorithm.

Algorithm 14.2 (Rejection from a dominating density).
Given a density function $f_{\mathbf{x}}(x)$ and a dominating density $g_{\mathbf{x}}(x)$ satisfying (14.9), this algorithm returns a random vector $\mathbf{x} \in \mathbb{R}^n$ with pdf $f_{\mathbf{x}}(x)$.

1. Generate a random vector $\mathbf{x} \in \mathbb{R}^n$ with pdf $g_{\mathbf{x}}(x)$;
2. Generate a random variable $\mathbf{w} \in \mathbb{R}$ uniform in $[0,1]$;
3. If $\eta \mathbf{w} g_{\mathbf{x}}(\mathbf{x}) \le f_{\mathbf{x}}(\mathbf{x})$ return $\mathbf{x}$ else goto 1.

The interpretation of the rejection algorithm is given in Figure 14.3. First, random samples

$$\begin{bmatrix} \mathbf{x}^{(i)} \\ \eta \mathbf{w}^{(i)} g_{\mathbf{x}}(\mathbf{x}^{(i)}) \end{bmatrix}$$

are generated uniformly inside the dominating set S_d. These uniform samples are obtained by generating $\mathbf{x}^{(i)}$ according to $g_{\mathbf{x}}(x)$, and $\mathbf{w}^{(i)}$ uniform in $[0,1]$.

Then, samples lying outside the set S (crosses) are rejected. The remaining ones (dots) are uniform in S and therefore, by Theorem 14.3, their projection onto the x space is distributed according to $f_\mathbf{x}(x)$.

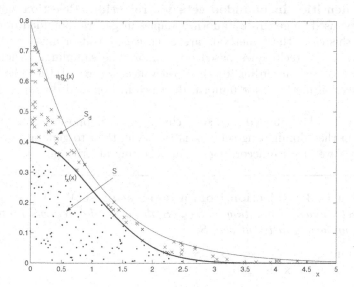

Fig. 14.3. Rejection from a dominating density.

It is worth noting that two features are essential for a rejection algorithm to work properly: first, the sample generation according to the dominating density $g_\mathbf{x}(x)$ should be feasible and simple. Second, the value of η should be known and not too large.

Remark 14.2 (Rejection rate). A critical parameter assessing the efficiency of a rejection algorithm is the *rejection rate*, defined as the expected value of the number of samples that have to be drawn from the dominating density $g_\mathbf{x}(x)$ in order to find one "good" sample. For Algorithm 14.2, the rejection rate coincides with the value of the constant η. The variance of the rejection rate is also related to η and it is given by $\eta(\eta - 1)$, see for instance [84].   ◁

Example 14.9 (A Gamma generator based on rejection).
An efficient algorithm for generation of samples distributed according to $G_{a,1}$, $a \in (0,1)$, can be derived via rejection from the Weibull density $W_a(x)$ given in (2.8). In fact

$$G_{a,1}(x) \leq \eta W_a(x), \quad x \geq 0$$

where the rejection constant η is given by

$$\eta = \frac{e^{(1-a)}a^{a/(1-a)}}{\Gamma(1+a)} < e.$$

Various other Gamma generators are discussed in detail in [84]. For instance, Gamma generators for $a > 1$ may be obtained via rejection from a so-called Burr XII density. $\qquad\qquad\qquad\qquad\qquad\qquad\qquad\qquad\qquad\qquad\qquad\qquad$ ⋆

Uniform densities in bounded sets via rejection. Rejection methods may also be used to generate uniform samples in generic bounded domains of $\mathbb{R}^n$. In this case, these methods are often called "hit-or-miss" and they are related to the techniques based on "importance sampling," where the samples are chosen according to their relevance, see e.g. [119]. The basic idea of rejection methods for uniform densities in bounded domains is now recalled.

Let S and S_d be bounded sets, such that $S \subseteq S_d$. Suppose that a uniform generator in the "dominating set" S_d is available, then to generate uniform samples in S we proceed according to the following algorithm.

Algorithm 14.3 (Rejection from a dominating set).
Given a set S and a dominating set $S_d \supseteq S$, this algorithm returns a random vector $\mathbf{x}$, uniformly distributed in S.

1. Generate a random vector $\mathbf{x} \in S_d$;
2. If $\mathbf{x} \in S$ return $\mathbf{x}$ else goto 1.

The rejection rate η of this method is given by the ratio of the volumes of the two sets

$$\eta = \frac{\text{Vol}(S_d)}{\text{Vol}(S)}. \qquad\qquad\qquad (14.10)$$

The effectiveness of this method is clearly related to three basic factors: first, the efficiency of the uniform generator in S_d, second the numerical evaluation of the membership of the sample $\mathbf{x}^{(i)}$ in S; and third, the ratio of volumes η, which is in general the most critical factor. Notice that, in principle, it is often possible to bound the set S with a suitable n-dimensional hyperrectangle, in which uniform generation is straightforward. However, this usually results in a dramatically large value of the rejection rate η, as shown in the following simple example.

Example 14.10 (Uniform generation in a sphere via rejection).
Suppose one is interested in generating uniform samples in the Euclidean sphere of unit radius $S = \mathcal{B}_{\|\cdot\|_2} = \{x \in \mathbb{R}^n : \|x\|_2 \le 1\}$. Clearly, the sphere is bounded by the n-dimensional hypercube $S_d = [-1, 1]^n$. The volume of the hypercube is 2^n, while the volume of the sphere is given by the formula

$$\text{Vol}(\mathcal{B}_{\|\cdot\|_2}(\mathbb{R}^n)) = \frac{2\pi^{n/2}}{n\Gamma(n/2)}.$$

The rejection rate grows with respect to the dimension n as

$$\eta(n) = \left(\frac{2}{\sqrt{\pi}}\right)^n \Gamma(n/2 + 1).$$

For small n, we have $\eta(1) = 1$, $\eta(2) = 1.2732$, $\eta(3) = 1.9099$, whereas for large n the method is not viable since we have for instance $\eta(20) = 4.06 \times 10^7$, $\eta(30) = 4.90 \times 10^{13}$. This "curse of dimensionality" turns out to be a common problem for set rejection methods, as discussed further in Chapter 15. $\star$

14.3.2 Conditional density method

The conditional density method is another standard method for random generation with multivariate distributions [84]. This is a recursive method in which the individual entries of the multivariate samples are generated according to their conditional probability density. In particular, let $f_{\mathbf{x}}(x)$ be the joint pdf of a vector of random variables $\mathbf{x} = [\mathbf{x}_1 \cdots \mathbf{x}_n]^T$. This pdf can be written as

$$f_{\mathbf{x}}(x_1, \ldots, x_n) = f_{\mathbf{x}_1}(x_1) f_{\mathbf{x}_2|x_1}(x_2|x_1) \cdots f_{\mathbf{x}_n|x_1 \cdots x_{n-1}}(x_n|x_1 \cdots x_{n-1})$$

where $f_{\mathbf{x}_i|x_1,\ldots,x_{i-1}}(x_i|x_1, \ldots, x_{i-1})$ are the conditional densities. These densities are defined, see (2.2), as the ratio of marginal densities

$$f_{\mathbf{x}_i|x_1,\ldots,x_{i-1}}(x_i|x_1, \ldots, x_{i-1}) = \frac{f_{\mathbf{x}_1,\ldots,\mathbf{x}_i}(x_1, \ldots, x_i)}{f_{\mathbf{x}_1,\ldots,\mathbf{x}_{i-1}}(x_1, \ldots, x_{i-1})}.$$

In turn, the marginal densities $f_{\mathbf{x}_1,\ldots,\mathbf{x}_i}(x_1, \ldots, x_i)$ are given by

$$f_{\mathbf{x}_1,\ldots,\mathbf{x}_i}(x_1, \ldots, x_i) = \int \cdots \int f_{\mathbf{x}}(x_1, \ldots, x_n) dx_{i+1} \cdots dx_n.$$

A random vector $\mathbf{x}$ with density $f_{\mathbf{x}}(x)$ can therefore be obtained by generating sequentially the $\mathbf{x}_i$, $i = 1, \ldots, n$, where $\mathbf{x}_i$ is distributed according to the univariate density $f_{\mathbf{x}_i|x_1,\ldots,x_{i-1}}(x_i)$.

The basic idea of this method, therefore, is to generate the first random variable according to $f_{\mathbf{x}_1}(x_1)$, then generate the next one conditional on the first one, and so forth. In other words, the conditional density method reduces an n-dimensional generation problem to n one-dimensional problems. The main difficulty in its application arises from the fact that the computation of the marginal densities is necessary. This is often a very difficult task, since it requires the computation of multiple integrals, see for instance [85]. The conditional density method is one of the basic tools used in Chapter 18 for generating random matrices uniformly distributed in norm bounded sets.

14.4 Asymptotic methods based on Markov chains

In this section we briefly account for a large body of literature on sampling methods generally known as Markov chain Monte Carlo (MCMC) methods. The basic idea behind these methods is to obtain the desired distribution by simulating a random walk on a graph. The sequence of randomly visited nodes on the graph constitutes a Markov chain, and the output distribution is the stationary distribution of the Markov chain (when such a distribution exists). Since the stationary distribution is only achieved after a certain "burn-in" period of time, these methods are here denoted as *asymptotic methods*. This makes a substantial difference with the "exact" or nonasymptotic methods previously studied in this chapter. The main difficulty in Markov chain methods is actually to prove bounds on the burn-in period (or mixing time), after which one has the guarantee that the chain has reached steady state. Some of these issues are discussed in the following sections.

14.4.1 Random walks on graphs

In this section, we discuss discrete distributions obtained as asymptotic distributions (among which the uniform distribution is a particular case) of random walks on the nodes of graphs. The exposition here is based on the surveys [45, 174], to which the reader is referred for a more extensive treatment and pointers to the existing literature.

Let $G(V, E)$ be a connected graph,[2] with vertex (or node) set V of cardinality n, and edge set E of cardinality m. The *degree* $d(v)$ of a vertex $v \in V$ is defined as the number of edges incident on v; if every vertex has degree d, then the graph is said d-regular. We now define a random walk on the nodes of $G(V, E)$, with initial state $\mathbf{v}_0$, as a sequence of random variables $\mathbf{v}_0, \mathbf{v}_1, \dots$ taking values in V, such that for $i, j \in V$ and $t \geq 0$

$$p_{ij} \doteq \mathrm{PR}\left\{\mathbf{v}_{t+1} = j | \mathbf{v}_t = i\right\} = \begin{cases} \frac{1}{d(i)} & \text{if } i, j \in E; \\ 0 & \text{otherwise.} \end{cases} \tag{14.11}$$

This means that, if at the tth step we are at node $v_t = i$, we move to some neighboring node with probability $1/d(i)$. Clearly, the sequence of random nodes $\mathbf{v}_0, \mathbf{v}_1, \dots$ is a Markov chain. Denoting with $\pi(t)$ the probability distribution of the nodes at time t (i.e. $\pi_i(t) \doteq \mathrm{PR}\{\mathbf{v}_t = i\}$, for $i = 1, \dots, n$), we have the Chapman-Kolmogorov recursion

$$\pi_i(t + 1) = P^T \pi_i(t)$$

where P is the transition probability matrix, having p_{ij} in the ith row and j^{th} column. A standard result from the elementary theory of Markov chains is that the chain admits a stationary distribution π, which solves $\pi = P^T \pi$, and

---

[2] A graph is connected if a path exists between all pairs of its vertices.

that this distribution is unique, if the graph is connected (we refer the reader to [190] for an in-depth treatment of the convergence properties of Markov chains). For the assigned transition probabilities, the stationary distribution is given by

$$\pi_i = \frac{d(i)}{2m}, \; i = 1, \ldots, n.$$

If the graph is d-regular, then this stationary distribution is the uniform distribution, since $d(i) = d$, for all nodes. Notice that for regular graphs the corresponding Markov chain is symmetric, meaning that the probability of moving to state j, given that the chain is at node i, is the same as the probability of moving to state i, given that the chain is at node j. Therefore, for regular graphs the transition matrix P is symmetric.

An important property of nonbipartite[3] (but non-necessarily regular) graphs is that the distribution of $\pi(t)$ tends to the stationary distribution as $t \to \infty$, regardless of the initial distribution $\pi(0)$. Therefore, for a d-regular nonbipartite graph with transition probabilities given by (14.11), the distribution $\pi(t)$ of v_t will tend to the uniform distribution $\pi_i = 1/n$, for $i = 1, \ldots, n$.

A natural question at this point is how many steps one has to wait before the distribution of v_t will be close to the stationary one, which leads to the fundamental notion of the *mixing rate* of the Markov chain. For a random walk starting at node i, we can express the deviation from stationarity in terms of the total variation distance $|\pi_j(t) - \pi_j|$. We then define the mixing rate as the smallest constant $\lambda > 0$ such that

$$|\pi_j(t) - \pi_j| \leq \sqrt{\frac{d(j)}{d(i)}} \lambda^t.$$

The following theorem establishes a key result for the mixing rate of a random walk on a graph, see [174] for a proof.

Theorem 14.4. *The mixing rate of a random walk on a nonbipartite graph is given by*

$$\lambda = \max\{|\lambda_2|, |\lambda_n|\}$$

where $\lambda_1 = 1 \geq \lambda_2 \geq \cdots \geq \lambda_n \geq -1$ are the eigenvalues of $M \doteq D^{-1/2} P D^{1/2}$, $D = \mathrm{diag}\left([1/d(1) \cdots 1/d(n)]\right)$.

---

[3] A bipartite graph is a graph whose nodes can be partitioned into two subsets, with no edge joining nodes that are in the same subset. A node in one of the subsets may be joined to all, some, or none of the nodes in the other. A bipartite graph is usually shown with the two subsets as top and bottom rows of nodes, or with the two subsets as left and right columns of nodes.

Very often the matrix M is positive semidefinite (or the graph can be transformed into one for which M is semidefinite, by adding self-loops to the nodes), in which case the mixing rate is simply given by $\lambda = \lambda_2$.

Unfortunately, however, in the applications where the random walk method is of interest, the "size" n of the underlying graph is exponentially large, and the estimation of the mixing rate via eigenvalue computation is hopeless. This is the case for instance in the "card deck shuffling" problem, where we wish to sample uniformly over all permutations of 52 elements, or if we are willing to sample uniformly in a convex body $K \subset \mathbb{R}^n$ (with large n), by defining a random walk on a fine lattice inside K, see for instance [99]. For this reason, a different approach is usually taken in the literature for proving the "fast mixing" property of a random walk. In this approach, further concepts such as the *conductance* of a graph and isoperimetric inequalities are introduced, which would lead us too far away from the scope of this chapter. The interested reader is referred to [45, 144, 174].

Metropolis random walk. The so-called Metropolis random walk [188] is a modification of the simple random walk described in the previous section. This modification is introduced in order to make the random walk converge asymptotically to an arbitrary desired probability distribution. Let $G(V, E)$ be a d-regular graph, let $f : V \to \mathbb{R}_+$, and let $\mathbf{v}_0$ be the initial node. Suppose that at time t we are at node $\mathbf{v}_t$, the modified random walk goes as follows: (i) select a random neighbor $\mathbf{u}$ of $\mathbf{v}_t$. (ii) if $f(u) \geq f(v_t)$, then move to $\mathbf{u}$, i.e. $\mathbf{v}_{t+1} = \mathbf{u}$; else move to $\mathbf{u}$ with probability $f(u)/f(v_t)$, or stay in $\mathbf{v}_t$ with probability $1 - f(u)/f(v_t)$. Clearly, this modified random walk is still a Markov chain. The fundamental fact that can be proven is that this chain admits the stationary distribution

$$\widetilde{\pi}_i = \frac{f(i)}{\sum_{j \in V} f(j)}.$$

Unlike the simple random walk, in general it is difficult to estimate the mixing rate of the Metropolis walk. Relevant results in this direction include [16, 86].

14.4.2 Asymptotic methods for continuous distributions

The idea of constructing random walks on graphs can be extended to the case when the node set of the graph is a "continuous," or dense set. In the following sections, we discuss some well-known methods to generate (asymptotically) random samples distributed according to a desired continuous distribution.

Metropolis–Hastings sampler. Suppose that the problem is to sample from a multivariate probability density $f(x)$, $x \in \mathbb{R}^n$, and we can evaluate $f(x)$ (up to a constant) as a function but have no means to generate a sample directly. As discussed previously, the rejection method is an option if a dominating density can be found with a computable bound η of moderate magnitude. However, this is very unlikely to be the case if the dimension n is high. We remark that rejection sampling, the inversion method, and the density transformation method all produce independent realizations from $f(x)$. If these methods are inefficient or difficult to implement, then we can adopt the weaker goal of generating a *dependent* sequence with marginal distribution equal to (or converging to) $f(x)$, using some variant of the random walk techniques describer earlier. It should be pointed out that this necessitates giving up independence of the samples, since the successive outcomes of a Markov chain (even when the chain is in steady state) are *not* independent.

The Metropolis–Hastings (MH) algorithm [127] builds the random walk (with continuous node space) according to the following rules. Suppose we have a transition density $g(y|x)$, such that

$$\text{PR}\{\mathbf{x}_{t+1} \in \mathcal{Y} | \mathbf{x}_t = x\} = \int_{\mathcal{Y}} g(y|x)\mathrm{d}y.$$

This density is sometimes called a "proposal density" in this context, and plays a role similar to the dominating density in the standard rejection method. Define the Metropolis–Hastings ratio by

$$\eta(x,y) \doteq \begin{cases} \min \left(\dfrac{f(y)g(x|y)}{f(x)g(y|x)}, 1 \right) & \text{if } f(x)g(y|x) > 0; \\ 1 & \text{otherwise.} \end{cases} \tag{14.12}$$

Notice that only density evaluations up to a constant are required, since unknown normalizing constants cancel out when forming the above ratio. Suppose that at time t the state is at $\mathbf{x}_t$, then choose the next state according to the following procedure: (i) draw a sample $\mathbf{y}$ from the proposal density $g(y|x_t)$. (ii) with probability $\eta(x_t, y)$ move to $\mathbf{y}$, i.e. set $\mathbf{x}_{t+1} = \mathbf{y}$, otherwise stay put, i.e. set $\mathbf{x}_{t+1} = \mathbf{x}_t$.

It can be proved that the resulting Markov chain will reach steady state, and that the stationary density is the target density $f(x)$. The analysis of the mixing rate of this algorithm is not, however, an easy task. We refer to [187] for a contribution in this latter direction, and to [119] for further references on variations and refinements of the basic MH algorithm. A recent survey on Metropolis–Hastings sampling and general Markov chain Monte Carlo methods for estimation of expectations is also presented in [251]. The explicit Metropolis–Hastings sampling algorithm is reported next.

Algorithm 14.4 (Metropolis–Hastings).

Given a target probability density $f(x)$, a "proposal density" $g(y|x)$, and a "burn-in" period $T \gg 0$, this algorithm returns a random variable $\mathbf{x}$ that is approximately distributed according to $f(x)$ (the approximating density asymptotically converging to $f(x)$ as T increases).

1. Select an arbitrary initial state $\mathbf{x}_0$, and set $k = 0$;
2. Generate a candidate point $\mathbf{y}$ according to $g(y|x_k)$;
3. Compute the ratio $\eta(x_k, y)$ in (14.12);
4. Generate $\mathbf{w}$ uniformly in $[0, 1]$;
5. If $\mathbf{w} \le \eta(x_k, y)$, set $\mathbf{x}_{k+1} = \mathbf{y}$, else set $\mathbf{x}_{k+1} = \mathbf{x}_k$;
6. Set $k = k + 1$;
7. If $k = T$, return $\mathbf{x}_k$ and end, else goto 2.

Remark 14.3. It is worth underlining that the result of Algorithm 14.4 is one single random sample, approximately distributed according to $f(x)$. Therefore, if another run of the algorithm is executed, and then a third run, and so on, we would obtain a sequence of samples that are all (approximately) distributed according to $f(x)$, and statistically *independent*, since after T steps the Markov chain approximately "forgets" its initial state. However, this procedure would be very inefficient, since at each run the algorithm should execute T burn-in iterations, with T that is typically very large.

Notice that this is not in contradiction with our initial statement, that Markov chain methods produce *dependent* samples, since the actual use of these algorithms is to produce not a single sample at step $k = T$, but the whole sequence of dependent samples that is generated by the chain *from step T on*. This is obtained by substituting step 7 in Algorithm 14.4 with "If $k \ge T$, return $\mathbf{x}_k$, goto 2; else goto 2."   ◁

14.4.3 Uniform sampling in a convex body

Random walk techniques have been successfully adopted to develop polynomial-time algorithms that produce approximately uniform (although not independent) samples in a convex body[4] $K \subset \mathbb{R}^n$. It is further commonly assumed that K contains the unit ball, and that it is contained in a ball of radius r.

In [99] an algorithm that is based on random walk on a sufficiently dense lattice inside K was originally proposed, and it was proved that the walk mixes in time polynomial in n (notice that this is far from obvious, since the number of nodes in the lattice grows exponentially with n). The motivation

---

[4] By convex body, we mean a closed, bounded convex set of nonzero volume.

of [99] was to design a polynomial-time algorithm to approximate the volume of a convex body, which is an "ancient" and extremely difficult problem. Indeed, several negative results on the hardness of this problem have appeared in the literature, stating for instance that any (deterministic) algorithm that approximates the volume within a factor of $n^{o(n)}$ necessarily takes exponential time [23, 153]. Therefore, the result of [99] has a particular importance, since it gives a breakthrough in the opposite direction, showing that randomization provably helps in solving efficiently those problems that are deterministically hard.

A similar method called "walk with ball steps" may be easily described. Let $\mathbf{x} \in K$ be the current point: (i) we generate a random point $\mathbf{z}$ in the Euclidean ball of radius δ centered in $\mathbf{x}$; (ii) if $\mathbf{z} \in K$, we move to $\mathbf{z}$, else we stay at $\mathbf{x}$. This procedure corresponds to a random walk on the graph whose vertex set is K, with two points x, z connected by an edge if and only if $|x-z| \le \delta$. It has been proved proved in [149] that (a slightly modified version of) this walk converges to an approximation of the uniform distribution in K which gets better as δ is reduced, and that the walk mixes in $O(nr^2/\delta^2)$ steps, provided that $\delta < 1/\sqrt{n}$.

Another popular method for generating (asymptotically) uniform samples in K is the so-called hit-and-run (HR) method introduced in [246]. In the HR method, the random walk is constructed as follows: if the current point is $\mathbf{x}$, then we generate the next point by selecting a random line through $\mathbf{x}$ (uniformly over all directions) and choosing the next point uniformly from the segment of the line in K. This walk has the uniform as the stationary distribution, and mixes in $O(n^2r^2)$ steps [175]. An explicit version of the HR algorithm is reported next.

Algorithm 14.5 (Hit-and-run).
Given a convex body $K \subset \mathbb{R}^n$ and a "burn-in" period $T \gg 0$, this algorithm returns a random variable $\mathbf{x}$ that is approximately uniform within K (the approximating density asymptotically converging to the uniform density in K, as T increases).

1. Select an arbitrary initial state $\mathbf{x}_0$ in the interior of K, and set $k = 0$;
2. Generate a random direction $\mathbf{v} = \mathbf{y}/\|\mathbf{y}\|$, where $\mathbf{y} \sim \mathcal{N}_{0,I_n}$;
3. Compute the extreme points $\mathbf{a}, \mathbf{b}$ (on the boundary of K) of the chord in K through $\mathbf{x}_k$ along direction $\mathbf{v}$;
4. Generate $\mathbf{w}$ uniformly in $[0, 1]$;
5. Set $\mathbf{x}_{k+1} = \mathbf{wa} + (1 - \mathbf{w})\mathbf{b}$;
6. Set $k = k + 1$;
7. If $k = T$, return $\mathbf{x}_k$ and end, else goto 2.

15. Statistical Theory of Radial Random Vectors

In this chapter we study the statistical properties of random vectors, for a certain class of symmetric probability distributions. In particular, we introduce the notions of ℓ_p radial and ℓ_2^W radial random vectors, and analyze the properties of these distributions, highlighting the connection among ℓ_p radial symmetry and the uniform distribution in ℓ_p norm balls. These properties will be used in the algorithms for vector sample generation presented in Chapter 16.

15.1 Radially symmetric densities

In this section we introduce a class of probability distributions with symmetry properties with respect to ℓ_p norms, and discuss some preliminary concepts that will be used in later parts of the book. We first formally define the notion of radially symmetric density functions (or simply radial densities).

Definition 15.1 (ℓ_p radial density). *A random vector* $\mathbf{x} \in \mathbb{F}^n$, *where* $\mathbb{F}$ *is either the real or complex field, is* ℓ_p *radial if its density function can be written as*

$$f_{\mathbf{x}}(x) = g(\rho), \ \rho = \|x\|_p$$

where $g(\rho)$ *is called the* defining function *of* $\mathbf{x}$.

In other words, for radially symmetric random vectors, the density function is uniquely determined by its radial shape, which is described by the defining function $g(\rho)$. When needed, we use the notation $g_{\mathbf{x}}(\rho)$ to specify the defining function of the random variable $\mathbf{x}$. For given ρ, the level set of the density function is an equal probability set represented by $\partial \mathcal{B}_{\|\cdot\|_p}(\rho)$. Examples of radial densities for real random vectors are, for instance, the normal density and the Laplace density. This is shown in the next example.

Example 15.1 (Radial densities).
   The classical multivariate normal density with identity covariance matrix and zero mean is an ℓ_2 radial density. In fact, setting $\rho = \|x\|_2$, we write

$$f_{\mathbf{x}}(x) = \frac{1}{\sqrt{(2\pi)^n}} e^{-\frac{1}{2}x^T x} = \frac{1}{\sqrt{(2\pi)^n}} e^{-\frac{1}{2}\rho^2} = g(\rho).$$

The multivariate Laplace density with zero mean is an ℓ_1 radial density. Indeed, setting $\rho = \|x\|_1$, we have

$$f_{\mathbf{x}}(x) = \frac{1}{2^n} e^{-\sum_{i=1}^n |x_i|} = \frac{1}{2^n} e^{-\rho} = g(\rho).$$

Furthermore, the generalized Gamma density, defined in (2.11), is radial with respect to the ℓ_p norm. The properties of this density are discussed in more detail in Section 16.2.

Another notable case of radial density is the uniform density over $\mathcal{B}_{\|\cdot\|_p}$. In fact, for a random vector $\mathbf{x}$ uniformly distributed in the ℓ_p norm ball of unit radius we have that

$$f_{\mathbf{x}}(x) = \mathcal{U}_{\mathcal{B}_{\|\cdot\|_p}}(x) = \begin{cases} \dfrac{1}{\mathrm{Vol}(\mathcal{B}_{\|\cdot\|_p})} & \text{if } \|x\|_p \le 1; \\ 0 & \text{otherwise.} \end{cases}$$

Clearly, this density can be expressed as a function of $\rho = \|x\|_p$, i.e. $f_{\mathbf{x}}(x) = g(\rho)$, with defining function

$$g(\rho) = \begin{cases} \dfrac{1}{\mathrm{Vol}(\mathcal{B}_{\|\cdot\|_p})} & \text{if } \rho \le 1; \\ 0 & \text{otherwise.} \end{cases} \tag{15.1}$$

$\star$

15.2 Statistical properties of ℓ_p radial real vectors

The next theorem provides a fundamental characterization of ℓ_p radial real random vectors, and relates them to the uniform distribution on the surface of the ℓ_p norm ball.

Theorem 15.1 (ℓ_p radial vectors in $\mathbb{R}^n$). *Let $\mathbf{x} \in \mathbb{R}^n$ be factored as $\mathbf{x} = \rho\mathbf{u}$, where $\rho > 0$, and $\mathbf{u} \in \mathbb{R}^n$, $\|\mathbf{u}\|_p = 1$. The following statements are equivalent:*

  1. *$\mathbf{x}$ is ℓ_p radial with defining function $g(\rho)$;*
  2. *ρ, $\mathbf{u}$ are independent, and $\mathbf{u}$ is uniformly distributed on $\partial\mathcal{B}_{\|\cdot\|_p}$.*

Moreover, we have that

$$f_\rho(\rho) = \frac{2^n \Gamma^n(1/p)}{p^{n-1}\Gamma(n/p)} g(\rho)\rho^{n-1};$$ (15.2)

$$f_{\mathbf{u}_1,\ldots,\mathbf{u}_{n-1}}(u_1,\ldots,u_{n-1}) = \frac{p^{n-1}\Gamma(n/p)}{2^{n-1}\Gamma^n(1/p)} \psi^{(1-p)/p}(u_1,\ldots,u_{n-1})$$ (15.3)

$$|u_i| < 1, \; i = 1,\ldots,n-1$$

$$\sum_{i=1}^{n-1} |u_i|^p < 1$$

where

$$\psi(u_1,\ldots,u_{n-1}) \doteq \left(1 - \sum_{i=1}^{n-1} |u_i|^p\right).$$ (15.4)

Proof. Consider $x \in \mathbb{R}^n$. If we assume $x_n > 0$, then x can be factored as $x = \rho u$, where ρ is positive, and the vector u lies on the unit surface $\partial \mathcal{B}_{\|\cdot\|_p}$ and is parameterized as

$$u = \left[u_1 \; u_2 \; \cdots \; u_{n-1} \; \left(1 - \sum_{i=1}^{n-1} |u_i|^p\right)^{1/p}\right]^T.$$

The transformation

$$x_i = \rho u_i, \; i = 1,\ldots,n-1;$$

$$x_n = \rho\left(1 - \sum_{i=1}^{n-1} |u_i|^p\right)^{1/p}, \; \rho > 0, \; |u_i| < 1, \; \sum_{i=1}^{n-1} |u_i|^p < 1$$

from the variables $x_1,\ldots,x_n$, $x_n > 0$, to the variables $u_1,\ldots,u_{n-1},\rho$ is then one-to-one. The Jacobian of this transformation is

$$J(x \to u,\rho) = \rho^{n-1}\left(1 - \sum_{i=1}^{n-1} |u_i|^p\right)^{(1-p)/p}.$$ (15.5)

Notice that, if we set $x_n < 0$, similar computations lead to the same Jacobian. Furthermore, the case $x_n = 0$ is neglected, since the event $\mathbf{x}_n = 0$ occurs with zero probability. For each of the two half-spaces (for $x_n > 0$ and for $x_n < 0$), the density $f_{\mathbf{x}}(x)$ restricted to the half-space is given by $2g(\rho)$, since $g(\rho)$ represents the density on the whole space. From the rule of change of variables, see Theorem 14.2, we then have

$$f_{\mathbf{u},\mathbf{r}}(u_1,\ldots,u_{n-1},\rho) = 2g(\rho)\rho^{n-1}\left(1 - \sum_{i=1}^{n-1} |u_i|^p\right)^{(1-p)/p}.$$ (15.6)

Then, it follows that $u_1,\ldots,u_{n-1}$ are independent from ρ. Since, see for instance [249]

$$\int_{\mathcal{D}} \left(1 - \sum_{i=1}^{n-1} |u_i|^p\right)^{(1-p)/p} du_1 \cdots du_{n-1} = \frac{2^{n-1}\Gamma^n(1/p)}{p^{n-1}\Gamma(n/p)} \tag{15.7}$$

where $\mathcal{D} = \{-1 < u_i < 1, i = 1,\ldots, n-1; \sum_{i=1}^{n-1} |u_i|^p < 1\}$, we obtain the marginal densities (15.2) and (15.3) by integration of (15.6) with respect to $u_1, \ldots, u_{n-1}$ and ρ respectively.   □

Remark 15.1 (Uniform density on $\partial\mathcal{B}_{\|\cdot\|_p}(\mathbb{R}^n)$). The density (15.3) is called the "ℓ_p norm uniform distribution" in [125, 249]. The interested reader is referred to those studies for further details on ℓ_p radial distributions. The pdf (15.3) is therefore a representation of the uniform distribution on the surface of the unit sphere $\partial\mathcal{B}_{\|\cdot\|_p}(\mathbb{R}^n)$ and is denoted by $\mathcal{U}_{\partial\mathcal{B}_{\|\cdot\|_p}(\mathbb{R}^n)}$. We also notice that Theorem 15.1 can be stated in the following alternative form: if $\mathbf{x} \in \mathbb{R}^n$ is ℓ_p radial, then the random vector $\mathbf{u} = \mathbf{x}/\|\mathbf{x}\|_p$ is uniformly distributed on $\partial\mathcal{B}_{\|\cdot\|_p}$, and $\mathbf{u}$ and $\|\mathbf{x}\|_p$ are independent.   ◁

Remark 15.2 (Uniform density on $\partial\mathcal{B}_{\|\cdot\|_2}(\mathbb{R}^n)$). For the case $p = 2$, a commonly used parameterization of u is obtained by means of the generalized polar coordinates

$$u_1 = \sin\phi_1;$$
$$u_2 = \cos\phi_1 \sin\phi_2;$$
$$u_3 = \cos\phi_1 \cos\phi_2 \sin\phi_3;$$
$$\vdots$$
$$u_{n-1} = \cos\phi_1 \cos\phi_2 \cdots \cos\phi_{n-2} \sin\phi_{n-1};$$
$$u_n = \pm\cos\phi_1 \cos\phi_2 \cdots \cos\phi_{n-2} \cos\phi_{n-1}$$

where $-\pi/2 < \phi_i \leq \pi/2$, for $i = 1,\ldots, n-1$. It is easy to verify that $u_n^2 = 1 - \sum_{i=1}^{n-1} u_i^2$. The Jacobian of the transformation from $u_1,\ldots, u_{n-1}$ to $\phi_1,\ldots, \phi_{n-1}$ is, see for instance [12]

$$J(u_1,\ldots, u_{n-1} \to \phi_1,\ldots, \phi_{n-1}) = \cos^{n-1}\phi_1 \cos^{n-2}\phi_2 \cdots \cos\phi_{n-1}.$$

In these coordinates, the uniform density on $\partial\mathcal{B}_{\|\cdot\|_2}$ is

$$f_{\phi_1,\ldots,\phi_{n-1}}(\phi_1,\ldots, \phi_{n-1}) = \frac{\Gamma(n/2)}{\Gamma^n(1/2)} \cos^{n-2}\phi_1 \cos^{n-3}\phi_2 \cdots \cos\phi_{n-2}.$$

◁

Remark 15.3 (ℓ_p norm density and volume of $\mathcal{B}_{\|\cdot\|_p}(r,\mathbb{R}^n)$). The *norm density* of an ℓ_p radial random vector $\mathbf{x} \in \mathbb{R}^n$ is defined as the probability density of its norm $\rho = \|\mathbf{x}\|_p$. The norm density of $\mathbf{x}$ is explicitly given in (15.2).

Notice that if $\mathbf{x}$ is uniformly distributed in $\mathcal{B}_{\|\cdot\|_p}(r,\mathbb{R}^n)$, then its defining function is given by (15.1). Therefore, substituting this $g(\rho)$ into (15.2) and integrating for ρ from 0 to r, we obtain a closed-form expression for the volume of the ℓ_p norm ball of radius r in $\mathbb{R}^n$

$$\text{Vol}\big(\mathcal{B}_{\|\cdot\|_p}(r,\mathbb{R}^n)\big) = 2^n \frac{\Gamma^n(1/p+1)}{\Gamma(n/p+1)} r^n. \tag{15.8}$$

$\triangleleft$

In the next section we study ℓ_p radial complex vectors.

15.3 Statistical properties of ℓ_p radial complex vectors

We now present an analogous result to Theorem 15.1 for the case of complex ℓ_p radial vectors.

Theorem 15.2 (ℓ_p radial vectors in $\mathbb{C}^n$). *Let $\mathbf{x} \in \mathbb{C}^n$ be factored as $\mathbf{x} = \rho\mathbf{v}$, where $\rho > 0$, $\mathbf{v} = e^{j\boldsymbol{\Phi}}\mathbf{u}$, with $\mathbf{u} \in \mathbb{R}_+^n$, $\|\mathbf{u}\|_p = 1$, $\boldsymbol{\Phi} = \text{diag}\,([\varphi_1 \cdots \varphi_n])$, $\varphi_i \in [0, 2\pi]$. The following statements are equivalent:*

1. $\mathbf{x}$ *is ℓ_p radial with defining function $g(\rho)$;*
2. ρ, $\mathbf{u}$, $\boldsymbol{\Phi}$ *are independent, and $\mathbf{v}$ is uniformly distributed on $\partial\mathcal{B}_{\|\cdot\|_p}$.*

Moreover, we have that

$$f_\rho(\rho) = \frac{(2\pi)^n \Gamma^n(2/p)}{p^{n-1}\Gamma(2n/p)} g(\rho)\rho^{2n-1}; \tag{15.9}$$

$$f_{\mathbf{u}_1,\ldots,\mathbf{u}_{n-1}}(u_1,\ldots,u_{n-1}) = \frac{p^{n-1}\Gamma(2n/p)}{\Gamma^n(2/p)}\tilde{\psi}(u_1,\ldots,u_{n-1}); \tag{15.10}$$

$$0 < u_i < 1,\, i = 1,\ldots,n-1$$

$$\sum_{i=1}^{n-1} |u_i|^p < 1$$

$$f_{\boldsymbol{\Phi}}(\varphi_1,\ldots,\varphi_n) = \frac{1}{(2\pi)^n} \tag{15.11}$$

where $\tilde{\psi}$ is defined as

$$\tilde{\psi}(u_1,\ldots,u_{n-1}) \doteq \left(\prod_{i=1}^{n-1} u_i\right)\psi^{2/p-1}(u_1,\ldots,u_{n-1})$$

and

$$\psi(u_1,\ldots,u_{n-1}) = \left(1 - \sum_{i=1}^{n-1} |u_i|^p\right).$$

Proof. Observe that, owing to the norm constraint, the vector u has $n-1$ free components, and it is expressed as $u = [u_1 \cdots u_{n-1}\, \psi^{1/p}(u_1, \ldots, u_{n-1})]^T$, $u_i \in [0, 1]$, $i = 1, \ldots, n-1$. Let $a_i = \mathrm{Re}(x_i)$, $b_i = \mathrm{Im}(x_i)$, then the change of variables $a_i + jb_i = \rho e^{j\varphi_i} u_i$ is one-to-one, and is explicitly written as

$$
\begin{aligned}
a_1 &= \rho u_1 \cos \varphi_1; \\
b_1 &= \rho u_1 \sin \varphi_1; \\
a_2 &= \rho u_2 \cos \varphi_2; \\
b_2 &= \rho u_2 \sin \varphi_2; \\
&\;\;\vdots \\
a_{n-1} &= \rho u_{n-1} \cos \varphi_{n-1}; \\
b_{n-1} &= \rho u_{n-1} \sin \varphi_{n-1}; \\
a_n &= \rho \psi^{1/p}(u_1, \ldots, u_{n-1}) \cos \varphi_n; \\
b_n &= \rho \psi^{1/p}(u_1, \ldots, u_{n-1}) \sin \varphi_n.
\end{aligned}
$$

To compute the Jacobian of this transformation, we construct the following table of derivatives, where $S\varphi_i$ and $C\varphi_i$ stand for $\sin \varphi_i$ and $\cos \varphi_i$.

	a_1	b_1	a_2	b_2	$\cdots$	a_{n-1}	b_{n-1}	a_n	b_n
φ_1	$-\rho u_1 S\varphi_1$	$\rho u_1 C\varphi_1$	0	0	$\cdots$	0	0	0	0
u_1	$\rho C\varphi_1$	$\rho S\varphi_1$	0	0	$\cdots$	0	0	$u_1^{p-1} C\varphi_n$	$u_1^{p-1} S\varphi_n$
φ_2	0	0	$-\rho u_2 S\varphi_2$	$\rho u_2 C\varphi_2$	$\cdots$	0	0	0	0
u_2	0	0	$\rho C\varphi_2$	$\rho S\varphi_2$	$\cdots$	0	0	$u_2^{p-1} C\varphi_n$	$u_2^{p-1} S\varphi_n$
$\vdots$	$\vdots$	$\vdots$	$\vdots$	$\vdots$	$\ddots$	$\vdots$	$\vdots$	$\vdots$	$\vdots$
φ_{n-1}	0	0	0	0	$\cdots$	$-\rho u_{n-1} S\varphi_{n-1}$	$\rho u_{n-1} C\varphi_{n-1}$	0	0
u_{n-1}	0	0	0	0	$\cdots$	$\rho C\varphi_{n-1}$	$\rho S\varphi_{n-1}$	$u_{n-1}^{p-1} C\varphi_n$	$u_{n-1}^{p-1} S\varphi_n$
φ_n	0	0	0	0	$\cdots$	0	0	$-\rho S\varphi_n$	$\rho C\varphi_n$
ρ	$u_1 C\varphi_1$	$u_1 S\varphi_1$	$u_2 C\varphi_2$	$u_2 S\varphi_2$	$\cdots$	$u_{n-1} C\varphi_{n-1}$	$u_{n-1} S\varphi_{n-1}$	$C\varphi_n$	$S\varphi_n$

Using the Schur determinant rule, and exploiting the block diagonal structure of the $2(n-1) \times 2(n-1)$ upper-left block in the above table, we obtain

$$
J(x \to \Phi, u, \rho) = \rho^{2n-1} \left(\prod_{i=1}^{n-1} u_i \right) \psi^{2/p-1}(u_1, \ldots, u_{n-1}).
$$

Therefore, we have that

$$
f_{\Phi, u, \rho}(\Phi, u, \rho) = g(\rho) J(x \to \Phi, u, \rho) = \rho^{2n-1} g(\rho) \tilde{\psi}(u_1, \ldots, u_{n-1}) \quad (15.12)
$$

which proves the statistical independence of Φ, u, ρ. The marginal density $f_\rho(\rho)$ in (15.9) is then obtained by integrating this joint pdf over Φ and u, and using the following facts

$$
\int_0^{2\pi} d\varphi_1 \cdots d\varphi_n = (2\pi)^n;
$$

$$
\int_D \tilde{\psi}(u_1, \ldots, u_{n-1}) du_1 \cdots du_{n-1} = \frac{\Gamma^n(2/p)}{p^{n-1} \Gamma(2n/p)}
$$

where $\mathcal{D} = \{0 < u_i < 1, i = 1, \ldots, n-1; \sum_{i=1}^{n-1} |u_i|^p < 1\}$. The marginal densities (15.10) and (15.11) are obtained in a similar way, integrating with respect to Φ, ρ and with respect to u, ρ respectively. Integrating (15.12) with respect to ρ, we obtain the pdf of $\mathbf{v}$, which is defined as the uniform density over $\partial \mathcal{B}_{\|\cdot\|_p}$, see Remark 15.4.   □

Remark 15.4 (Uniform pdf on $\partial \mathcal{B}_{\|\cdot\|_p}(\mathbb{C}^n)$). Similar to the development for the real case in [63, 108, 249], the marginal density $f_{\Phi,\mathbf{u}}(\Phi, u)$, obtained by integrating (15.12) with respect to ρ, is defined as the "(complex) ℓ_p norm uniform distribution" on the surface of the unit sphere $\partial \mathcal{B}_{\|\cdot\|_p}(\mathbb{C}^n)$, and it is denoted by $\mathcal{U}_{\partial \mathcal{B}_{\|\cdot\|_p}(\mathbb{C}^n)}$. In addition, notice that Theorem 15.2 could be stated in the following form: if $\mathbf{x} \in \mathbb{C}^n$ is ℓ_p radial, then $\mathbf{v} = \mathbf{x}/\|\mathbf{x}\|_p$ is uniformly distributed on $\partial \mathcal{B}_{\|\cdot\|_p}$, and $\mathbf{v}$ and $\|\mathbf{x}\|_p$ are independent.   ◁

Remark 15.5 (ℓ_p norm density and volume of $\mathcal{B}_{\|\cdot\|_p}(r, \mathbb{C}^n)$). The norm density (see Remark 15.3 for its definition) of an ℓ_p radial random vector $\mathbf{x} \in \mathbb{C}^n$ is explicitly given in (15.9). We also notice that if $\mathbf{x}$ is uniformly distributed in $\mathcal{B}_{\|\cdot\|_p}(r, \mathbb{C}^n)$, its defining function is given by (15.1), and hence substituting this $g(\rho)$ into (15.9) and integrating for ρ from 0 to r, we obtain a closed-form expression for the volume of the ℓ_p norm ball of radius r in $\mathbb{C}^n$

$$\text{Vol}\big(\mathcal{B}_{\|\cdot\|_p}(r, \mathbb{C}^n)\big) = \pi^n \frac{\Gamma^n(2/p+1)}{\Gamma(2n/p+1)} r^{2n}. \tag{15.13}$$

◁

Finally, we observe that the expressions for the norm density (15.9) and (15.2) can be unified in a single formula that is valid for both the real and complex cases. This is stated in the next lemma.

Lemma 15.1 (ℓ_p norm density). *If $\mathbf{x} \in \mathbb{F}^n$ is ℓ_p radial, then the random variable $\rho = \|\mathbf{x}\|_p$ has density function $f_\rho(\rho)$ given by*

$$f_\rho(\rho) = \text{Vol}\big(\mathcal{B}_{\|\cdot\|_p}\big) d\rho^{d-1} g(\rho) \tag{15.14}$$

where $d = n$ if $\mathbb{F} \equiv \mathbb{R}$ or $d = 2n$ if $\mathbb{F} \equiv \mathbb{C}$.

15.4 ℓ_p radial vectors and uniform distribution in $\mathcal{B}_{\|\cdot\|_p}$

The results of the previous sections provide a connection between ℓ_p radial distributions and uniform distributions within ℓ_p norm balls, for real and complex random vectors. This connection is analyzed in [60] and it is stated in the next corollary.

Corollary 15.1 (Uniform vectors in $\mathcal{B}_{\|\cdot\|_p}$). *Let $d = n$ if $\mathbb{F}$ is the real field, and $d = 2n$ if $\mathbb{F}$ is the complex field. The following two conditions are equivalent:*

1. $\mathbf{x} \in \mathbb{F}^n$ is ℓ_p radial, with norm density function $f_\rho(\rho) = \rho^{d-1}d$, $\rho \in [0,1]$;
2. $\mathbf{x} \in \mathbb{F}^n$ is uniformly distributed in $\mathcal{B}_{\|\cdot\|_p}$.

Proof.
1.→ 2. Since $\mathbf{x}$ is ℓ_p radial, its norm density is given by (15.14), then the defining function of $\mathbf{x}$ is

$$f_\mathbf{x}(x) = g(\|x\|_p) = \frac{1}{\text{Vol}(\mathcal{B}_{\|\cdot\|_p})}, \quad \|x\|_p \le 1$$

which implies that the pdf of $\mathbf{x}$ is constant on its domain, i.e. $\mathbf{x}$ is uniformly distributed in $\mathcal{B}_{\|\cdot\|_p}$.
2.→1. Since $\mathbf{x}$ is uniform in $\mathcal{B}_{\|\cdot\|_p}$ then

$$f_\mathbf{x}(x) = \begin{cases} \dfrac{1}{\text{Vol}(\mathcal{B}_{\|\cdot\|_p})} & \text{if } \|x\|_p \le 1; \\ 0 & \text{otherwise.} \end{cases}$$

Notice that $f_\mathbf{x}(x)$ depends only on $\|x\|_p$. Therefore, $\mathbf{x}$ is ℓ_p radial, with defining function $g(\rho) = f_\mathbf{x}(x)$, $\rho = \|x\|_p$. Substituting $g(\rho)$ in (15.14), we obtain the norm density as claimed. □

The next corollary shows how to obtain an ℓ_p radial distribution with given defining function, or with given norm density, starting from any arbitrary ℓ_p radial distribution.

Corollary 15.2. *Let $\mathbf{x} \in \mathbb{F}^n$ be ℓ_p radial and let $\mathbf{z} \in \mathbb{R}^+$ be an independent random variable with density $f_\mathbf{z}(z)$. Then, the random vector*

$$\mathbf{y} = \mathbf{z} \frac{\mathbf{x}}{\|\mathbf{x}\|_p}$$

is ℓ_p radial with norm density function $f_\rho(\rho) = f_\mathbf{z}(\rho)$, $\rho = \|y\|_p$. Moreover, the defining function of y, $g_\mathbf{y}(\rho)$, is given by

$$g_\mathbf{y}(\rho) = \frac{1}{\text{Vol}(\mathcal{B}_{\|\cdot\|_p})\rho^{d-1}d} f_\rho(\rho). \tag{15.15}$$

Proof. Clearly, $\mathbf{y}$ is ℓ_p radial, and $\|\mathbf{y}\|_p = \mathbf{z}$. Therefore, the norm density function $f_\rho(\rho)$ of $\mathbf{y}$ coincides with the density function $f_\mathbf{z}(z)$ of $\mathbf{z}$. Relation (15.15) follows immediately from (15.14). □

The previous corollary may be used to generate ℓ_p radial random vectors with a given defining function. In the next corollary, we specialize this result to uniform distributions.

Corollary 15.3 (Uniform vectors in $\mathcal{B}_{\|\cdot\|_p}(r, \mathbb{F}^n)$). *Let $\mathbf{x} \in \mathbb{F}^n$ be ℓ_p radial and let $\mathbf{w} \in \mathbb{R}$ be an independent random variable uniformly distributed in $[0, 1]$. Then*

$$\mathbf{y} = r\mathbf{z}\frac{\mathbf{x}}{\|\mathbf{x}\|_p}, \quad \mathbf{z} = \mathbf{w}^{1/d}$$

where $d = n$ if $\mathbb{F} \equiv \mathbb{R}$ or $d = 2n$ if $\mathbb{F} \equiv \mathbb{C}$, is uniformly distributed in $\mathcal{B}_{\|\cdot\|_p}(r)$.

Proof. By the inversion method, see Corollary 14.1, it follows that the distribution of $\mathbf{z}$ is $F_{\mathbf{z}}(z) = z^d$, therefore $f_{\mathbf{z}}(z) = dz^{d-1}$. For $r = 1$, the statement is proved by means of Theorem 15.1. With a rescaling, it is immediate to show that $\mathbf{y}$ is uniformly distributed in $\mathcal{B}_{\|\cdot\|_p}(r)$.   □

This result can be interpreted as follows: first, an ℓ_p radial random vector $\mathbf{x}$ is normalized to obtain a uniform distribution on the surface $\partial\mathcal{B}_{\|\cdot\|_p}(r)$ of the set $\mathcal{B}_{\|\cdot\|_p}(r)$, then each sample is projected into $\mathcal{B}_{\|\cdot\|_p}(r)$ by the random *volumetric factor* $\mathbf{z}$. Therefore, the problem of uniform generation is reduced to that of generation of ℓ_p radially symmetric random vectors. This is discussed in more detail in Chapter 16, where explicit algorithms for uniform generation are provided.

15.5 Statistical properties of ℓ_2^W radial vectors

In this section we introduce the notion of real random vectors with ℓ_2^W radial distribution (or ℓ_2^W radial vectors) and discuss their statistical properties. Then, we study how the uniform distribution on the unit ball in the ℓ_2 norm is transformed by a linear mapping. For further details and references on distributions with ellipsoidal support, the reader is referred to [124].

We recall that the ℓ_2^W norm of a vector $x \in \mathbb{R}^n$ is defined in (3.4) as

$$\|x\|_2^W = \left(x^T W^{-1} x\right)^{1/2}$$

where $W \succ 0$ is a given weighting matrix. We first introduce the notion of ℓ_2^W radially symmetric density.

Definition 15.2 (ℓ_2^W radial density). *A random vector $\mathbf{x} \in \mathbb{R}^n$ is ℓ_2^W radial if its probability density function $f_{\mathbf{x}}(x)$ can be expressed in the form*

$$f_{\mathbf{x}}(x) = g(\rho), \quad \rho = \|x\|_2^W.$$

Example 15.2 (Normal density as ℓ_2^W radial density).
The multivariate normal density $\mathcal{N}_{0,W}$ is ℓ_2^W radial since

$$\mathcal{N}_{0,W}(x) = (2\pi)^{-n/2}|W|^{-1/2}e^{-\frac{1}{2}g(\rho)}, \quad \rho = \|x\|_2^W.$$

★

We now state a well-known result, see e.g. [12], regarding the linear mapping of a vector with multivariate normal density.

Lemma 15.2. *Let* $\mathbf{x} \in \mathbb{R}^n \sim \mathcal{N}_{\bar{x},W}$, *and let* $\mathbf{y} = T\mathbf{x}$, *where* $\mathbf{y} \in \mathbb{R}^m$, *and* T *may be rank deficient. Then,* $\mathbf{y}$ *is also normally distributed, with* $E(\mathbf{y}) = T\bar{x}$ *and* $\mathrm{Cov}(\mathbf{y}) = TWT^T$.

The following lemma states the relation between linear transformations of ℓ_2 radial vectors and ℓ_2^W radial vectors.

Lemma 15.3. *Let* $\mathbf{x} \in \mathbb{R}^n$ *be* ℓ_2 *radial with defining function* $g_{\mathbf{x}}(\|x\|_2)$, *and let* $\mathbf{y} = T\mathbf{x}$, *with* $T \in \mathbb{R}^{n,n}$ *invertible. Then,* $\mathbf{y}$ *is* ℓ_2^W *radial with* $W = TT^T$, *and has defining function*

$$g_{\mathbf{y}}(\rho) = |W|^{-1/2}g_{\mathbf{x}}(\rho), \quad \rho = \|y\|_2^W.$$

Proof. By assumption, the random vector $\mathbf{x}$ is ℓ_2 radial, therefore $f_{\mathbf{x}}(x) = g_{\mathbf{x}}(\|x\|_2)$. Since $x = T^{-1}y$, from the rule of change of variables, see Theorem 14.2, we have

$$f_{\mathbf{y}}(y) = f_{\mathbf{x}}(T^{-1}y)J(x \to y)$$

where $J(x \to y) = |T|^{-1}$. Letting $W = TT^T$, this expression becomes

$$f_{\mathbf{y}}(y) = g_{\mathbf{x}}(\|T^{-1}y\|_2)|W|^{-1/2} = g_{\mathbf{x}}(\rho)|W|^{-1/2}$$

for $\rho = (y^T(TT^T)^{-1}y)^{1/2}$. □

An immediate consequence of this lemma is that if $\mathbf{x}$ is uniform in $\mathcal{B}_{\|\cdot\|_2}$, then the vector $\mathbf{y} = T\mathbf{x}$, with T invertible, is uniformly distributed inside the ellipsoid $\mathcal{E}(0, TT^T)$. The subsequent result states the relation between the defining function $g_{\mathbf{y}}(\rho)$ of an ℓ_2^W radial vector, and its ℓ_2^W norm density $f_{\boldsymbol{\rho}}(\rho)$, defined as the pdf of the random variable $\boldsymbol{\rho} = \|\mathbf{y}\|_2^W$. This result is analogous to Lemma 15.1, which was stated for ℓ_p radial densities.

Lemma 15.4 (ℓ_2^W **norm density**). *Let* $\mathbf{y} \in \mathbb{R}^n$ *be* ℓ_2^W *radial with defining function* $g_{\mathbf{y}}(\rho)$, *where* $\rho = \|y\|_2^W$. *Then, the pdf of* $\boldsymbol{\rho}$ *is called the* ℓ_2^W *norm density and is given by*

$$f_{\boldsymbol{\rho}}(\rho) = |W|^{1/2}\mathrm{Vol}(\mathcal{B}_{\|\cdot\|_2})\, n\rho^{n-1}g_{\mathbf{y}}(\rho). \tag{15.16}$$

Proof. We first notice that any ℓ_2^W radial vector $\mathbf{y}$ can be factored as $\mathbf{y} = Tr\mathbf{u}$, where T is such that $W = TT^T$, $\rho > 0$, and $\|\mathbf{u}\|_2 = 1$. We then compute the joint pdf in the new variables $\mathbf{u}, \rho$, in a way similar to Theorem 15.1

$$f_{\mathbf{u},\boldsymbol{\rho}}(u_1, \ldots, u_{n-1}, \rho) = 2f_{\mathbf{y}}(y)J(y \to u, \rho).$$

Now, introducing the slack variable $x = \rho u$, and applying the chain rule for the computation of Jacobians, see Rule A.1 in Appendix A.2, we have that

$$J(y \to u, \rho) = J(y \to x)J(x \to u, \rho).$$

The first factor is equal to $J(y \to x) = |T| = |W|^{1/2}$, since $y = Tx$. The second factor has been computed in Theorem 15.1, and is equal to

$$J(x \to u, \rho) = \rho^{n-1}\psi^{-1/2}(u_1, \ldots, u_{n-1})$$

where ψ is defined in (15.4). Therefore

$$f_{\mathbf{u},\mathbf{r}}(u_1, \ldots, u_{n-1}, \rho) = 2|W|^{1/2}\rho^{n-1}g_{\mathbf{y}}(\rho)\psi^{-1/2}(u_1, \ldots, u_{n-1})$$

where $\rho = \|y\|_2^W$. Integrating over $u_1, \ldots, u_{n-1}$, and using (15.7), we obtain

$$f_\rho(\rho) = 2|W|^{1/2}\frac{\pi^{n/2}}{\Gamma(n/2)}\rho^{n-1}g_{\mathbf{y}}(\rho).$$

The statement of the lemma then follows noticing that

$$\frac{2\pi^{n/2}}{\Gamma(n/2)} = n\mathrm{Vol}(\mathcal{B}_{\|\cdot\|_2}).$$

$\square$

We now extend the previous results to the case when the transformation matrix T is rectangular. In particular, we address the problem of determining the distribution on the image of a linear transformation, when the distribution on the domain is uniform. More precisely, given a random variable $\mathbf{x} \in \mathbb{R}^n \sim \mathcal{U}_{\mathcal{B}_{\|\cdot\|_2}}$, and given a full-rank matrix $T \in \mathbb{R}^{m,n}$, $m \leq n$, we derive the distribution of $\mathbf{y} = T\mathbf{x}$. Of course, when $m = n$ and $|T| \neq 0$, the answer follows from Lemma 15.3, i.e. a nonsingular linear mapping transforms uniform distributions into uniform distributions. The general case when $m < n$ is addressed in the next lemma.

Lemma 15.5. *Let* $\mathbf{x} \in \mathbb{R}^n$ *be* ℓ_2 *radial with defining function* $g_{\mathbf{x}}(\|x\|_2)$, *and let* $T \in \mathbb{R}^{m,n}$ *be full-rank, with* $m < n$. *Then, the random vector* $\mathbf{y} = T\mathbf{x}$ *is* ℓ_2^W *radial with* $W = TT^T$, *and in particular the pdf* $f_{\mathbf{y}}(y)$ *is given by*

$$f_{\mathbf{y}}(y) = g_{\mathbf{y}}(\rho) = |\Sigma|^{-1}\mathrm{Surf}\big(\mathcal{B}_{\|\cdot\|_2}(1, \mathbb{R}^{n-m})\big)$$

$$\int_0^\infty g_{\mathbf{x}}((\tilde{\rho}^2 + \rho^2)^{1/2})\tilde{\rho}^{n-m-1}d\tilde{\rho}, \quad \rho = \|y\|_2^W \quad (15.17)$$

where Σ *is a diagonal matrix containing the singular values of* T.

Proof. Consider the singular value decomposition $T = U\Sigma V^T$, where $U \in \mathbb{R}^{m,m}$, $V \in \mathbb{R}^{n,m}$ are orthogonal and $\Sigma \in \mathbb{R}^{m,m}$ is the diagonal matrix of the singular values of T. Take $\widetilde{V} \in \mathbb{R}^{n,n-m}$, such that $\widetilde{V}^T\widetilde{V} = I_{n-m}$, $V^T\widetilde{V} = 0_{m,n-m}$, and define

$$\widetilde{T} = \begin{bmatrix} T \\ \widetilde{V}^T \end{bmatrix} \in \mathbb{R}^{n,n}.$$

Then, $\widetilde{T}$ is invertible, and $\widetilde{T}^{-1} = [V\Sigma^{-1}U^T \ \widetilde{V}], |\widetilde{T}| = |\Sigma|$. Next, consider the change of variables $w = \widetilde{T}x$, where $w^T = [y^T \ \widetilde{y}^T]$. Hence, it follows that

$$f_\mathbf{w}(w) \doteq f_{\mathbf{y},\widetilde{\mathbf{y}}}(y,\widetilde{y}) = f_\mathbf{x}(\widetilde{T}^{-1}w)J(x \to w) = g_\mathbf{x}(\|\widetilde{T}^{-1}w\|_2)|\widetilde{T}|^{-1}.$$

Since $\|\widetilde{T}^{-1}w\|_2^2 = \widetilde{y}^T\widetilde{y} + y^T(TT^T)^{-1}y$, we have that

$$f_{\mathbf{y},\widetilde{\mathbf{y}}}(y,\widetilde{y}) = g_\mathbf{x}((\widetilde{\rho}^2 + \rho^2)^{1/2})|\Sigma|^{-1}$$

where $\widetilde{\rho} = \|\widetilde{y}\|_2$, and $\rho = \|y\|_2^W$, with $W = TT^T$. The marginal density $f_\mathbf{y}(y)$ can be derived by integrating $f_{\mathbf{y},\widetilde{\mathbf{y}}}(y,\widetilde{y})$ over $\widetilde{y}$. This integration can be performed using a radial element of volume, obtaining

$$f_\mathbf{y}(y) = |\Sigma|^{-1}\mathrm{Surf}(\mathcal{B}_{\|\cdot\|_2}(\mathbb{R}^{n-m}))\int_0^\infty g_\mathbf{x}((\widetilde{\rho}^2 + \rho^2)^{1/2})\widetilde{\rho}^{n-m-1}\mathrm{d}\widetilde{\rho} = g_\mathbf{y}(\rho)$$

which proves that $\mathbf{y}$ is ℓ_2^W radial with $W = TT^T$.   □

We now explicitly determine the distribution of $\mathbf{y}$, under the assumption that $\mathbf{x}$ is uniformly distributed in the unit ball.

Theorem 15.3. *Let $\mathbf{x} \in \mathbb{R}^n \sim \mathcal{U}_{\mathcal{B}_{\|\cdot\|_2}}$ and let $T \in \mathbb{R}^{m,n}$ be full-rank, with $m < n$. Then, the pdf of the random vector $\mathbf{y} = T\mathbf{x}$ is given by*

$$f_\mathbf{y}(y) = |W|^{-1/2}\frac{\Gamma(n/2 + 1)}{\Gamma^m(1/2)\Gamma((n-m)/2 + 1)}(1 - \rho^2)^{(n-m)/2} \qquad (15.18)$$

where $\rho = \|y\|_2^W$, with $W = TT^T$. Moreover, the ℓ_2^W norm density of $\mathbf{y}$ is given by

$$f_\rho(\rho) = \frac{2\Gamma(n/2 + 1)}{\Gamma(m/2)\Gamma((n-m)/2 + 1)}\rho^{m-1}(1 - \rho^2)^{(n-m)/2}. \qquad (15.19)$$

Proof. Using (15.17) of Lemma 15.5, we first observe that $|\Sigma| = |TT^T|^{1/2}$. Then, we introduce the change of variables $s = (\widetilde{\rho}^2 + \rho^2)^{1/2}$, from which $\mathrm{d}\widetilde{\rho} = (s^2 - \rho^2)^{-1/2}s\,\mathrm{d}s$, and the integral in (15.17) becomes

$$g_\mathbf{y}(\rho) = |TT^T|^{-1/2}\mathrm{Surf}(\mathcal{B}_{\|\cdot\|_2}(\mathbb{R}^{n-m}))\int_\rho^\infty g_\mathbf{x}(s)(s^2 - \rho^2)^{(n-m)/2-1}s\,\mathrm{d}s.$$

Since $\mathbf{x}$ is uniform in $\mathcal{B}_{\|\cdot\|_2}$, we have that

$$g_{\mathbf{x}}(s) = \begin{cases} \dfrac{1}{\text{Vol}(\mathcal{B}_{\|\cdot\|_2}(\mathbb{R}^n))} & \text{if } s < 1; \\ 0 & \text{otherwise.} \end{cases}$$

Therefore, $g_{\mathbf{y}}(\rho)$ can be written as

$$g_{\mathbf{y}}(\rho) = |TT^T|^{-1/2} \frac{\text{Surf}(\mathcal{B}_{\|\cdot\|_2}(\mathbb{R}^{n-m}))}{\text{Vol}(\mathcal{B}_{\|\cdot\|_2}(\mathbb{R}^n))} \int_\rho^1 (s^2 - \rho^2)^{(n-m)/2-1} s\, ds.$$

Using the fact that

$$\int (s^2 - \rho^2)^{q+1/2} s\, ds = \frac{(s^2 - \rho^2)^{q+3/2}}{2q+3}$$

with $q = (n-m)/2 - 3/2$, and substituting the values for $\text{Surf}(\mathcal{B}_{\|\cdot\|_2}(\mathbb{R}^{n-m}))$ and $\text{Vol}(\mathcal{B}_{\|\cdot\|_2}(\mathbb{R}^n))$, we obtain $g_{\mathbf{y}}(\rho)$ as stated in (15.18). Now, substituting the expression of $g_{\mathbf{y}}(\rho)$ into (15.16), we finally derive the ℓ_2^W norm density as given in (15.19). $\qquad\square$

Remark 15.6 (Rectangular transformation of uniform densities). An important consequence of Theorem 15.3 is that if a uniform distribution is transformed by a rectangular full-rank linear mapping, then the resulting image density is no longer uniform. A linear (rectangular) transformation therefore changes the nature of the uniform distribution on a ball. In particular, the transformed vector $\mathbf{y} = T\mathbf{x}$ tends to concentrate towards the center of the image of the support set (the ellipsoid $\mathcal{E}(0, TT^T)$), rather than to its surface, see Figure 15.1 for an illustration of this phenomenon. An extension of Lemma 15.5 and Theorem 15.3 to the complex case is reported in [58]. $\quad\lhd$

Remark 15.7 (Probabilistic predictors). The result of Theorem 15.3 has been proved in [220], using an alternative derivation. This result is then exploited to determine probabilistic confidence ellipsoids for random vectors, and to construct *probabilistic predictors* of certain sets in $\mathbb{R}^n$. The predictors are subsequently applied to various systems and control problems. $\quad\lhd$

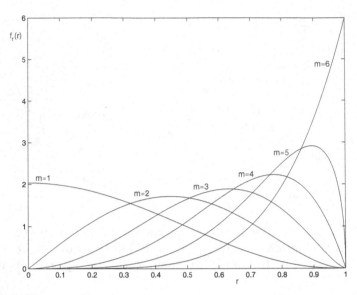

Fig. 15.1. ℓ_2^W norm density of $\mathbf{y} = T\mathbf{x}$, when $\mathbf{x} \sim \mathcal{U}_{\mathcal{B}_{\|\cdot\|_2}}$ and $T \in \mathbb{R}^{m,n}$ is full-rank. The plot shows the case $n = 6$ and $m = 1, \ldots, 6$.

16. Vector Randomization Methods

In this chapter we address the issue of real and complex vector randomization in ℓ_p norm balls, according to the uniform distribution. We present efficient algorithms based upon the theoretical developments of Chapter 15 regarding the statistical properties of random vectors. The presentation is partly based on the results reported in [60].

We recall that the uniform density in the ℓ_p norm ball $\mathcal{B}_{\|\cdot\|_p}(r)$ is a special case of the more general ℓ_p radial densities. Hence, random vector generation for ℓ_p radial densities is an immediate extension of uniform generation and follows from the application of Corollary 15.2. We also observe that the algorithms for vector generation presented in this chapter are based on simple algebraic transformations on samples obtained from the *univariate* generalized Gamma density $\overline{G}_{a,c}$ defined in (2.11). Therefore, whenever a generator for $\overline{G}_{a,c}$ is available (such as the one presented in Example 14.4), uniform generation in ℓ_p norm balls can be readily performed without the need of the asymptotic techniques of Section 14.4.

16.1 Rejection methods for uniform vector generation

We first discuss an application of the rejection technique for uniform generation of n-dimensional vector samples in ℓ_p norm balls. As already observed in Section 14.3.1, we expect these methods to be inefficient for large n, due to the curse of dimensionality. These rejection techniques are discussed here mainly to motivate the need for more efficient methods such as those presented in Section 16.3 for real vectors and in Section 16.4 for complex vectors.

To show the application of the rejection technique to the problem at hand, we first construct an outer bounding set for the ball $\mathcal{B}_{\|\cdot\|_p}$. Notice that the norm inequality

$$\|x\|_{p_2} \geq \|x\|_{p_1}, \text{ for } p_2 > p_1$$

implies the set inclusion

$$\mathcal{B}_{\|\cdot\|_{p_1}} \subseteq \mathcal{B}_{\|\cdot\|_{p_2}}, \text{ for } p_2 > p_1.$$

Assuming that generation in $\mathcal{B}_{\|\cdot\|_{p_2}}$ is easier to perform than generation in $\mathcal{B}_{\|\cdot\|_{p_1}}$ (see Example 16.1), the idea is therefore to generate samples uniformly in the outer bounding set $\mathcal{B}_{\|\cdot\|_{p_2}}$ and then to reject those which fall outside the set $\mathcal{B}_{\|\cdot\|_{p_1}}$. The rejection rate of such an algorithm is equal to the ratio of volumes, see (14.10), and is given by

$$\eta(n) = \frac{\mathrm{Vol}\big(\mathcal{B}_{\|\cdot\|_{p_2}}(\mathbb{F}^n)\big)}{\mathrm{Vol}\big(\mathcal{B}_{\|\cdot\|_{p_1}}(\mathbb{F}^n)\big)}.$$

In the subsequent examples, we compute rejection rates for real and complex balls.

Example 16.1 (Rejection method for real random vectors).
    Observe that the norm ball $\mathcal{B}_{\|\cdot\|_{p_2}}(\mathbb{R}^n)$ with $p_2 = \infty$ contains all the other norm balls. Clearly, generation of uniform samples in $\mathcal{B}_{\|\cdot\|_\infty}(\mathbb{R}^n)$ is straightforward, since

$$\mathcal{B}_{\|\cdot\|_\infty}(\mathbb{R}^n) = \{x \in \mathbb{R}^n : \|x\|_\infty \leq 1\}$$

is an n-dimensional hypercube whose edges have length equal to two. Hence, a random vector $\mathbf{x} \in \mathbb{R}^n$ uniformly distributed in $\mathcal{B}_{\|\cdot\|_\infty}$ can be immediately obtained generating independently the n components of $\mathbf{x}$, each uniform in the interval $[-1, 1]$. Using the volume formulae (15.8), the rejection rate of an algorithm for uniform generation in $\mathcal{B}_{\|\cdot\|_p}(\mathbb{R}^n)$ based on samples in $\mathcal{B}_{\|\cdot\|_\infty}(\mathbb{R}^n)$ is given by

$$\eta(n) = \frac{\Gamma(n/p + 1)}{\Gamma^n(1/p + 1)}.$$

Table 16.1 presents the rejection rate $\eta(n)$ for different values of n and p.

Table 16.1. Rejection rates for generating samples uniformly in $\mathcal{B}_{\|\cdot\|_p}(\mathbb{R}^n)$ from uniform samples in $\mathcal{B}_{\|\cdot\|_\infty}(\mathbb{R}^n)$.

	$p = 1$	$p = 1.5$	$p = 2$
$n = 2$	2	1.4610	1.2732
$n = 3$	6	2.7185	1.9099
$n = 4$	24	6.0412	3.2423
$n = 5$	120	15.446	6.0793
$n = 10$	3.62×10^6	7.21×10^3	401.54
$n = 20$	2.43×10^{18}	1.15×10^{11}	4.06×10^7
$n = 30$	2.65×10^{32}	5.24×10^{19}	4.90×10^{13}

The values of the rejection rates reported in this table clearly show that this rejection technique may be useful only for small values of n, whereas it becomes extremely inefficient for large values of n.   ⋆

Example 16.2 (Rejection method for complex random vectors).

As with the previous case, a rejection method may be based on the complex hypercube

$$\{x \in \mathbb{C}^n : \|x\|_\infty \le 1\}.$$

A random vector $\mathbf{x}$ uniformly distributed in this set is obtained by generating independently each component $\mathbf{x}_i$ uniformly in the complex disk of radius one. The rejection rate of this method can be easily computed by means of the formula for the volume (15.13), obtaining

$$\eta(n) = \frac{\Gamma(2n/p + 1)}{\Gamma^n(2/p + 1)}.$$

Table 16.2 reports several values of the rejection rate, showing the extreme inefficiency of this method for large values of n.

Table 16.2. Rejection rates for generating samples uniformly in $\mathcal{B}_{\|\cdot\|_p}(\mathbb{C}^n)$ from uniform samples in $\mathcal{B}_{\|\cdot\|_\infty}(\mathbb{C}^n)$.

	$p = 1$	$p = 1.5$	$p = 2$
$n = 2$	6	2.8302	2
$n = 3$	90	14.219	6
$n = 4$	2.52×10^3	106.50	24
$n = 5$	1.13×10^5	1.08×10^3	120
$n = 10$	2.38×10^{15}	2.60×10^9	3.63×10^6
$n = 20$	7.78×10^{41}	1.10×10^{26}	2.43×10^{18}
$n = 30$	7.75×10^{72}	4.35×10^{45}	2.65×10^{32}

$\star$

16.2 The generalized Gamma density

The main technical tool used in the algorithms for random vector generation presented in this chapter is the generalized Gamma density, defined in (2.11) as

$$\overline{G}_{a,c}(x) = \frac{c}{\Gamma(a)} x^{ca-1} e^{-x^c}, \quad x \ge 0.$$

We notice that $\overline{G}_{a,c}(x)$ is a *unilateral* density, since $x \ge 0$. A *bilateral* generalized Gamma density is defined accordingly as

$$f_{\mathbf{x}}(x) = \frac{c}{2\Gamma(a)} |x|^{ca-1} e^{-|x|^c}.$$

We recall that the generalized Gamma density coincides with classical density functions, such as Gaussian and Laplace, for a specific choice of the parameters a and c. This is further discussed in the next example.

Example 16.3 (Generalized Gamma density).
    We illustrate the behavior of $\overline{G}_{a,c}$ for specific values of a, c. First, we set $a = 1/p$ and $c = p$. Taking $p = 1$, we obtain the unilateral Laplace density with maximum value equal to one

$$f_{\mathbf{x}}(x) = e^{-x}, \quad x \geq 0. \tag{16.1}$$

For $p = 2$ we have the unilateral normal density with mean value zero and variance equal to $1/2$

$$f_{\mathbf{x}}(x) = \frac{2}{\sqrt{\pi}} e^{-x^2}, \quad x \geq 0.$$

In addition, it can be shown that

$$\lim_{p \to \infty} \frac{p}{\Gamma(1/p)} e^{-x^p} = \begin{cases} 1 \text{ if } x \in (0,1); \\ 0 \text{ if } x > 1. \end{cases}$$

Thus, the generalized Gamma density approaches the uniform density on $[0,1]$ for large values of p. Figure 16.1 shows a plot of the unilateral generalized Gamma density for various values of p.

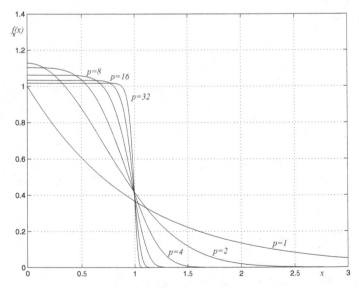

Fig. 16.1. Generalized Gamma density $\overline{G}_{1/p,p}$ for different values of p.

To conclude this example, we discuss the relation between $\overline{G}_{a,c}$ and the unilateral Gamma density $G_{a,b}$. The unilateral Gamma density, see (2.10), is

$$G_{a,b}(x) = \frac{1}{\Gamma(a)b^a} x^{a-1} e^{-x/b}, \quad x \geq 0.$$

In Example 14.4 it is shown that a random variable $\mathbf{x} \sim \overline{G}_{a,c}$ is obtained from a random variable with Gamma density $\mathbf{z} \sim G_{a,1}$ by means of the change of variables $\mathbf{x} = \mathbf{z}^{1/c}$. Hence, it is straightforward to generate samples according to $\overline{G}_{a,c}$ if a random generator for $G_{a,b}$ is available.

Finally, we remark that samples distributed according to a bilateral univariate density $\mathbf{x} \sim f_\mathbf{x}(x)$ can be immediately obtained from a unilateral univariate density $\mathbf{z} \sim f_\mathbf{z}(z)$ taking $\mathbf{x} = \mathbf{sz}$, where $\mathbf{s}$ is an independent random sign that takes the values ± 1 with equal probability.   ⋆

16.3 Uniform sample generation of real vectors

In this section we present an efficient algorithm for uniform generation of samples in $\mathcal{B}_{\|\cdot\|_p}(r, \mathbb{R}^n)$, based on the results of Corollary 15.3. Let each component of $\mathbf{x}$ be independently distributed according to the (bilateral) generalized Gamma density with parameters $1/p, p$. Then, the joint density $f_\mathbf{x}(x)$ is

$$f_\mathbf{x}(x) = \prod_{i=1}^{n} \frac{p}{2\Gamma(1/p)} e^{-|x_i|^p} = \frac{p^n}{2^n \Gamma^n(1/p)} e^{-\|x\|_p^p}.$$

Hence, it follows immediately that $f_\mathbf{x}(x)$ is ℓ_p radial with defining function

$$g(\rho) = \frac{p^n}{2^n \Gamma^n(1/p)} e^{-\rho^p}, \quad \rho = \|x\|_p.$$

Further, we have from Corollary 15.3 that if $\mathbf{w}$ is uniform in $[0,1]$ then $\mathbf{y} = r\mathbf{w}^{1/n}\mathbf{x}/\|\mathbf{x}\|_p$ is uniform in $\mathcal{B}_{\|\cdot\|_p}(r, \mathbb{R}^n)$.

The algorithm for uniform sample generation in real ℓ_p balls is hence summarized next.

Algorithm 16.1 (Uniform generation in real ℓ_p norm ball).
Given n, p and r, this algorithm returns a real random vector $\mathbf{y}$ uniformly distributed in $\mathcal{B}_{\|\cdot\|_p}(r, \mathbb{R}^n)$.

1. Generate n independent random real scalars $\boldsymbol{\xi}_i \sim \overline{G}_{1/p,p}$;
2. Construct the vector $\mathbf{x} \in \mathbb{R}^n$ of components $\mathbf{x}_i = \mathbf{s}_i \boldsymbol{\xi}_i$, where $\mathbf{s}_i$ are independent random signs;
3. Generate $\mathbf{z} = \mathbf{w}^{1/n}$, where $\mathbf{w}$ is uniform in $[0,1]$;
4. Return $\mathbf{y} = r\mathbf{z}\frac{\mathbf{x}}{\|\mathbf{x}\|_p}$.

We remark that this algorithm is an extension of the method proposed in [132, 194] for generating random points uniformly distributed on n-dimensional spheres starting from normally distributed real vectors.

Example 16.4 (Uniform generation in $\mathcal{B}_{\|\cdot\|_p}(\mathbb{R}^2)$).
For illustrative purposes, in this example we consider the case $n = 2$, $p = 1.5$ and $r = 1$. Figure 16.2 shows the three steps involved in uniform generation:

1. Each vector sample $\mathbf{x}^{(i)}$ is generated according to a generalized Gamma density with parameters $1/p, p$;
2. Each sample is normalized taking $\mathbf{x}^{(i)}/\|\mathbf{x}^{(i)}\|_p$, obtaining a uniform distribution on the contour $\partial\mathcal{B}_{\|\cdot\|_p}$;
3. Each normalized sample is scaled by the *volumetric factor* $\sqrt{\mathbf{w}^{(i)}}$, where $\mathbf{w}^{(i)} \sim \mathcal{U}_{[0,1]}$, which smudges the samples uniformly inside $\mathcal{B}_{\|\cdot\|_p}$.

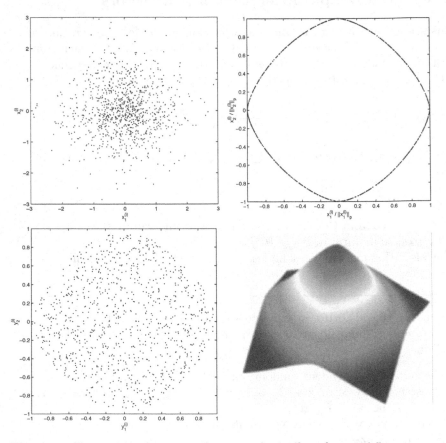

Fig. 16.2. Generation of $1,000$ uniform samples in $\mathcal{B}_{\|\cdot\|_p}$, for $p = 1.5$.

We remark that Algorithm 16.1 can be also used when $p \in (0,1)$. However, in this case $\|x\|_p^p = \sum_{i=1}^{n} |x_i|^p$ is not a norm and the set

$$\mathcal{B}_{\|\cdot\|_p}(r) = \{x \in \mathbb{R}^n : \|x\|_p \leq r\}$$

is not convex. Figure 16.3 shows $1,000$ samples of real two-dimensional vectors uniformly distributed in $\mathcal{B}_{\|\cdot\|_p}$ for $p = 0.7$ and $p = 1$. ★

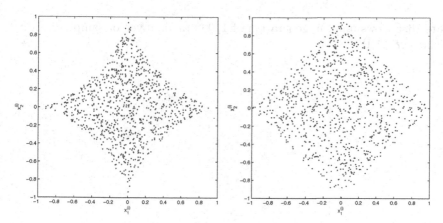

Fig. 16.3. Generation of $1,000$ uniform samples in $\mathcal{B}_{\|\cdot\|_p}$, for $p = 0.7$ (left) and $p = 1$ (right).

Next, we discuss sample generation within an ellipsoid

$$\mathcal{E}(\bar{x}, W) = \{x \in \mathbb{R}^n : x^T W^{-1} x \leq 1\}$$

and show that it can be easily performed using the techniques discussed in Section 15.5. This is described in the following algorithm.

Algorithm 16.2 (Uniform generation in an ellipsoid).
Given $n, \bar{x}, W$, this algorithm returns a random vector $\mathbf{y} \in \mathbb{R}^n$ uniformly distributed in the ellipsoid $\mathcal{E}(\bar{x}, W)$.

1. Compute a matrix T such that $W = TT^T$;
2. Generate a random vector $\mathbf{x} \sim \mathcal{U}_{\mathcal{B}_{\|\cdot\|_2}}$ using Algorithm 16.1;
3. Return $\mathbf{y} = T\mathbf{x} + \bar{x}$.

Example 16.5 (Uniform generation in an ellipse).
   Consider the ellipse

$$\mathcal{E}\,(\bar{x}, W) = \{x \in \mathbb{R}^2 : x^T W^{-1} x \leq 1\} \tag{16.2}$$

where

$$\bar{x} = \begin{bmatrix} 1 \\ 2 \end{bmatrix}, \quad W = \begin{bmatrix} 5 & 6 \\ 6 & 8 \end{bmatrix}.$$

Figure 16.4 shows uniform generation of $1,000$ random vector samples inside the ellipse $\mathcal{E}\,(\bar{x}, W)$. ⋆

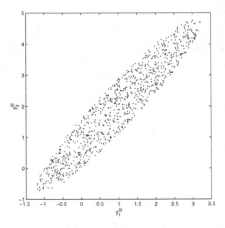

Fig. 16.4. Uniform generation in the ellipse (16.2).

Example 16.6 (Uniform sample generation in a simplex).
   We describe in this example a simple method for generating uniform points in the standard simplex, based on the generalized Gamma density $\overline{G}_{1,1}$, which corresponds to the unilateral Laplace density (16.1). The standard unit simplex, also known as *probability simplex*, is defined as

$$\mathrm{PS}(n) \doteq \left\{ x \in \mathbb{R}^n : \sum_{i=1}^{n} x_i = 1,\ x_i \geq 0,\ i = 1, \ldots, n \right\}.$$

Notice that the probability simplex is the intersection between the nonnegative orthant of $\mathbb{R}^n$ and the surface of the unit ball in the ℓ_1 norm, i.e.

$$\mathrm{PS}(n) = \mathbb{R}_+^n \cap \mathcal{B}_{\|\cdot\|_1}.$$

As discussed in Remark 15.1, if a random vector $\mathbf{x}$ is ℓ_1 radial, then $\mathbf{y} = \mathbf{x}/\|\mathbf{x}\|_1$ is uniformly distributed on the surface of the unit ℓ_1 norm ball. Therefore, the uniform distribution on the unit simplex can be obtained by projecting the $\mathbf{y}$ samples onto the nonnegative orthant. The explicit algorithm is given next.

Algorithm 16.3 (Uniform generation in the unit simplex).
Given n, this algorithm returns a real random vector $\mathbf{y}$ uniformly distributed in $\mathrm{PS}(n)$.

1. Generate n independent random real scalars $\mathbf{x}_i \sim \overline{G}_{1,1}$;
2. Return $\mathbf{y} = \frac{\mathbf{x}}{\|\mathbf{x}\|_1}$.

Uniform samples in the probability simplex can be used to generate uniform samples in generic simplices. A generic $(k-1)$-dimensional simplex in $\mathbb{R}^n$ is defined as the convex hull of k affinely independent vectors

$$\mathrm{Simplex}(v_1,\ldots,v_k) = \{\xi = x_1 v_1 + \cdots + x_k v_k, x = [x_1 \cdots x_k]^T \in \mathrm{PS}(k)\}$$

where $v_i \in \mathbb{R}^n$, $i = 1,\ldots,k$, and $[v_2 - v_1 \; v_3 - v_1 \; \cdots \; v_k - v_1]$ is full-rank (affine independence). Therefore, defining the simplex vertex matrix $V \doteq [v_1 \; \cdots \; v_k] \in \mathbb{R}^{n,k}$, it can be verified that if the random vector $\mathbf{x}$ is uniformly distributed in $\mathrm{PS}(k)$, then the random vector

$$\mathbf{y} = V\mathbf{x}$$

is uniformly distributed in $\mathrm{Simplex}(v_1,\ldots,v_k)$.   ⋆

16.4 Uniform sample generation of complex vectors

In this section we present an algorithm based on Corollary 15.3 for generating complex vectors uniformly distributed in ℓ_p norm balls. Let

$$\mathbf{x} = [\mathbf{x}_1 \; \cdots \; \mathbf{x}_n]^T \in \mathbb{C}^n$$

be a complex random vector. Clearly, each component $\mathbf{x}_i$ of $\mathbf{x}$ can be considered as a two-dimensional real vector $\mathbf{z}_i = [\mathrm{Re}(\mathbf{x}_i) \; \mathrm{Im}(\mathbf{x}_i)]^T \in \mathbb{R}^2$, so that the absolute value of $\mathbf{x}_i$ coincides with the ℓ_2 norm of $\mathbf{z}_i$. Let the components $\mathbf{z}_i$ be independent ℓ_2 radial vectors, with defining function

$$f_{\mathbf{z}_i}(z_i) = \frac{p}{2\pi\Gamma(2/p)} e^{-\rho_i^p} = g_i(\rho_i), \quad \rho_i = \|z_i\|_2.$$

Therefore, using (15.14) the corresponding norm density function is

$$f_{\rho_i}(\rho_i) = \frac{p}{\Gamma(2/p)} \rho_i e^{-\rho_i^p}, \quad \rho_i \geq 0.$$

This density is a (unilateral) generalized Gamma density with parameters $(2/p, p)$. Since the components of $\mathbf{x}$ are independent, the density function $f_{\mathbf{x}}(x)$ is obtained as the (bilateral) joint density function

$$f_{\mathbf{x}}(x) = \prod_{i=1}^{n} \frac{p}{2\pi\Gamma(2/p)} e^{-|x_i|^p} = \frac{p^n}{2^n\pi^n\Gamma^n(2/p)} e^{-\|x\|_p^p}.$$

As in the real case, from this expression it follows that the random vector $\mathbf{x}$ is ℓ_p radial, with defining function

$$g(\rho) = \frac{p^n}{2^n\pi^n\Gamma^n(2/p)} e^{-\rho^p}, \quad \rho = \|x\|_p$$

and the results of Corollary 15.3 can be applied. The algorithm for uniform generation in complex ℓ_p balls is now presented.

Algorithm 16.4 (Uniform generation in complex ℓ_p norm ball).
Given n, p and r, this algorithm returns a complex random vector $\mathbf{y}$ uniformly distributed in $\mathcal{B}_{\|\cdot\|_p}(r, \mathbb{C}^n)$.

1. Generate n independent random complex numbers $\mathbf{s}_i = e^{j\theta}$, where θ is uniform in $[0, 2\pi]$;
2. Construct the vector $\mathbf{x} \in \mathbb{C}^n$ of components $\mathbf{x}_i = \mathbf{s}_i \boldsymbol{\xi}_i$, where the $\boldsymbol{\xi}_i$ are independent random variables $\boldsymbol{\xi}_i \sim \bar{G}_{2/p,p}$;
3. Generate $\mathbf{z} = \mathbf{w}^{1/(2n)}$, where $\mathbf{w}$ is uniform in $[0, 1]$;
4. Return $\mathbf{y} = r\mathbf{z}\frac{\mathbf{x}}{\|\mathbf{x}\|_p}$.

---

17. Statistical Theory of Radial Random Matrices

In this chapter we study the statistical properties of random matrices whose probability density belongs to the class of matrix radial densities. The results presented in this chapter constitute the theoretical foundations of the algorithms for generation of random matrices uniformly distributed in norm bounded sets presented in Chapter 18.

This chapter has the following structure. In Section 17.1 we introduce the notion of ℓ_p induced radial matrix density, which is a direct extension of the concept of radial vector densities introduced in Chapter 15. In Sections 17.2 and 17.3 we state specific results for ℓ_p induced radial densities for the cases $p = 1, \infty$ and $p = 2$ respectively. For the case $p = 2$, we first study symmetric real matrices, and then extend these results to real and complex rectangular matrices. The contents of Section 17.2 are based on concepts of multivariate statistical analysis. The reader interested in this topic may refer to [12] for an introductory presentation. More specific material on the theory of random matrices can be found in [100, 101, 124, 186]. Additional material is available in [120, 140]. The results presented in this chapter are largely based on [62].

17.1 Radial matrix densities

In Chapter 15 we defined a particular class of vector densities which depend only on the norm of the random vector. This definition is now extended to random matrices. We first discuss Hilbert–Schmidt ℓ_p radial densities.

17.1.1 Hilbert–Schmidt ℓ_p radial matrix densities

Definition 17.1 (Hilbert–Schmidt ℓ_p radial matrix densities). *A random matrix $\mathbf{X} \in \mathbb{F}^{n,m}$ is radial in the Hilbert–Schmidt ℓ_p norm if its density function can be written as*

$$f_{\mathbf{X}}(X) = g(\rho), \quad \rho = \|X\|_p$$

where $g(\rho)$ is the defining function of $\mathbf{X}$.

Using the vectorization operator $\text{vec}(\cdot)$ introduced in (3.8), the Hilbert–Schmidt ℓ_p norm of X can be written as

$$\|X\|_p = \|\text{vec}(X)\|_p.$$

Then, the statistical properties of an ℓ_p radial matrix $\mathbf{X}$ are equivalent to the properties of the ℓ_p radial random vector $\mathbf{x} = \text{vec}(\mathbf{X})$ studied in Chapter 15.

17.1.2 ℓ_p induced radial matrix densities

In this chapter we mainly concentrate on the properties of the class of random matrices whose densities depend only on the ℓ_p induced norm, and refer to this class as ℓ_p induced radial densities. A formal definition is given below.

Definition 17.2 (ℓ_p induced radial matrix densities). *A random matrix* $\mathbf{X} \in \mathbb{F}^{n,m}$ *is radial in the ℓ_p induced norm if its density function can be written as*

$$f_{\mathbf{X}}(X) = g(\rho), \quad \rho = \|X\|_p$$

where $g(\rho)$ is the defining function of $\mathbf{X}$.

Example 17.1 (Uniform matrices in $\mathcal{B}_{\|\cdot\|_p}$ are ℓ_p induced radial).
    Consider the definition of uniform density in the ℓ_p induced norm unit ball

$$f_{\mathbf{X}}(X) = \mathcal{U}_{\mathcal{B}_{\|\cdot\|_p}}(X) = \begin{cases} \dfrac{1}{\text{Vol}(\mathcal{B}_{\|\cdot\|_p})} & \text{if } \|X\|_p \leq 1; \\ 0 & \text{otherwise.} \end{cases}$$

It is easy to verify that this pdf depends only on the norm of X. That is,

$$f_{\mathbf{X}}(X) = g(\rho), \quad \rho = \|X\|_p$$

where the defining function $g(\rho)$ is given by

$$g(\rho) = \begin{cases} \dfrac{1}{\text{Vol}(\mathcal{B}_{\|\cdot\|_p})} & \text{if } \rho \leq 1; \\ 0 & \text{otherwise.} \end{cases} \tag{17.1}$$

The fact that the uniform density is ℓ_p induced radial turns out to be fundamental for the development of efficient algorithms for random matrix generation in ℓ_p induced norm ball presented in Chapter 18. $\star$

17.2 Statistical properties of ℓ_1 and ℓ_∞ induced densities

In this section we study ℓ_1 and ℓ_∞ radial densities for real and complex random matrices.

17.2.1 Real matrices with ℓ_1 and ℓ_∞ induced densities

The ℓ_1 induced norm of a given matrix $X \in \mathbb{F}^{n,m}$ is equal to the maximum of the ℓ_1 norms of its columns, see (3.10), that is

$$\|X\|_1 = \max_{i=1,\ldots,m} \|\xi_i\|_1$$

where $\xi_1, \ldots, \xi_m$ are the columns of X. The pdf of an ℓ_p induced radial matrix $\mathbf{X} \in \mathbb{F}^{n,m}$ can therefore be written as

$$f_{\mathbf{X}}(X) = g(\bar{\rho}), \quad \bar{\rho} \doteq \max_{i=1,\ldots,m} \rho_i$$

where $\rho_i \doteq \|\xi_i\|_1$, for $i = 1, \ldots, m$.

The following theorem defines a decomposition of an ℓ_1 induced radial real matrix $\mathbf{X}$ in two terms: a normalized matrix $\mathbf{U}$ and a diagonal matrix $\mathbf{R}$ containing the norms of the columns, and provides a closed-form expression for their probability densities. This theorem is the counterpart, for ℓ_1 induced radial matrices, of Theorem 15.1.

Theorem 17.1 (ℓ_1 induced radial matrices in $\mathbb{R}^{n,m}$). *Let the random matrix* $\mathbf{X} \in \mathbb{R}^{n,m}$ *be factored as* $\mathbf{UR}$, *where*

$$\mathbf{U} = [\mathbf{u}_1 \cdots \mathbf{u}_m], \quad \|\mathbf{u}_i\|_1 = 1;$$
$$\mathbf{R} = \text{diag}\left([\rho_1 \cdots \rho_m]\right), \quad \rho_i > 0$$

being $\mathbf{u}_i \in \mathbb{R}^n$ *the* i^{th} *column of* $\mathbf{U}$. *The following statements are equivalent:*

1. $\mathbf{X}$ *is* ℓ_1 *induced radial with defining function* $g(\bar{\rho})$, $\bar{\rho} = \max_{i=1,\ldots,m} \rho_i$;
2. $\mathbf{U}$ *and* $\mathbf{R}$ *are independent, and their densities are given by*

$$f_{\mathbf{U}}(U) = \prod_{i=1}^{m} f_{\mathbf{u}_i}(u_i), \quad f_{\mathbf{u}_i}(u_i) = \mathcal{U}_{\partial \mathcal{B}_{\|\cdot\|_1}(\mathbb{R}^n)};$$

$$f_{\mathbf{R}}(R) = \left[\frac{2^n}{(n-1)!}\right]^m g(\bar{\rho}) \prod_{i=1}^{m} \rho_i^{n-1}.$$

Proof. The proof follows the same lines as Theorem 15.1. For each column ξ_i of the matrix variable X we write

$$\xi_i = \rho_i u_i, \quad i = 1, \ldots, m.$$

Then, we assume $[X]_{n,i} > 0$, $i = 1, \ldots, n$, and observe that in this case the transformation from X to U, R is one-to-one. The Jacobian of this transformation is easily computed as

$$J(X \to U, R) = \prod_{i=1}^{m} J(\xi_i \to u_i, \rho_i) = \prod_{i=1}^{m} \rho_i^{n-1}$$

where the last equality follows from equation (15.5) with $p = 1$. From Theorem A.1 (see Appendix) on the transformation of random matrices, we write

$$f_{\mathbf{U},\mathbf{R}}(U, R) = 2^m g(\bar{\rho}) \prod_{i=1}^{m} \rho_i^{n-1}$$

where the factor 2^m is introduced to consider all possible combinations of signs in the terms $[X]_{n,i}$, and thus to remove the condition $[X]_{n,i} > 0$. From this equation, it follows that $\mathbf{U}$ and $\mathbf{R}$ are independent, and that the pdf of each $\mathbf{u}_i$ is a constant that can be computed by setting $p = 1$ in equation (15.3). The density $f_{\mathbf{R}}(R)$ is then obtained integrating the joint density $f_{\mathbf{U},\mathbf{R}}(U, R)$ with respect to U. $\square$

The following corollary, based on the results of Theorem 17.1, provides a statistical characterization of real matrices uniformly distributed in the norm ball $\mathcal{B}_{\|\cdot\|_1}$.

Corollary 17.1 (Uniform real matrices in $\mathcal{B}_{\|\cdot\|_1}$). *Let $\mathbf{X} \in \mathbb{R}^{n,m}$ and let $\xi_1, \ldots, \xi_m$ be the columns of $\mathbf{X}$. The following statements are equivalent:*

1. *$\mathbf{X}$ is uniformly distributed in the set $\mathcal{B}_{\|\cdot\|_1}(\mathbb{R}^{n,m})$;*
2. *$\xi_1, \ldots, \xi_m$ are independent and uniformly distributed in $\mathcal{B}_{\|\cdot\|_1}(\mathbb{R}^n)$.*

Proof. The defining function of $\mathbf{X}$ is given by

$$g(\bar{\rho}) = \begin{cases} \dfrac{1}{\mathrm{Vol}(\mathcal{B}_{\|\cdot\|_1})} & \text{if } \bar{\rho} \leq 1; \\ 0 & \text{otherwise} \end{cases}$$

where $\bar{\rho} = \max_{i=1,\ldots,m} \rho_i$ and $\rho_i = \|\xi_i\|_1$, for $i = 1, \ldots, m$.

From Theorem 17.1, we immediately obtain

$$f_{\mathbf{U}}(U) = \prod_{i=1}^{m} f_{\mathbf{u}_i}(u_i), \quad f_{\mathbf{u}_i}(u_i) = \mathcal{U}_{\partial \mathcal{B}_{\|\cdot\|_1}(\mathbb{R}^n)};$$

$$f_{\mathbf{R}}(R) = \left[\frac{2^n}{(n-1)!} \right]^m \frac{1}{\mathrm{Vol}(\mathcal{B}_{\|\cdot\|_1})} \prod_{i=1}^{m} \rho_i^{n-1}, \quad \bar{\rho} \leq 1.$$

Since $f_{\mathbf{R}}(R)$ is a density function, we impose

$$\int f_{\mathbf{R}}(R)\mathrm{d}R = \int_{\bar{\rho} \leq 1} \left[\frac{2^n}{(n-1)!} \right]^m \frac{1}{\mathrm{Vol}(\mathcal{B}_{\|\cdot\|_1})} \prod_{i=1}^{m} \rho_i^{n-1} \mathrm{d}\rho_1 \cdots \mathrm{d}\rho_m = 1$$

obtaining

$$\mathrm{Vol}(\mathcal{B}_{\|\cdot\|_1}) = \left[\frac{2^n}{n!}\right]^m.$$

Moreover, substituting this expression in $f_{\mathbf{R}}(R)$, we have

$$f_{\mathbf{R}}(R) = \prod_{i=1}^{m} n\rho_i^{n-1}, \quad \bar{\rho} \le 1.$$

The joint density of $\mathbf{R}$, $\mathbf{U}$ is then given by

$$f_{\mathbf{U},\mathbf{R}}(U, R) = f_{\mathbf{R}}(R)f_{\mathbf{U}}(U) = \prod_{i=1}^{m} n\rho_i^{n-1} f_{\mathbf{u}_i}(u_i) = \prod_{i=1}^{m} f_{\mathbf{u}_i,\rho_i}(u_i, \rho_i)$$

being $f_{\mathbf{u}_i}(u_i) = \mathcal{U}_{\partial \mathcal{B}_{\|\cdot\|_1}}$. The statement then follows from Theorem 15.1, observing that $\xi_i = \rho_i u_i$, $i = 1, \dots, m$.   □

Remark 17.1 (Volume of the ℓ_1 induced norm ball in $\mathbb{R}^{n,m}$). From the proof of Corollary 17.1, we obtain a closed-form expression for the volume of the ℓ_1 induced norm ball of radius r in $\mathbb{R}^{n,m}$

$$\mathrm{Vol}(\mathcal{B}_{\|\cdot\|_1}(r, \mathbb{R}^{n,m})) = \left[\frac{2^n}{n!}\right]^m r^{nm}.$$

◁

Remark 17.2 (ℓ_∞ induced radial densities in $\mathbb{R}^{n,m}$). The case of ℓ_∞ induced radial matrices may be treated in a similar way. Indeed, the ℓ_∞ induced norm is the maximum of the ℓ_1 norms of the rows of X, i.e.

$$\|X\|_\infty = \max_{i=1,\dots,n} \|\eta_i\|_1$$

where $\eta_1^T, \dots, \eta_n^T$ are the rows of X.

The statistical properties of a real ℓ_∞ induced radial random matrix $\mathbf{X}$ can be immediately deduced, noticing that $\|\mathbf{X}\|_\infty = \|\mathbf{X}^T\|_1$. Therefore, if a random matrix $\mathbf{X}$ has an ℓ_∞ induced radial density, then its transpose $\mathbf{X}^T$ is ℓ_1 induced radial. In particular, the volume of an ℓ_∞ induced norm ball in $\mathbb{R}^{n,m}$ is given by

$$\mathrm{Vol}(\mathcal{B}_{\|\cdot\|_\infty}(r, \mathbb{R}^{n,m})) = \left[\frac{2^m}{m!}\right]^n r^{nm}.$$

◁

17.2.2 Complex matrices with ℓ_1 and ℓ_∞ induced densities

The previous results for ℓ_1 and ℓ_∞ induced radial real matrices are immediately extended to the complex case. The results are reported without proof.

Theorem 17.2 (ℓ_1 induced radial matrices in $\mathbb{C}^{n,m}$). *Let the random matrix* $\mathbf{X} \in \mathbb{C}^{n,m}$ *be factored as* $\mathbf{UR}$*, where*

$$\mathbf{U} = [\mathbf{u}_1 \cdots \mathbf{u}_m], \quad \|\mathbf{u}_i\|_1 = 1;$$
$$\mathbf{R} = \operatorname{diag}([\rho_1 \cdots \rho_m]), \quad \rho_i > 0$$

being $\mathbf{u}_i \in \mathbb{C}^n$ *the* i^{th} *column of* $\mathbf{U}$*. The following statements are equivalent:*

1. $\mathbf{X}$ *is* ℓ_1 *induced radial with defining function* $g(\bar{\rho})$, $\bar{\rho} \doteq \max_{i=1,\dots,m} \rho_i$;
2. $\mathbf{U}$ *and* $\mathbf{R}$ *are independent, and their densities are given by*

$$f_{\mathbf{U}}(U) = \prod_{i=1}^{m} f_{\mathbf{u}_i}(u_i), \quad f_{\mathbf{u}_i}(u_i) = \mathcal{U}_{\partial \mathcal{B}_{\|\cdot\|_1}(\mathbb{C}^n)};$$

$$f_{\mathbf{R}}(R) = \left[\frac{(2\pi)^n}{(2n-1)!} \right]^m g(\bar{\rho}) \prod_{i}^{m} \rho_i^{2n-1}.$$

Next, we present a corollary which gives a characterization of uniformly distributed complex matrices in ℓ_1 induced norm balls.

Corollary 17.2 (Uniform complex matrices in $\mathcal{B}_{\|\cdot\|_1}$). *Let* $\mathbf{X} \in \mathbb{C}^{n,m}$ *and let* $\boldsymbol{\xi}_1, \dots, \boldsymbol{\xi}_m$ *be the columns of* $\mathbf{X}$*. The following statements are equivalent:*

1. $\mathbf{X}$ *is uniformly distributed in the set* $\mathcal{B}_{\|\cdot\|_1}(\mathbb{C}^{n,m})$;
2. $\boldsymbol{\xi}_1, \dots, \boldsymbol{\xi}_m$ *are independent and uniformly distributed in* $\mathcal{B}_{\|\cdot\|_1}(\mathbb{C}^n)$.

Remark 17.3 (ℓ_∞ induced radial densities in $\mathbb{C}^{n,m}$). Similar to the real case, the statistical properties of a complex ℓ_∞ induced radial random matrix $\mathbf{X}$ can be immediately derived. In fact, if a complex random matrix $\mathbf{X}$ has an ℓ_∞ induced radial density, then $\mathbf{X}^*$ is ℓ_1 induced radial.

Remark 17.4 (Volume of the ℓ_1 and ℓ_∞ induced norm balls in $\mathbb{C}^{n,m}$). The volume of the ℓ_1 and ℓ_∞ induced norm balls of radius r in $\mathbb{C}^{n,m}$ are given by

$$\operatorname{Vol}(\mathcal{B}_{\|\cdot\|_1}(r, \mathbb{C}^{n,m})) = \left[\frac{(2\pi)^n}{(2n)!} \right]^m r^{2nm};$$

$$\operatorname{Vol}(\mathcal{B}_{\|\cdot\|_\infty}(r, \mathbb{C}^{n,m})) = \left[\frac{(2\pi)^m}{(2m)!} \right]^n r^{2nm}.$$

$\triangleleft$

Next, we study in detail the important case of ℓ_2 induced radial matrices, also denoted as σ radial matrices.

17.3 Statistical properties of σ radial densities

In this section we consider random matrices with radial distribution with respect to the ℓ_2 induced norm. We recall that the ℓ_2 induced norm of a matrix $X \in \mathbb{F}^{n,m}$ is usually referred to as the spectral norm (or σ norm), which is defined in (3.11) as

$$\|X\|_2 = \bar{\sigma}(X)$$

where $\bar{\sigma}(X)$ is the largest singular value of X. We now define σ radial matrix density functions.

Definition 17.3 (σ radial matrix densities). *A random matrix* $\mathbf{X} \in \mathbb{F}^{n,m}$ *is radial in the ℓ_2 induced norm, or σ radial, if its density can be written as*

$$f_{\mathbf{X}}(X) = g(\bar{\sigma}), \quad \bar{\sigma} = \bar{\sigma}(X)$$

where $g(\bar{\sigma})$ is the defining function of $\mathbf{X}$.

17.3.1 Positive definite matrices with σ radial densities

We first consider the case of a real positive definite matrix $X \succ 0$ and define a *normalized* singular value decomposition (SVD) of X.

Definition 17.4 (Normalized SVD of positive definite matrices). *Any positive definite matrix $X \in \mathbb{S}^n$, $X \succ 0$, can be written in the form*

$$X = U \Sigma U^T \tag{17.2}$$

where $\Sigma = \mathrm{diag}\,(\sigma)$, $\sigma = [\sigma_1 \cdots \sigma_n]^T$, with $\sigma_1 \geq \cdots \geq \sigma_n \geq 0$, and $U \in \mathbb{R}^{n,n}$ has orthonormal columns, normalized so that the first nonvanishing component of each column is positive.

Remark 17.5 (The orthogonal group). The set of orthogonal matrices in $\mathbb{R}^{n,n}$ forms a group. This group is generally referred to as the *orthogonal group* and is denoted by

$$\mathcal{G}_\mathcal{O}^n \doteq \{U \in \mathbb{R}^{n,n} : U^T U = I\}. \tag{17.3}$$

Moreover, the real manifold

$$\mathcal{R}^{m,n} \doteq \{U \in \mathbb{R}^{m,n} : U^T U = I;\ [U]_{1,i} > 0,\ i = 1, \ldots, n\} \tag{17.4}$$

represents the set of matrices $U \in \mathbb{R}^{m,n}$, $m \geq n$, whose columns are orthonormal with positive first component.   ◁

Remark 17.6 (Haar invariant distributions). In the literature, see e.g. [12], the uniform distribution over the orthogonal group $\mathcal{G}_{\mathcal{O}}^n$ is known as the Haar invariant distribution, which is here denoted as $\mathcal{U}_{\mathcal{G}_{\mathcal{O}}^n}$. Similarly, the uniform distribution over the manifold $\mathcal{R}^{n,n}$ of normalized orthogonal matrices is known as the conditional Haar invariant distribution, denoted here as $\mathcal{U}_{\mathcal{R}^{n,n}}$. The Haar invariant distribution is the only distribution with the property that if $\mathbf{U} \in \mathcal{G}_{\mathcal{O}}^n$ is distributed according to Haar, then $Q\mathbf{U} \sim \mathbf{U}$ for any fixed orthogonal matrix Q.   ◁

The objective of this section is to relate the density function of the positive definite random matrix $\mathbf{X}$ to the pdfs of its SVD factors $\mathbf{U}$ and $\mathbf{\Sigma}$, using the mapping defined in (17.2). However, we notice that this mapping presents some ambiguity, in the sense that X may not be uniquely defined by a U, Σ pair, i.e. the mapping is not one-to-one. This is discussed in the next example.

Example 17.2 (Non uniqueness of the SVD).
     Consider the identity matrix $X \equiv I$. In this case, for any $U_1 \in \mathcal{G}_{\mathcal{O}}^n$, $U_2 \in \mathcal{G}_{\mathcal{O}}^n$ such that $U_1 \neq U_2$, we can write the SVD of X either as

$$X = U_1 \Sigma U_1^T, \quad \Sigma = I$$

or as

$$X = U_2 \Sigma U_2^T.$$

Furthermore, consider a positive definite matrix $X \succ 0$ with singular value decomposition $X = U_1 \Sigma U_1^T$, $U_1 \in \mathcal{G}_{\mathcal{O}}^n$. It can be easily seen that

$$X = U_2 \Sigma U_2^T, \quad U_2 = -U_1.$$

                                                                    ⋆

Remark 17.7 (One-to-one SVD). The matrices with at least two coincident singular values are not uniquely represented by (17.2). However, this mapping may be made one-to-one by considering strict inequalities in the ordering of the singular values. That is, we consider $\sigma \in \mathcal{D}_\sigma$, where the singular value *ordered domain* $\mathcal{D}_\sigma$ is defined as

$$\mathcal{D}_\sigma \doteq \{\sigma \in \mathbb{R}^n : 1 \geq \sigma_1 > \cdots > \sigma_n > 0\}. \tag{17.5}$$

For a similar reason, to avoid the possible ambiguities shown in Example 17.2, a normalization condition may be imposed on every *first* element of the columns of U. That is, in (17.2) we fix the signs of the rows of the matrix U, taking $U \in \mathcal{R}^{n,n}$.
     We notice that these normalizations do not affect the probabilistic results developed in the following. In fact, we are excluding a set of measure zero from the decomposition (17.2). In other words, for the class of densities under

study, the probability of two singular values being equal is zero and the probability of $[\mathbf{U}]_{1,i} = 0$ is also zero. Finally, we remark that these normalizations are in agreement with classical literature on this topic, see e.g. [12], where strict inequalities in the ordering of the eigenvalues of symmetric matrices and a normalization condition on the columns of the eigenvector matrices are considered.   ◁

The next result relates the density function of a σ radial positive definite matrix to the densities of its SVD factors Σ and U.

Theorem 17.3 (Positive definite σ radial matrices). *Let the positive definite random matrix* $\mathbf{X} \in \mathbb{S}^n$, $\mathbf{X} \succ 0$, *be factored as* $\mathbf{U}\Sigma\mathbf{U}^T$ *according to Definition 17.4, with* $\boldsymbol{\sigma} \in \mathcal{D}_\sigma$ *and* $U \in \mathcal{R}^{n,n}$. *The following statements are equivalent:*

1. $\mathbf{X}$ *is* σ *radial with defining function* $g(\bar{\sigma})$;
2. $\mathbf{U}$ *and* Σ *are independent, and their densities are given by*

$$f_{\mathbf{U}}(U) = \mathcal{U}_{\mathcal{R}^{n,n}}; \tag{17.6}$$

$$f_{\Sigma}(\Sigma) = \Upsilon_{\mathbb{S}}\, g(\bar{\sigma}) \prod_{1 \leq i < k \leq n} (\sigma_i - \sigma_k) \tag{17.7}$$

where the normalization constant $\Upsilon_{\mathbb{S}}$ *is*

$$\Upsilon_{\mathbb{S}} = \pi^{\frac{n}{4}(n+1)} \prod_{i=1}^{n} \frac{1}{\Gamma\left(\frac{n-i+1}{2}\right)}. \tag{17.8}$$

Proof. Consider the transformation $X = U\Sigma U^T$. The strict inequalities in the ordering of the singular values and the normalization conditions on the columns of U make the mapping between X and U, Σ one-to-one. The joint pdf in the new variables U, Σ may be obtained applying Theorem A.1 (see Appendix) on the transformation of random variables

$$f_{\mathbf{U},\boldsymbol{\Sigma}}(U, \Sigma) = g(\bar{\sigma})J(X \rightarrow U, \Sigma). \tag{17.9}$$

To compute the Jacobian $J(X \rightarrow U, \Sigma)$, we make use of the rules stated in Appendix A.2. In particular, using Rule A.2 on the Jacobian of the differentials, we have that $J(X \rightarrow U, \Sigma) = J(\mathrm{d}X \rightarrow \mathrm{d}U, \mathrm{d}\Sigma)$. The differential of X is given by

$$\mathrm{d}X = \mathrm{d}U\,\Sigma U^T + U\,\mathrm{d}\Sigma\,U^T + U\,\Sigma\,\mathrm{d}U^T.$$

Multiplying this equation by U^T on the left and by U on the right, we obtain

$$Z \doteq U^T\mathrm{d}XU = U^T\mathrm{d}U\Sigma + \mathrm{d}\Sigma + \Sigma\mathrm{d}U^T U. \tag{17.10}$$

Applying the chain rule for Jacobians (Rule A.1), we have

$$J(\mathrm{d}X \rightarrow \mathrm{d}U, \mathrm{d}\Sigma) = J(\mathrm{d}X \rightarrow Z)J(Z \rightarrow \mathrm{d}U, \mathrm{d}\Sigma).$$

Since, by Rule A.4, $J(\mathrm{d}X \to Z) = 1$, it follows that

$$J(X \to U, \Sigma) = J(Z \to \mathrm{d}U, \mathrm{d}\Sigma).$$

Next, we rewrite (17.10) in the form

$$Z = S_u \Sigma + \mathrm{d}\Sigma + \Sigma S_u^T$$

where $S_u \doteq U^T \mathrm{d}U \in \mathbb{R}^{n,n}$. We notice that the Jacobian $J(S_u, \mathrm{d}\Sigma \to \mathrm{d}U, \mathrm{d}\Sigma)$ is equal to one by Rule A.4. Then, applying the chain rule again, we have

$$J(Z \to \mathrm{d}U, \mathrm{d}\Sigma) = J(Z \to S_u, \mathrm{d}\Sigma)\, J(S_u, \mathrm{d}\Sigma \to \mathrm{d}U, \mathrm{d}\Sigma) = J(Z \to S_u, \mathrm{d}\Sigma).$$

We now concentrate on the evaluation of $J(Z \to S_u, \mathrm{d}\Sigma)$. First, notice that S_u is skew-symmetric. This is easily seen by differentiating the identity $U^T U = I$, obtaining $\mathrm{d}U^T U + U^T \mathrm{d}U = 0$. The matrix Z may therefore be rewritten in the form

$$Z = S_u \Sigma - \Sigma S_u + \mathrm{d}\Sigma.$$

Then, we examine the number of free variables that describe the quantities of interest. The symmetric matrix X is described by means of $\frac{n}{2}(n + 1)$ real variables. The orthogonal matrix U is described by $n_u \doteq \frac{n}{2}(n - 1)$ real variables, Σ is described by means of its n diagonal entries. The differentials $\mathrm{d}U$ and $\mathrm{d}\Sigma$ are described by the same number of free variables of U and Σ respectively. Therefore, S_u is described by n_u variables. Since $\mathrm{d}\Sigma$ is diagonal, we choose its n free variables as the diagonal entries $\eta_i = \mathrm{d}\sigma_i$, $1 \le i \le n$. Since S_u is skew-symmetric, we choose the n_u free variables μ_{ik} as the coefficients of the standard orthonormal basis of the space of $n \times n$ skew-symmetric matrices. In particular

$$S_u = \sum_{1 \le i < k \le n} \mu_{ik} B_{ik}^{\mathbb{R}}$$

where $B_{ik}^{\mathbb{R}}$ are the elements of the basis. Denoting by E_{ik} an $n \times n$ matrix having one in position (i, k) and zero otherwise, the elements of the basis are defined as

$$B_{ik}^{\mathbb{R}} \doteq \frac{1}{\sqrt{2}}(E_{ik} - E_{ki}), \quad 1 \le i < k \le n. \tag{17.11}$$

The (i, k) entry of Z may now be expressed as

$$[Z]_{i,k} = \begin{cases} \mu_{ik}(\sigma_k - \sigma_i); & 1 \le i < k \le n; \\ \eta_i; & i = k,\ 1 \le i \le n. \end{cases}$$

To compute the Jacobian $J(Z \to S_u, \mathrm{d}\Sigma)$, we construct the following scheme of partial derivatives

	$\begin{matrix}[Z]_{i,i} \\ 1 \le i \le n\end{matrix}$	$\begin{matrix}[Z]_{i,k} \\ 1 \le i < k \le n\end{matrix}$
$\begin{matrix}\eta_i; \\ 1 \le i \le n\end{matrix}$	I	0
$\begin{matrix}\mu_{ik}; \\ 1 \le i < k \le n\end{matrix}$	0	$C - D$

where

$$C = \frac{1}{\sqrt{2}}\text{bdiag}(\text{diag}\left([\sigma_2 \cdots \sigma_n]\right), \text{diag}\left([\sigma_3 \cdots \sigma_n]\right), \ldots, \sigma_n);$$

$$D = \frac{1}{\sqrt{2}}\text{bdiag}(\sigma_1 I_{n-1}, \sigma_2 I_{n-2}, \ldots, \sigma_{n-2}I_2, \sigma_{n-1}).$$

The matrix of partial derivatives is block diagonal and, therefore, its determinant is given by

$$J(Z \to S_u, d\Sigma) = |C - D| = 2^{\frac{n}{4}(1-n)} \prod_{1 \le i < k \le n} (\sigma_i - \sigma_k).$$

Now, from (17.9) it follows that

$$f_{\mathbf{U},\mathbf{\Sigma}}(U, \Sigma) = g(\bar{\sigma}) \, 2^{\frac{n}{4}(1-n)} \prod_{1 \le i < k \le n} (\sigma_i - \sigma_k). \tag{17.12}$$

From this equation we immediately conclude that $\mathbf{U}$ and $\mathbf{\Sigma}$ are statistically independent. It also follows that $f_{\mathbf{U}}(U)$ is constant over its domain, and this proves (17.6). Finally, integrating (17.12) with respect to U we get

$$f_{\mathbf{\Sigma}}(\Sigma) = \int f_{\mathbf{U},\mathbf{\Sigma}}(U, \Sigma)dU = \Upsilon_{\mathbb{S}} \, g(\bar{\sigma}) \prod_{1 \le i < k \le n} (\sigma_i - \sigma_k).$$

The constant $\Upsilon_{\mathbb{S}}$ is given by

$$\Upsilon_{\mathbb{S}} = 2^{\frac{n}{4}(1-n)} \int_{\mathcal{R}^{n,n}} dU$$

where (see for instance [140])

$$\int_{\mathcal{R}^{n,n}} dU = \frac{1}{2^n} \int_{\mathcal{G}_{\mathcal{O}}^n} dU = \frac{(8\pi)^{\frac{n}{4}(n-1)}}{2^n} \prod_{i=1}^{n} \frac{\Gamma\left(\frac{i-1}{2}\right)}{\Gamma(i-1)}.$$

Simple computations finally lead to equation (17.8). □

This theorem is closely related to a classical result, given in [12], on the density function of the eigenvalues of a symmetric matrix whose density depends only on its eigenvalues.

Remark 17.8 (Symmetric σ radial matrices). A result analogous to Theorem 17.3 can be easily obtained for symmetric (not necessarily positive definite) σ radial matrices $\mathbf{X} \in \mathbb{S}^n$. In this case we consider the factorization

$\mathbf{X} = \mathbf{U\Sigma S U}^T$, where $\mathbf{U}$ and $\mathbf{\Sigma}$ are given in Definition 17.4, and $\mathbf{S}$ is a diagonal matrix of signs. In this case, the densities of $\mathbf{U}$ and $\mathbf{\Sigma}$ are as in Theorem 17.3 and the random signs in $\mathbf{S}$ are uniform. Therefore, the joint density of the factors is given by

$$f_{\mathbf{U,\Sigma,S}}(U, \Sigma, S) = \frac{1}{2^n} f_{\mathbf{U}}(U) f_{\mathbf{\Sigma}}(\Sigma)$$

where the constant $1/2^n$ takes into account all possible combinations of signs.

◁

The following corollary, based on the results of Theorem 17.3, provides a characterization of positive definite matrices uniformly distributed in the σ norm ball

$$\mathcal{B}_\sigma(\mathbb{S}_+^n) \doteq \{X \in \mathbb{S}^n, X \succ 0 : \bar{\sigma}(X) \leq 1\}.$$

Corollary 17.3 (Uniform positive definite matrices in $\mathcal{B}_\sigma(\mathbb{S}_+^n)$). *Let the positive definite random matrix* $\mathbf{X} \in \mathbb{S}^n$, $\mathbf{X} \succ 0$, *be factored as* $\mathbf{U\Sigma U}^T$ *according to Definition 17.4, with* $\boldsymbol{\sigma} \in \mathcal{D}_\sigma$ *and* $\mathbf{U} \in \mathcal{R}^{n,n}$. *The following statements are equivalent:*

1. $\mathbf{X}$ *is uniformly distributed in* $\mathcal{B}_\sigma(\mathbb{S}_+^n)$;
2. $\mathbf{U}$ *and* $\mathbf{\Sigma}$ *are independent, and their densities are given by*

$$f_{\mathbf{U}}(U) = \mathcal{U}_{\mathcal{R}^{n,n}};$$
$$f_{\mathbf{\Sigma}}(\Sigma) = K_{\mathbb{S}} \prod_{1 \leq i < k \leq n} (\sigma_i - \sigma_k) \tag{17.13}$$

where the normalization constant $K_{\mathbb{S}}$ *is*

$$K_{\mathbb{S}} = \pi^{\frac{n}{2}} \prod_{i=1}^n \frac{\Gamma\left(\frac{n+i}{2} + 1\right)}{\Gamma\left(\frac{i}{2}\right) \Gamma^2\left(\frac{i+1}{2}\right)}.$$

Proof. To obtain (17.13) we substitute the defining function (17.1) of the uniform pdf in $\mathcal{B}_\sigma(\mathbb{S}_+^n)$ into equation (17.7) of Theorem 17.3. The normalization constant $K_{\mathbb{S}}$ can be computed by imposing

$$\int_{\mathcal{D}_\sigma} f_{\mathbf{\Sigma}}(\Sigma) d\Sigma = 1.$$

To solve this integral we use a standard technique (see e.g. [186]) that consists in removing the ordering condition on the singular values, introducing an absolute value sign, and dividing the resulting integral by $n!$. That is, we obtain

$$\int_{\mathcal{D}_\sigma} K_{\mathbb{S}} \prod_{1 \leq i < k \leq n} (\sigma_i - \sigma_k) d\sigma = \frac{1}{n!} \int_0^1 \cdots \int_0^1 K_{\mathbb{S}} \prod_{1 \leq i < k \leq n} |\sigma_i - \sigma_k| d\sigma.$$

The corollary is then proved, noticing that the right-hand side of this equation is a Selberg integral with parameters $\gamma = 1/2$, $\alpha = \beta = 1$, see Appendix A.3.

□

Remark 17.9 (Volumes of $\mathcal{B}_\sigma(r, \mathbb{S}_+^n)$ and $\mathcal{B}_\sigma(r, \mathbb{S}^n)$). From the proof of Corollary 17.3, comparing equations (17.7) and (17.13), we derive a closed-form expression for the volume of the σ norm ball of radius r in the space of positive definite matrices

$$\mathrm{Vol}\big(\mathcal{B}_\sigma(r, \mathbb{S}_+^n)\big) = \frac{\Upsilon_\mathbb{S}}{K_\mathbb{S}} = \pi^{\frac{n}{4}(n-1)} \prod_{i=1}^n \frac{\Gamma^2\left(\frac{i+1}{2}\right)}{\Gamma\left(\frac{n+i}{2}+1\right)} r^{n^2}.$$

Similarly, uniform symmetric matrices in $\mathcal{B}_\sigma(\mathbb{S}^n)$ (not necessarily positive definite) have the same singular values density (17.13), but the volume of the norm ball is in this case given by

$$\mathrm{Vol}(\mathcal{B}_\sigma(r, \mathbb{S}^n)) = 2^n \pi^{\frac{n}{4}(n-1)} \prod_{i=1}^n \frac{\Gamma^2\left(\frac{i+1}{2}\right)}{\Gamma\left(\frac{n+i}{2}+1\right)} r^{n^2}.$$

★

Next, we consider the general case of rectangular real matrices with σ radial density.

17.3.2 Real σ radial matrix densities

Consider the normalized singular value decomposition given below.

Definition 17.5 (Normalized SVD of real matrices). *Any matrix $X \in \mathbb{R}^{n,m}$ can be written in the form*

$$X = U\Sigma V^T$$

where $\Sigma = \mathrm{diag}\,(\sigma)$, $\sigma = [\sigma_1 \cdots \sigma_n]^T$, with $\sigma_1 \geq \cdots \geq \sigma_n \geq 0$, $U \in \mathbb{R}^{n,n}$ and $V \in \mathbb{R}^{m,n}$ have orthonormal columns, and V is normalized so that the first nonvanishing component of each column is positive.

We now state a result that relates the pdf of a σ radial real matrix to the pdfs of its SVD factors. In order to make the SVD mapping one-to-one, the next theorem requires the additional conditions $V \in \mathcal{R}^{m,n}$ and $\sigma \in \mathcal{D}_\sigma$ (see Definitions (17.4) and (17.5)). This issue is also discussed in Remark 17.7 for positive definite matrices, and a similar reasoning applies here.

Theorem 17.4 (σ radial matrices in $\mathbb{R}^{n,m}$). *Let the real random matrix $X \in \mathbb{R}^{n,m}$, $m \geq n$, be factored as $U\Sigma V^T$ according to Definition 17.5, with $\sigma \in \mathcal{D}_\sigma$ and $V \in \mathcal{R}^{m,n}$. The following statements are equivalent:*

1. $\mathbf{X}$ *is* σ *radial with defining function* $g(\bar{\sigma})$;
2. $\mathbf{U}, \mathbf{\Sigma}$ *and* $\mathbf{V}$ *are independent, and their densities are given by*

$$f_{\mathbf{U}}(U) = \mathcal{U}_{\mathcal{G}_{\mathbb{O}}^n};$$ (17.14)

$$f_{\mathbf{\Sigma}}(\Sigma) = \Upsilon_{\mathbb{R}} g(\bar{\sigma}) \prod_{i=1}^{n} \sigma_i^{m-n} \prod_{1 \leq i < k \leq n} (\sigma_i^2 - \sigma_k^2);$$ (17.15)

$$f_{\mathbf{V}}(V) = \mathcal{U}_{\mathcal{R}^{m,n}}$$ (17.16)

where the normalization constant $\Upsilon_{\mathbb{R}}$ *is*

$$\Upsilon_{\mathbb{R}} = 2^n \pi^{\frac{n}{2}(m+1)} \prod_{i=1}^{n} \frac{1}{\Gamma\left(\frac{n-i+1}{2}\right) \Gamma\left(\frac{m-i+1}{2}\right)}.$$ (17.17)

Proof. This proof follows the lines of that of Theorem 17.3. The SVD $X = U\Sigma V^T$, with the strict ordering of the singular values $\sigma \in \mathcal{D}_\sigma$ and the normalization condition on the columns of $V \in \mathcal{R}^{m,n}$, is one-to-one. Using Theorem A.1 (see Appendix), the joint pdf of the random matrices $\mathbf{U}, \mathbf{\Sigma}, \mathbf{V}$ is

$$f_{\mathbf{U},\mathbf{\Sigma},\mathbf{V}}(U, \Sigma, V) = g(\bar{\sigma}) J(X \to U, \Sigma, V).$$ (17.18)

The differential of X is given by

$$\mathrm{d}X = \mathrm{d}U \, \Sigma \, V^T + U \, \mathrm{d}\Sigma \, V^T + U \, \Sigma \, \mathrm{d}V^T.$$ (17.19)

If $m > n$, then let $V_1 \in \mathbb{R}^{m,m-n}$ be such that $\bar{V} \doteq [V \quad V_1]$ is orthogonal; otherwise, if $m = n$, then let $\bar{V} = V$. Then, multiplying (17.19) by U^T on the left and by $\bar{V}$ on the right, we obtain

$$Z \doteq U^T \mathrm{d}X \bar{V} = [U^T \mathrm{d}U \Sigma \quad 0] + [\mathrm{d}\Sigma \quad 0] + \Sigma \mathrm{d}V^T \bar{V}.$$

Proceeding as in the proof of Theorem 17.3, we write

$$\begin{aligned} J(X \to U, \Sigma, V) &= J(\mathrm{d}X \to \mathrm{d}U, \mathrm{d}\Sigma, \mathrm{d}V) \\ &= J(\mathrm{d}X \to Z) J(Z \to \mathrm{d}U, \mathrm{d}\Sigma, \mathrm{d}V) \\ &= J(Z \to \mathrm{d}U, \mathrm{d}\Sigma, \mathrm{d}V) \end{aligned}$$ (17.20)

since $J(\mathrm{d}X \to Z) = 1$. Next, we rewrite this equation in the form

$$Z = [S_u \Sigma \quad 0] + [\mathrm{d}\Sigma \quad 0] + \Sigma \bar{S}_v^T$$

where

$$S_u \doteq U^T \mathrm{d}U \in \mathbb{R}^{n,n} \text{ and } \bar{S}_v \doteq \bar{V}^T \mathrm{d}V \in \mathbb{R}^{m,n}.$$

By Rule A.4, the Jacobian $J(S_u, \mathrm{d}\Sigma, \bar{S}_v \to \mathrm{d}U, \mathrm{d}\Sigma, \mathrm{d}V)$ is equal to one and, applying the chain rule, we have

$$J(Z \to dU, d\Sigma, dV) = J(Z \to S_u, d\Sigma, \bar{S}_v)J(S_u, d\Sigma, \bar{S}_v \to dU, d\Sigma, dV)$$
$$= J(Z \to S_u, d\Sigma, \bar{S}_v). \qquad (17.21)$$

We now concentrate on the evaluation of $J(Z \to S_u, d\Sigma, \bar{S}_v)$. We notice that S_u is skew-symmetric and, if $m > n$, $\bar{S}_v$ can be partitioned as

$$\bar{S}_v = \begin{bmatrix} S_v \\ Q^T \end{bmatrix}$$

where $S_v \doteq V^T dV \in \mathbb{R}^{n,n}$ is skew-symmetric and $Q \doteq dV^T V_1 \in \mathbb{R}^{n,m-n}$. The matrix Z is finally rewritten in the form

$$Z = [S_u \Sigma - \Sigma S_v + d\Sigma \quad \Sigma Q].$$

Clearly, if $m = n$, then $\bar{S}_v \equiv S_v$ and $Z = S_u \Sigma - \Sigma S_v + d\Sigma$.

We now examine the number of free variables that describe the quantities of interest. The matrix X is described by means of nm real variables. The orthogonal matrix U is described by $n_u \doteq \frac{n}{2}(n-1)$ real variables and Σ is described by means of its n diagonal entries; therefore, V is described by the remaining $n_v \doteq nm - \frac{n}{2}(n-1) - n = n(m-n) + \frac{n}{2}(n-1)$ real variables.[1] The differentials dU, dV, and $d\Sigma$ are described by the same number of free variables as, respectively, U, V, and Σ. Therefore, S_u and $\bar{S}_v^T = [S_v^T \quad Q]$ are described by n_u and n_v variables respectively. Since $d\Sigma$ is diagonal, we choose its n free variables as the diagonal entries $\eta_i = d\sigma_i$, $1 \le i \le n$. Since S_u is skew-symmetric, we choose the n_u free variables μ_{ik} as the coefficients of the standard orthonormal basis of the space of $n \times n$ skew-symmetric matrices. Therefore, using the notation introduced in the proof of Theorem 17.3, we write

$$S_u = \sum_{1 \le i < k \le n} \mu_{ik} B_{ik}^{\mathbb{R}}.$$

Similarly, considering that the matrix $\bar{S}_v = [S_v^T \quad Q]^T$ is the first (block) column of the $m \times m$ skew-symmetric matrix $\bar{V}^T d\bar{V}$, we choose $n(n-1)/2$ free variables ν_{ik} such that

$$S_v = \sum_{1 \le i < k \le n} \nu_{ik} B_{ik}^{\mathbb{R}}$$

where $B_{ik}^{\mathbb{R}}$ are defined in (17.11). The remaining $n_v - n(n-1)/2 = n(m-n)$ free variables are needed to describe Q. Hence, we write

---

[1] The number $n_v = nm - \frac{n}{2}(n+1)$ of free variables needed to represent an $m \times n$ matrix $V = [v_1 \cdots v_n]$, $m \ge n$ with orthonormal columns can be constructed as follows: the first column v_1 can be chosen in $m - 1$ different ways (m free variables with one norm constraint), v_2 can be chosen in $m - 2$ different ways (m free variables with one norm constraint and one orthogonality constraint), ..., v_n can be chosen in $m - n$ different ways.

$$[Q]_{i,k} = \frac{1}{\sqrt{2}} q_{ik}, \quad 1 \le i \le n; 1 \le k \le m - n.$$

The (i, k) entry of Z may now be expressed as

$$[Z]_{i,k} = \begin{cases} \mu_{ik}\sigma_k - \nu_{ik}\sigma_i; & 1 \le i < k \le n; \\ -\mu_{ki}\sigma_k + \nu_{ki}\sigma_i; & 1 \le k < i \le n; \\ \eta_i; & i = k, 1 \le i \le n; \\ q_{ir}\sigma_i; & r = k - n; 1 \le i \le n; n < k \le m. \end{cases}$$

To compute the Jacobian $J(S_u, \mathrm{d}\Sigma, \bar{S}_v; Z)$, we construct the following scheme of partial derivatives

	$[Z]_{i,i}$ $1 \le i \le n$	$[Z]_{i,k}$ $1 \le i < k \le n$	$[Z]_{i,k}$ $1 \le k < i \le n$	$[Z]_{i,k}$ $1 \le i \le n; n < k \le m$
$i;$ $1 \le i \le n$	I	0	0	0
$\mu_{ik};$ $1 \le i < k \le n$	0	C	$-D$	0
$\nu_{ik};$ $1 \le i < k \le n$	0	$-D$	C	0
$q_{ik};$ $1 \le i \le n;$ $1 \le k \le m - m$	0	0	0	F

where

$$C = \frac{1}{\sqrt{2}} \mathrm{bdiag}(\mathrm{diag}([\sigma_2 \cdots \sigma_n]), \mathrm{diag}([\sigma_3 \cdots \sigma_n]), \dots, \sigma_n);$$

$$D = \frac{1}{\sqrt{2}} \mathrm{bdiag}(\sigma_1 I_{n-1}, \sigma_2 I_{n-2}, \dots, \sigma_{n-2} I_2, \sigma_{n-1});$$

$$F = \frac{1}{\sqrt{2}} \mathrm{bdiag}(\sigma_1 I_{m-n}, \sigma_2 I_{m-n}, \dots, \sigma_{n-1} I_{m-n}, \sigma_n I_{m-n}).$$

The matrix of partial derivatives is block diagonal and, therefore, its determinant is given by

$$J(Z \to S_u, \mathrm{d}\Sigma, \bar{S}_v) = |F| \begin{vmatrix} C & -D \\ -D & C \end{vmatrix}.$$

Using the Schur complement, we have

$$J(Z \to S_u, \mathrm{d}\Sigma, \bar{S}_v) = |F||C^2 - D^2| = 2^{\frac{n}{2}(1-m)} \prod_{i=1}^{n} \sigma_i^{m-n} \prod_{1 \le i < k \le n} (\sigma_i^2 - \sigma_k^2).$$

Now, from (17.18), (17.20), and (17.21) it follows that

$$f_{\mathbf{U},\boldsymbol{\Sigma},\mathbf{V}}(U, \Sigma, V) = g(\bar{\sigma}) 2^{\frac{n}{2}(1-m)} \prod_{i=1}^{n} \sigma_i^{m-n} \prod_{1 \le i < k \le n} (\sigma_i^2 - \sigma_k^2). \tag{17.22}$$

From this equation we immediately obtain that $\mathbf{U}, \mathbf{\Sigma}, \mathbf{V}$ are statistically independent. It also follows that $f_{\mathbf{U}}(U)$ and $f_{\mathbf{V}}(V)$ are constant over their respective domains, which proves (17.14) and (17.16). Finally, integrating (17.22) with respect to U and V we get the marginal density (17.15)

$$
f_{\mathbf{\Sigma}}(\Sigma) = \int \cdots \int f_{\mathbf{U}, \mathbf{\Sigma}, \mathbf{V}}(U, \Sigma, V) \mathrm{d}U \mathrm{d}V
$$

$$
= \Upsilon_{\mathbb{R}} g(\bar{\sigma}) \prod_{i=1}^{n} \sigma_i^{m-n} \prod_{1 \leq i < k \leq n} (\sigma_i^2 - \sigma_k^2)
$$

where $\Upsilon_{\mathbb{R}}$ is a constant computed as

$$
\Upsilon_{\mathbb{R}} = 2^{\frac{n}{2}(1-m)} \int_{\mathcal{G}_{\mathcal{O}}^n} \mathrm{d}U \int_{\mathcal{R}^{m,n}} \mathrm{d}V.
$$

The above integrals evaluate to (see for instance [140])

$$
\int_{\mathcal{G}_{\mathcal{O}}^n} \mathrm{d}U = (8\pi)^{\frac{n}{4}(n-1)} \prod_{i=1}^{n} \frac{\Gamma\left(\frac{i-1}{2}\right)}{\Gamma(i-1)};
$$

$$
\int_{\mathcal{R}^{m,n}} \mathrm{d}V = \frac{(8\pi)^{\frac{mn}{2} - \frac{n}{4}(n+1)}}{2^n} \prod_{i=m-n+1}^{m} \frac{\Gamma\left(\frac{i-1}{2}\right)}{\Gamma(i-1)}
$$

where, for continuity, we take $\frac{\Gamma\left(\frac{i-1}{2}\right)}{\Gamma(i-1)} = 2$ for $i = 1$. Finally, simple computations lead to equation (17.17). $\qquad\square$

The following corollary, based on the results of Theorem 17.3, provides a characterization of real matrices uniformly distributed in $\mathcal{B}_\sigma(\mathbb{R}^{n,m})$.

Corollary 17.4 (Uniform real matrices in $\mathcal{B}_\sigma(\mathbb{R}^{n,m})$). *Let the real random matrix $\mathbf{X} \in \mathbb{R}^{n,m}$, $m \geq n$, be factored as $\mathbf{U}\mathbf{\Sigma}\mathbf{V}^T$ according to Definition 17.5, with $\sigma \in \mathcal{D}_\sigma$ and $\mathbf{V} \in \mathcal{R}^{m,n}$. The following statements are equivalent:*

1. *$\mathbf{X}$ is uniformly distributed in $\mathcal{B}_\sigma(\mathbb{R}^{n,m})$;*
2. *$\mathbf{U}, \mathbf{\Sigma}$ and $\mathbf{V}$ are independent, and their densities are given by*

$$
f_{\mathbf{U}}(U) = \mathcal{U}_{\mathcal{G}_{\mathcal{O}}^n};
$$

$$
f_{\mathbf{\Sigma}}(\Sigma) = K_{\mathbb{R}} \prod_{i=1}^{n} \sigma_i^{m-n} \prod_{1 \leq i < k \leq n} (\sigma_i^2 - \sigma_k^2); \tag{17.23}
$$

$$
f_{\mathbf{V}}(V) = \mathcal{U}_{\mathcal{R}^{m,n}}
$$

where the normalization constant $K_{\mathbb{R}}$ is

$$
K_{\mathbb{R}} = (4\pi)^{\frac{n}{2}} \prod_{i=1}^{n} \frac{\Gamma\left(\frac{m+i+1}{2}\right)}{\Gamma\left(\frac{i}{2}\right) \Gamma\left(\frac{i+1}{2}\right) \Gamma\left(\frac{m-n+i}{2}\right)}. \tag{17.24}
$$

Proof. The proof is similar that of Corollary 17.3. To obtain (17.23) we substitute the defining function (17.1) of the uniform pdf in equation (17.15).The normalization constant $K_{\mathbb{R}}$ can be computed by imposing

$$\int f_{\Sigma}(\Sigma)\mathrm{d}\Sigma = 1.$$

With some algebraic manipulations, we notice that this integral is a Selberg integral with parameters $\gamma = 1/2$, $\alpha = \frac{m-n+1}{2}$, $\beta = 1$, see Appendix A.3. Therefore, we obtain $K_{\mathbb{R}}$ as given in (17.24). $\qquad\square$

Remark 17.10 (Volume of $\mathcal{B}_{\sigma}(r, \mathbb{R}^{n,m})$). From the proof of Corollary 17.4, comparing equations (17.15) and (17.23), we derive a closed-form expression for the volume of the σ norm real ball of radius r

$$\mathrm{Vol}(\mathcal{B}_{\sigma}(r, \mathbb{R}^{n,m})) = \pi^{\frac{nm}{2}} \prod_{i=1}^{n} \frac{\Gamma\left(\frac{i+1}{2}\right)\Gamma\left(\frac{m-n+i}{2}\right)}{\Gamma\left(\frac{m+i+1}{2}\right)\Gamma\left(\frac{m-i+1}{2}\right)} r^{nm}. \qquad (17.25)$$

$\triangleleft$

Next, we derive similar results for complex matrices with σ radial density.

17.3.3 Complex σ radial matrix densities

To analyze the complex case, we first introduce the following normalized SVD decomposition.

Definition 17.6 (Normalized SVD of complex matrices). *Any matrix $X \in \mathbb{C}^{n,m}$ can be written in the form*

$$X = U\Sigma V^{*}$$

where $\Sigma = \mathrm{diag}(\sigma)$, $\sigma = [\sigma_1 \cdots \sigma_n]^{T}$, with $\sigma_1 \geq \cdots \geq \sigma_n \geq 0$, $U \in \mathbb{C}^{n,n}$ and $V \in \mathbb{C}^{m,n}$ have orthonormal columns, and V is normalized so that the first nonvanishing component of each column is real and positive.

Remark 17.11 (The unitary group). The set of unitary matrices in $\mathbb{C}^{n,n}$ forms a group, called the *unitary group*, which is denoted by

$$\mathcal{G}_{\mathcal{U}}^{n} \doteq \{U \in \mathbb{C}^{n,n} : U^{*}U = I\}. \qquad (17.26)$$

Similar to the real case, see Remark 17.5, we also define the following complex manifold

$$\mathcal{C}^{m,n} \doteq \{V \in \mathbb{C}^{m,n} : V^{*}V = I; \ \mathrm{Re}([V]_{1,i}) > 0, \mathrm{Im}([V]_{1,i}) = 0, i = 1, \ldots, n\}. \qquad (17.27)$$

$\triangleleft$

Remark 17.12 (Complex Haar invariant distributions). The uniform distribution over the unitary group $\mathcal{G}_{\mathcal{U}}^n$ is the (complex) Haar invariant distribution, here denoted as $\mathcal{U}_{\mathcal{G}_{\mathcal{U}}^n}$. Similarly, the uniform distribution over the complex manifold $\mathcal{C}^{n,n}$ of normalized unitary matrices is denoted as $\mathcal{U}_{\mathcal{C}^{n,n}}$.   ◁

The next result relates the pdf of a σ radial complex matrix to the pdfs of its SVD factors. In order to make the SVD mapping one-to-one, the theorem requires the additional conditions $V \in \mathcal{C}^{m,n}$ and $\sigma \in \mathcal{D}_\sigma$, see Remark 17.7 for a discussion regarding the case of positive definite matrices. A similar reasoning also applies to complex matrices.

Theorem 17.5 (σ radial matrices in $\mathbb{C}^{n,m}$). *Let the complex random matrix $\mathbf{X} \in \mathbb{C}^{n,m}$, $m \geq n$, be factored as $\mathbf{U\Sigma V}^*$ according to Definition 17.6, with $\boldsymbol{\sigma} \in \mathcal{D}_\sigma$ and $\mathbf{V} \in \mathcal{C}^{m,n}$. The following statements are equivalent:*

1. *$\mathbf{X}$ is σ radial with defining function $g(\bar{\sigma})$;*
2. *$\mathbf{U}$, $\mathbf{\Sigma}$ and $\mathbf{V}$ are independent, and their densities are given by*

$$f_\mathbf{U}(U) = \mathcal{U}_{\mathcal{G}_{\mathcal{U}}^n};\tag{17.28}$$

$$f_\mathbf{\Sigma}(\Sigma) = \Upsilon_C g(\bar{\sigma}) \prod_{i=1}^{n} \sigma_i^{2(m-n)+1} \prod_{1 \leq i < k \leq n} (\sigma_i^2 - \sigma_k^2)^2;\tag{17.29}$$

$$f_\mathbf{V}(V) = \mathcal{U}_{\mathcal{C}^{m,n}}\tag{17.30}$$

where the normalization constant Υ_C is

$$\Upsilon_C = 2^n \pi^{nm} \prod_{i=1}^{n} \frac{1}{\Gamma(n-i+1)\Gamma(m-i+1)}.\tag{17.31}$$

Proof. This proof follows the lines of that of Theorem 17.4. Indeed, the derivation up to the expression of Z as

$$Z = [S_u \Sigma - \Sigma S_v + \mathrm{d}\Sigma \quad \Sigma Q]$$

is identical to the real case, considering that all the quantities involved are now complex (and, therefore, matrix transpose should be treated as conjugate transpose), and S_v, S_u are skew-Hermitian. In particular, we have that

$$J(X \to U, \Sigma, V) = J(Z \to S_u, \mathrm{d}\Sigma, \bar{S}_v).\tag{17.32}$$

We now examine the number of free variables that describe the quantities of interest. The matrix X is described by means of $2nm$ real variables. The unitary matrix U is described by $n_u \doteq n^2$ real variables and Σ is described by means of its n diagonal entries; therefore, V is described by the remaining $n_v \doteq 2nm - n^2 - n$ real variables. Since an $m \times n$ complex matrix with orthonormal columns is described by $2nm - n^2$ variables, see e.g. [140], we notice that the normalization imposed by Definition 17.6 on the columns of V fixes n of the free variables. The differentials $\mathrm{d}U$, $\mathrm{d}V$, $\mathrm{d}\Sigma$ are described by the same number of free variables as U, V, and Σ. Therefore, S_u and $\bar{S}_v^* = \begin{bmatrix} S_v^* & Q \end{bmatrix}$ are described by n_u and n_v variables respectively. Since $d\Sigma$ is real diagonal, we choose its n free variables as the diagonal entries $\eta_i = d\sigma_i$, $1 \le i \le n$. Since S_u is skew-Hermitian, we choose the n_u free variables $\mu_{ik}^{\mathbb{R}}$, $\mu_{ik}^{\mathbb{I}}$ as the coefficients of the standard orthonormal basis of the space of $n \times n$ skew-Hermitian matrices. In particular, we write

$$S_u = \sum_{1 \le i < k \le n} (\mu_{ik}^{\mathbb{R}} B_{ik}^{\mathbb{R}} + \mu_{ik}^{\mathbb{I}} B_{ik}^{\mathbb{I}}) + \sum_{i=1}^{n} \mu_{ii}^{\mathbb{I}} D_i$$

where $B_{ik}^{\mathbb{R}}, B_{ik}^{\mathbb{I}}$, and D_i are the elements of the basis. Denoting by E_{ik} an $n \times n$ matrix having one in position (i, k) and zero otherwise, the elements of the basis are defined as

$$B_{ik}^{\mathbb{R}} \doteq \frac{1}{\sqrt{2}}(E_{ik} - E_{ki}), \quad 1 \le i < k \le n;$$

$$B_{ik}^{\mathbb{I}} \doteq \frac{j}{\sqrt{2}}(E_{ik} + E_{ki}), \quad 1 \le i < k \le n;$$

$$D_i \doteq j E_{ii}, \; 1 \le i \le n.$$

Similarly, considering that the matrix $\bar{S}_v = \begin{bmatrix} S_v^* & Q \end{bmatrix}$ is the first (block) column of the $m \times m$ skew-Hermitian matrix $\bar{V}^* d\bar{V}$, we choose $n^2 - n$ free variables $\nu_{ik}^{\mathbb{R}}, \nu_{ik}^{\mathbb{I}}$ such that

$$S_v = \sum_{1 \le i < k \le n} (\nu_{ik}^{\mathbb{R}} B_{ik}^{\mathbb{R}} + \nu_{ik}^{\mathbb{I}} B_{ik}^{\mathbb{I}}) + \sum_{i=1}^{n} h_i(\nu^{\mathbb{R}}, \nu^{\mathbb{I}}) D_i$$

where $h_i(\nu^{\mathbb{R}}, \nu^{\mathbb{I}})$ is a function of the variables $\nu^{\mathbb{R}}, \nu^{\mathbb{I}}$. The remaining $n_v - (n^2 - n) = 2n(m - n)$ free variables $q_{ik}^{\mathbb{R}}$, $q_{ik}^{\mathbb{I}}$ are needed to describe Q

$$[Q]_{i,k} = \frac{1}{\sqrt{2}}(q_{ik}^{\mathbb{R}} + j q_{ik}^{\mathbb{I}}), \; 1 \le i \le n; 1 \le k \le m - n.$$

The (i, k) entry of Z may now be expressed as

$$[Z]_{i,k} = \begin{cases} \frac{1}{\sqrt{2}}(\mu_{ik}^{\mathbb{R}} + j\mu_{ik}^{\mathbb{I}})\sigma_k - \frac{1}{\sqrt{2}}(\nu_{ik}^{\mathbb{R}} + j\nu_{ik}^{\mathbb{I}})\sigma_i; & 1 \le i < k \le n; \\ -\frac{1}{\sqrt{2}}(\mu_{ki}^{\mathbb{R}} + j\mu_{ki}^{\mathbb{I}})\sigma_k + \frac{1}{\sqrt{2}}(\nu_{ki}^{\mathbb{R}} - j\nu_{ki}^{\mathbb{I}})\sigma_i; & 1 \le k < i \le n; \\ \eta_i + j\mu_{ii}^{\mathbb{I}}\sigma_i - jh_i(\nu^{\mathbb{R}}, \nu^{\mathbb{I}})\sigma_i; & i = k, 1 \le i \le n; \\ \frac{1}{\sqrt{2}}(q_{ir}^{\mathbb{R}} + j q_{ir}^{\mathbb{I}})\sigma_i; & r = k - n; \\ & 1 \le i \le n; n < k \le m. \end{cases}$$

To compute the Jacobian $J(Z \to S_u, d\Sigma, \bar{S}_v)$, we construct the following scheme of partial derivatives

	$\mathrm{Re}([Z]_{i,i})$ $1 \leq i \leq n$	$\mathrm{Im}([Z]_{i,i})$ $1 \leq i \leq n$	$\mathrm{Re}([Z]_{i,k})$ $1 \leq i < k \leq n$	$\mathrm{Re}([Z]_{i,k})$ $1 \leq k < i \leq n$	$\mathrm{Im}([Z]_{i,k})$ $1 \leq i < k \leq n$	$\mathrm{Im}([Z]_{i,k})$ $1 \leq k < i \leq n$	$\mathrm{Re}([Z]_{i,k})$ $1 \leq i \leq n;\ n < k \leq m$	$\mathrm{Im}([Z]_{i,k})$ $1 \leq i \leq n;\ n < k \leq m$
i	I	0	0	0	0	0	0	0
μ^I_{ii} $1 \leq i \leq n$	0	Σ	0	0	0	0	0	0
μ^R_{ik}	0	0	C	$-D$	0	0	0	0
ν^R_{ik} $1 \leq i < k \leq n$	0	$H^{\mathbb{R}}$	$-D$	C	0	0	0	0
μ^I_{ik}	0	0	0	0	C	D	0	0
ν^I_{ik} $1 \leq i < k \leq n$	0	$H^{\mathbb{I}}$	0	0	$-D$	$-C$	0	0
q^R_{ik}	0	0	0	0	0	0	F	0
q^I_{ik} $1 \leq i \leq n$ $1 \leq k \leq m-n$	0	0	0	0	0	0	0	F

where

$$C = \frac{1}{\sqrt{2}}\mathrm{bdiag}(\mathrm{diag}\left([\sigma_2 \cdots \sigma_n]\right), \mathrm{diag}\left([\sigma_3 \cdots \sigma_n]\right), \ldots, \sigma_n);$$

$$D = \frac{1}{\sqrt{2}}\mathrm{bdiag}(\sigma_1 I_{n-1}, \sigma_2 I_{n-2}, \ldots, \sigma_{n-2}I_2, \sigma_{n-1});$$

$$F = \frac{1}{\sqrt{2}}\mathrm{bdiag}(\sigma_1 I_{m-n}, \sigma_2 I_{m-n}, \ldots, \sigma_{n-1}I_{m-n}, \sigma_n I_{m-n}).$$

Since the matrix of partial derivatives is block triangular, the matrices $H^{\mathbb{R}}, H^{\mathbb{I}}$ do not affect the value of its determinant. Therefore, we obtain

$$J(Z \to S_u, \mathrm{d}\Sigma, \bar{S}_v) = |\Sigma||F|^2 \begin{vmatrix} C & -D \\ -D & C \end{vmatrix} \begin{vmatrix} C & D \\ -D & -C \end{vmatrix}.$$

Using the Schur complement, we have that

$$J(Z \to S_u, \mathrm{d}\Sigma, \bar{S}_v) = |\Sigma||F|^2|C^2 - D^2|^2$$

$$= 2^{n(1-m)} \prod_{i=1}^{n} \sigma_i^{2(m-n)+1} \prod_{1 \leq i < k \leq n} (\sigma_i^2 - \sigma_k^2)^2. \quad (17.33)$$

Now, by means of Theorem A.1 (see Appendix), we write the joint pdf of the random matrices $\mathbf{U}, \boldsymbol{\Sigma}, \mathbf{V}$ as

$$f_{\mathbf{U},\boldsymbol{\Sigma},\mathbf{V}}(U, \Sigma, V) = g(\bar{\sigma})J(X \to U, \Sigma, V).$$

Using equations (17.32) and (17.33) we immediately obtain

$$f_{\mathbf{U},\boldsymbol{\Sigma},\mathbf{V}}(U, \Sigma, V) = g(\bar{\sigma})2^{n(1-m)} \prod_{i=1}^{n} \sigma_i^{2(m-n)+1} \prod_{1 \leq i < k \leq n} (\sigma_i^2 - \sigma_k^2)^2. \quad (17.34)$$

From this equation we conclude that $\mathbf{U}, \mathbf{\Sigma}, \mathbf{V}$ are statistically independent. It also follows that $f_{\mathbf{U}}(U)$ and $f_{\mathbf{V}}(V)$ are constant over their respective domains, which proves (17.28) and (17.30). Finally, integrating (17.34) with respect to U and V, we get the marginal density (17.29) as

$$f_{\mathbf{\Sigma}}(\Sigma) = \int \cdots \int f_{\mathbf{U},\mathbf{\Sigma},\mathbf{V}}(U, \Sigma, V) dU dV$$

$$= \Upsilon_{\mathbb{C}} g(\bar{\sigma}) \prod_{i=1}^{n} \sigma_i^{2(m-n)+1} \prod_{1 \le i < k \le n} (\sigma_i^2 - \sigma_k^2)^2$$

where the constant $\Upsilon_{\mathbb{C}}$ is given by

$$\Upsilon_{\mathbb{C}} = 2^{n(1-m)} \int_{\mathcal{G}_{\mathcal{U}}^n} dU \int_{\mathcal{C}^{m,n}} dV$$

and the measure of the unitary group $\mathcal{G}_{\mathcal{U}}^n$ and of the complex manifold $\mathcal{C}^{m,n}$ are given by (see for instance [140])

$$\int_{\mathcal{G}_{\mathcal{U}}^n} dU = (2\pi)^{n(n+1)/2} \prod_{i=1}^{n} \frac{1}{(n-i)!};$$

$$\int_{\mathcal{C}^{m,n}} dV = \frac{(2\pi)^{mn-n(n-1)/2}}{(2\pi)^n} \prod_{i=1}^{n} \frac{1}{(m-i)!}.$$

$\square$

The following corollary, based on the results of Theorem 17.5, provides a characterization of complex matrices uniformly distributed in $\mathcal{B}_\sigma$.

Corollary 17.5 (Uniform complex matrices in $\mathcal{B}_\sigma(\mathbb{C}^{n,m})$). *Let the complex random matrix $\mathbf{X} \in \mathbb{C}^{n,m}$, $m \ge n$, be factored as $\mathbf{U\Sigma V}^*$ according to Definition 17.6, with $\sigma \in \mathcal{D}_\sigma$ and $\mathbf{V} \in \mathcal{C}^{m,n}$. The following statements are equivalent:*

1. *$\mathbf{X}$ is uniformly distributed in $\mathcal{B}_\sigma(\mathbb{C}^{n,m})$;*
2. *$\mathbf{U}, \mathbf{\Sigma}$ and $\mathbf{V}$ are independent, and their densities are given by*

$$f_{\mathbf{U}}(U) = \mathcal{U}_{\mathcal{G}_{\mathcal{U}}^n};$$

$$f_{\mathbf{\Sigma}}(\Sigma) = K_{\mathbb{C}} \prod_{i=1}^{n} \sigma_i^{2(m-n)+1} \prod_{1 \le i < k \le n} (\sigma_i^2 - \sigma_k^2)^2; \tag{17.35}$$

$$f_{\mathbf{V}}(V) = \mathcal{U}_{\mathcal{C}^{m,n}}$$

where the normalization constant $K_{\mathbb{C}}$ is

$$K_{\mathbb{C}} = 2^n \prod_{i=1}^{n} \frac{\Gamma(m+i)}{\Gamma^2(i)\Gamma(m-n+i)}. \tag{17.36}$$

Proof. The proof is similar to that of Corollaries 17.3 and 17.4. To obtain (17.35) we substitute the defining function (17.1) of the uniform pdf in equation (17.29). The normalization constant $K_{\mathbb{C}}$ can be computed by imposing

$$\int f_{\Sigma}(\Sigma)\mathrm{d}\Sigma = 1.$$

With some algebraic manipulations, we notice that this is a Selberg integral with parameters $\gamma = 1$, $\alpha = m - n + 1$, $\beta = 1$, see Appendix A.3.   □

Remark 17.13 (Volume of $\mathcal{B}_{\sigma}(r, \mathbb{C}^{n,m})$). From the proof of the above corollary, comparing equations (17.29) and (17.35), we derive a closed-form expression for the volume of the σ norm complex ball of radius r

$$\mathrm{Vol}(\mathcal{B}_{\sigma}(r, \mathbb{C}^{n,m})) = \pi^{nm} \prod_{i=1}^{n} \frac{\Gamma(i)\Gamma(m - n + i)}{\Gamma(m + i)\Gamma(m - i + 1)} r^{2nm}. \tag{17.37}$$

◁

17.4 Statistical properties of unitarily invariant matrices

The results presented in the previous two sections can be immediately extended to the more general class of unitarily invariant random matrices defined below.

Definition 17.7 (Unitarily invariant matrices). *A random matrix $\mathbf{X} \in \mathbb{F}^{n,m}$ is unitarily invariant if its density function can be written as*

$$f_{\mathbf{X}}(X) = g(\Sigma) \tag{17.38}$$

where $g(\Sigma)$ is the defining function of $\mathbf{X}$ and Σ is the (diagonal) singular values matrix of X.

The name unitarily invariant follows from the fact that if $\mathbf{X} \in \mathbb{C}^{n,m}$ has a unitarily invariant density, then $Q\mathbf{X} \sim \mathbf{X}$ and $\mathbf{X}W \sim \mathbf{X}$, for any given unitary matrices Q and W. Clearly, any σ radial matrix is also unitarily invariant. However, there are some important examples of unitarily invariant matrices that are not σ radial. One of these examples is the so-called Wishart density, which is discussed next.

Example 17.3 (Wishart density). Let $\mathbf{X} = [\mathbf{x}_1 \cdots \mathbf{x}_m]$, where $\mathbf{x}_1, \ldots, \mathbf{x}_m \in \mathbb{R}^n$, $m \geq n$, are iid random vectors normally distributed, i.e. $\mathbf{x}_i \sim \mathcal{N}_{0,W}$, $i = 1, \ldots, m$. Construct the random matrix $\mathbf{Y} = \mathbf{X}\mathbf{X}^T$, $\mathbf{Y} \in \mathbb{S}^m$. It can be shown, see e.g. [12], that $\mathbf{Y}$ is positive definite (with probability one) and has the Wishart density, defined as

$$\mathcal{W}_W \doteq K_W |Y|^{(m-n-1)/2} e^{-\frac{1}{2}\operatorname{Tr} W^{-1}Y}, \quad Y \succ 0$$

where the normalization constant K_W is given by

$$K_W = \left(2^{nm/2} |W|^{m/2} \pi^{n(n-1)/4} \prod_{i=1}^{n} \Gamma\left((m-i+1)/2\right) \right)^{-1}.$$

When $W = I$ the Wishart density $\mathcal{W}_I$ is unitarily invariant. In fact, $|Y| = |\Sigma|$ and $\operatorname{Tr} Y = \operatorname{Tr} \Sigma$, where Σ is the singular values matrix of Y. Therefore

$$\mathcal{W}_I = g(\Sigma)$$

where the defining function $g(\Sigma)$ is given by

$$g(\Sigma) = K_W |\Sigma|^{(m-n-1)/2} e^{-\frac{1}{2}\operatorname{Tr}\Sigma}.$$

When $W \neq I$, the distribution of $\mathbf{Y}$ is no longer unitarily invariant, but it belongs to the more general class of elliptically contoured matrix distributions, which are studied for instance in [124].   ⋆

Example 17.4 (Radial densities in the Frobenius norm). For $p = 2$, the Hilbert–Schmidt ℓ_p matrix norm is also known as the Frobenius norm. For a matrix $X \in \mathbb{F}^{n,m}$, the Frobenius norm is given by

$$\|X\|_2 = \operatorname{Tr} XX^T = \|\Sigma\|_2$$

where Σ is the singular values matrix of X. Therefore, the Hilbert–Schmidt ℓ_2 radial densities studied in Section 17.1.1 are also unitarily invariant.   ⋆

The three theorems stated next, without proof, provide the extensions of Theorems 17.3, 17.4 and 17.5 to the case of random matrices with unitarily invariant distribution.

Theorem 17.6 (Unitarily invariant positive definite matrices). *Let the positive definite random matrix* $\mathbf{X} \in \mathbb{S}^n$, $\mathbf{X} \succ 0$, *be factored as* $\mathbf{U}\Sigma\mathbf{U}^T$ *according to Definition 17.4, with* $\boldsymbol{\sigma} \in \mathcal{D}_\sigma$ *and* $\mathbf{U} \in \mathcal{R}^{n,n}$. *The following statements are equivalent:*

1. $\mathbf{X}$ *is unitarily invariant with defining function* $g(\Sigma)$;
2. $\mathbf{U}$ *and* Σ *are independent, and their densities are given by*

$$f_{\mathbf{U}}(U) = \mathcal{U}_{\mathcal{R}^{n,n}};$$

$$f_{\Sigma}(\Sigma) = \Upsilon_{\mathbb{S}}\, g(\Sigma) \prod_{1 \leq i < k \leq n} (\sigma_i - \sigma_k)$$

where the normalization constant $\Upsilon_{\mathbb{S}}$ *is given in (17.8).*

Theorem 17.7 (Unitarily invariant real matrices). *Let the real random matrix* $\mathbf{X} \in \mathbb{R}^{n,m}$, $m \geq n$, *be factored as* $\mathbf{U\Sigma V}^T$ *according to Definition 17.5, with* $\boldsymbol{\sigma} \in \mathcal{D}_\sigma$ *and* $\mathbf{V} \in \mathcal{R}^{m,n}$. *The following statements are equivalent:*

1. $\mathbf{X}$ *is unitarily invariant with defining function* $g(\Sigma)$;
2. $\mathbf{U}$, $\boldsymbol{\Sigma}$ *and* $\mathbf{V}$ *are independent, and their densities are given by*

$$f_{\mathbf{U}}(U) = \mathcal{U}_{\mathcal{G}_{\mathcal{O}}^n};$$

$$f_{\boldsymbol{\Sigma}}(\Sigma) = \Upsilon_{\mathbb{R}} g(\Sigma) \prod_{i=1}^{n} \sigma_i^{m-n} \prod_{1 \leq i < k \leq n} (\sigma_i^2 - \sigma_k^2);$$

$$f_{\mathbf{V}}(V) = \mathcal{U}_{\mathcal{R}^{m,n}}$$

where the normalization constant $\Upsilon_{\mathbb{R}}$ *is given in (17.17).*

Theorem 17.8 (Unitarily invariant complex matrices). *Let the complex random matrix* $\mathbf{X} \in \mathbb{C}^{n,m}$, $m \geq n$, *be factored as* $\mathbf{U\Sigma V}^*$ *according to Definition 17.6, with* $\boldsymbol{\sigma} \in \mathcal{D}_\sigma$ *and* $\mathbf{V} \in \mathcal{C}^{m,n}$. *The following statements are equivalent:*

1. $\mathbf{X}$ *is unitarily invariant with defining function* $g(\Sigma)$;
2. $\mathbf{U}$, $\boldsymbol{\Sigma}$ *and* $\mathbf{V}$ *are independent, and their densities are given by*

$$f_{\mathbf{U}}(U) = \mathcal{U}_{\mathcal{G}_{\mathcal{U}}^n};$$

$$f_{\boldsymbol{\Sigma}}(\Sigma) = \Upsilon_{\mathbb{C}} g(\Sigma) \prod_{i=1}^{n} \sigma_i^{2(m-n)+1} \prod_{1 \leq i < k \leq n} (\sigma_i^2 - \sigma_k^2)^2;$$

$$f_{\mathbf{V}}(V) = \mathcal{U}_{\mathcal{C}^{m,n}}$$

where the normalization constant $\Upsilon_{\mathbb{C}}$ *is given in (17.31).*

18. Matrix Randomization Methods

In this chapter we present algorithms for uniform matrix sample generation in norm bounded sets. First, we discuss the simple case of matrix sampling in sets defined by ℓ_p Hilbert–Schmidt norm, which reduces to the vector ℓ_p norm randomization problem, and subsequently we present an efficient solution to the more challenging problem of uniform generation in sets defined by the spectral norm.

18.1 Uniform sampling in ℓ_p Hilbert–Schmidt norm balls

As discussed in Section 17.1.1, the vectorization operation $x = \text{vec}(X)$ defined in (3.8) reduces the ℓ_p Hilbert–Schmidt matrix sample generation problem into an equivalent problem concerning vector samples. Therefore, the vector randomization algorithms presented in Chapter 16 can be used directly for the ℓ_p Hilbert–Schmidt matrix sampling problem. A simple illustration of this idea is reported next for matrix generation in the Frobenius norm ball.

Example 18.1 (Uniform matrices in the Frobenius norm ball).
    Suppose we are interested in generating samples of a real matrix $\mathbf{X} \in \mathbb{R}^{2,3}$ uniformly distributed in the unit Frobenius norm ball, i.e. the ℓ_2 Hilbert–Schmidt norm ball

$$\mathcal{B}_{\|\cdot\|_2}(\mathbb{R}^{2,3}) = \left\{ X \in \mathbb{R}^{2,3} : \|X\|_2 \leq 1 \right\}.$$

Using the vectorization operator $x = \text{vec}(X)$, we observe that this problem is equivalent to the generation of uniform random vectors in the ℓ_2 norm ball $\mathcal{B}_{\|\cdot\|_2}(\mathbb{R}^6)$, which can be easily performed by means of Algorithm 16.1.   ⋆

    Next, we discuss uniform generation of matrix samples in the ℓ_p induced norm balls, for $p = 1$ and $p = \infty$.

18.2 Uniform sampling in ℓ_1 and ℓ_∞ induced norm balls

The algorithm for matrix sample generation in the ℓ_1 induced norm ball follows directly from Corollary 17.1. That is, to generate a matrix $\mathbf{X} \in \mathbb{R}^{n,m}$

with uniform distribution in $\mathcal{B}_{\|\cdot\|_1}(\mathbb{R}^{n,m})$, it suffices to generate its columns independently and uniformly in the ℓ_1 vector norm ball $\mathcal{B}_{\|\cdot\|_1}(\mathbb{R}^m)$ using Algorithm 16.1. The complex case follows from direct application of Corollary 17.2, and hence Algorithm 16.4 can be used for random generation. Similarly, to generate a matrix sample $\mathbf{X}$ (real or complex) uniformly distributed in the ℓ_∞ induced norm ball $\mathcal{B}_{\|\cdot\|_\infty}(\mathbb{F}^{n,m})$, it suffices to generate $\mathbf{X}^T$ (or $\mathbf{X}^*$, in the complex case) uniformly in the ℓ_1 induced norm ball $\mathcal{B}_{\|\cdot\|_1}(\mathbb{F}^{m,n})$.

We now discuss the problem of uniform generation in the spectral norm ball, which turns out to be technically more difficult than the cases previously considered. This difficulty is mainly due to the special structure of the spectral norm, which depends on the entries of the matrix only through an implicit relation given by the singular value decomposition. In Section 18.3 we present several sampling schemes based on rejection, and show that these methods become rapidly inefficient as the dimension of the matrix increases. This motivates the development of the algorithms subsequently studied in Sections 18.4 and 18.5.

18.3 Rejection methods for uniform matrix generation

A simple algorithm for generating uniform matrix samples in the ℓ_2 induced (spectral) norm unit ball $\mathcal{B}_\sigma$ is given by the rejection method from a bounding set discussed in Section 14.3.1. We notice that for a matrix $X \in \mathbb{F}^{n,m}$, $m \geq n$, the well-known norm inequalities hold

$$\|X\|_2 \leq \sqrt{n}\|X\|_2;$$
$$\|X\|_\infty \leq \|X\|_2$$

where $\|X\|_2 = \bar{\sigma}(X)$. These inequalities in turn imply the set inclusions

$$\mathcal{B}_\sigma(1, \mathbb{F}^{n,m}) \subseteq \mathcal{B}_{\|\cdot\|_2}(\sqrt{n}, \mathbb{F}^{n,m}); \tag{18.1}$$
$$\mathcal{B}_\sigma(1, \mathbb{F}^{n,m}) \subseteq \mathcal{B}_{\|\cdot\|_\infty}(1, \mathbb{F}^{n,m}). \tag{18.2}$$

A tighter set inclusion can also be obtained by considering the inequality

$$\|X\|_2 = \max_{\|y\|_2=1} \|Xy\|_2 = \max_{\|y\|_2=1} \left\| \sum_{i=1}^m \xi_i y_i \right\|_2 \geq \max_{i=1,\dots,m} \|\xi_i\|_2$$

where ξ_i is the ith column of X. In this case, it follows that

$$\mathcal{B}_\sigma(1, \mathbb{F}^{n,m}) \subseteq \mathcal{B}_{\text{col}}(1, \mathbb{F}^{n,m}) \doteq \left\{ X \in \mathbb{F}^{n,m} : \max_{i=1,\dots,m} \|\xi_i\|_2 \leq 1 \right\}.$$

Uniform generation in the bounding sets $\mathcal{B}_{\|\cdot\|_2}(\sqrt{n}, \mathbb{F}^{n,m})$ and $\mathcal{B}_{\|\cdot\|_\infty}(1, \mathbb{F}^{n,m})$ can be easily performed by means of the methods described in Section 18.1. Uniform sample generation in the set $\mathcal{B}_{\text{col}}(1, \mathbb{F}^{n,m})$ can also be easily obtained by generating independent columns ξ_i, $i = 1, \ldots, m$, uniformly distributed in the ℓ_2 norm ball $\mathcal{B}_{\|\cdot\|_2}(1, \mathbb{F}^n)$.

The efficiency of the rejection method is dictated by the rejection rate, defined in Section 14.3.1 as the expected number of samples that should be generated in the outer set in order to have one sample in the set of interest $\mathcal{B}_\sigma$. In the case of uniform densities, the rejection rate is given by the ratio of the volumes of the outer bounding set and the set of interest.

The volume of the spectral norm ball has been derived in (17.25) and (17.37) for real and complex matrices respectively. The volumes of the ℓ_2 and ℓ_∞ Hilbert–Schmidt norm balls may be computed using equations (15.8) and (15.13) for real and complex matrices respectively. As for the set $\mathcal{B}_{\text{col}}$, its volume may be derived as the product of the volumes of the m unit balls in the ℓ_2 norm in $\mathbb{F}^n$, obtaining

$$\text{Vol}(\mathcal{B}_{\text{col}}(1, \mathbb{R}^{n,m})) = \frac{(\pi)^{nm/2}}{\Gamma^m(\frac{n}{2} + 1)};$$

$$\text{Vol}(\mathcal{B}_{\text{col}}(1, \mathbb{C}^{n,m})) = \frac{(\pi)^{nm}}{\Gamma^m(n + 1)}.$$

Therefore, we can compute in closed form the rejection rates when different bounding sets are used. These rejection rates are reported in Table 18.1.

Table 18.1. Rejection rates for generating samples uniformly in $\mathcal{B}_\sigma(1, \mathbb{F}^{n,n})$ with overbounding sets given by $\mathcal{B}_{\|\cdot\|_\infty}(1, \mathbb{F}^{n,n})$, $\mathcal{B}_{\|\cdot\|_2}(\sqrt{n}, \mathbb{F}^{n,n})$ and $\mathcal{B}_{\text{col}}(1, \mathbb{F}^{n,n})$.

	$\mathcal{B}_\sigma(\mathbb{R}^{n,n})$			$\mathcal{B}_\sigma(\mathbb{C}^{n,n})$		
n	$\mathcal{B}_{\|\cdot\|_\infty}$	$\mathcal{B}_{\|\cdot\|_2}$	$\mathcal{B}_{\text{col}}$	$\mathcal{B}_{\|\cdot\|_\infty}$	$\mathcal{B}_{\|\cdot\|_2}$	$\mathcal{B}_{\text{col}}$
2	2.432	3	1.5	12	8	3
3	29.57	26.72	4.244	8640	468.6	40
4	2720	640	24.61	8.71×10^{08}	1.79×10^{05}	2625
5	2.53×10^{06}	3.95×10^{04}	305	2.21×10^{16}	4.25×10^{08}	8.89×10^{05}
6	2.99×10^{10}	6.14×10^{06}	8290	2.23×10^{26}	6.17×10^{12}	1.60×10^{09}
7	5.38×10^{15}	2.38×10^{09}	5.03×10^{05}	1.28×10^{39}	5.41×10^{17}	1.55×10^{13}
8	1.72×10^{22}	2.28×10^{12}	6.88×10^{07}	5.75×10^{54}	2.84×10^{23}	8.23×10^{17}
9	1.12×10^{30}	5.38×10^{15}	2.14×10^{10}	2.63×10^{73}	8.92×10^{29}	2.41×10^{23}
10	1.67×10^{39}	3.12×10^{19}	1.53×10^{13}	1.56×10^{95}	1.67×10^{37}	3.93×10^{29}

Table 18.1 shows that, using $\mathcal{B}_{\text{col}}$ as the bounding set, we can construct "good" rejection schemes up to $n = 4$ for the real case and $n = 3$ for the complex case. For larger values of n, the rejection method becomes highly inefficient.

Next, we report two simple algorithms for generating uniform (real or complex) samples in the spectral norm ball by rejection from $\mathcal{B}_{\text{col}}(1, \mathbb{F}^{n,m})$.

Algorithm 18.1 (Uniform generation in $\mathcal{B}_\sigma(r, \mathbb{R}^{n,m})$ by rejection).
Given n, m and r this algorithm returns a random matrix $\mathbf{X} \in \mathbb{R}^{n,m}$ uniformly distributed in the (real) spectral norm ball of radius r.

1. Generate m independent random columns $\boldsymbol{\xi}_i \sim \mathcal{U}_{\mathcal{B}_{\|\cdot\|_2}(\mathbb{R}^n)}$, $i = 1, \ldots, m$, using Algorithm 16.1;
2. Construct matrix $\mathbf{X} = [\boldsymbol{\xi}_1 \cdots \boldsymbol{\xi}_m]$;
3. If $\bar{\sigma}(\mathbf{X}) \leq 1$ return $r\mathbf{X}$ else goto 1.

Algorithm 18.2 (Uniform generation in $\mathcal{B}_\sigma(r, \mathbb{C}^{n,m})$ by rejection).
Given n, m and r this algorithm returns a random matrix $\mathbf{X} \in \mathbb{C}^{n,m}$ uniformly distributed in the (complex) spectral norm ball of radius r.

1. Generate m independent random columns $\boldsymbol{\xi}_i \sim \mathcal{U}_{\mathcal{B}_{\|\cdot\|_2}(\mathbb{C}^n)}$, $i = 1, \ldots, m$, using Algorithm 16.4;
2. Construct matrix $\mathbf{X} = [\boldsymbol{\xi}_1 \cdots \boldsymbol{\xi}_m]$;
3. If $\bar{\sigma}(\mathbf{X}) \leq 1$ return $r\mathbf{X}$ else goto 1.

The inefficiency of the rejection method for large dimension motivates the need for more sophisticated techniques for direct generation of uniform samples, which are discussed in the next section. We first concentrate on the complex case, which turns out to be easier than the real one.

18.4 Uniform sample generation of complex matrices

In this section we show how to generate uniform matrix samples $X \in \mathbb{C}^{n,m}$, by first generating the samples of the SVD factors U, Σ, V according to their respective densities, and then constructing $X = U\Sigma V^*$. We analyze the generation of Σ in the next section, and subsequently discuss a technique for generating U and V in Section 18.4.2.

18.4.1 Sample generation of singular values

We recall that if the random matrix $\mathbf{X} \in \mathbb{C}^{n,m}$, $m \geq n$, is uniformly distributed over the set $\mathcal{B}_\sigma(\mathbb{C}^{n,m})$, then from Corollary 17.5 the pdf of Σ is

$$f_\Sigma(\Sigma) = K_{\mathbb{C}} \prod_{i=1}^{n} \sigma_i^{2(m-n)+1} \prod_{1 \leq i < k \leq n} (\sigma_i^2 - \sigma_k^2)^2, \quad \sigma \in \mathcal{D}_\sigma \tag{18.3}$$

where the (ordered) domain $\mathcal{D}_\sigma$ is defined in (17.5) and the constant $K_{\mathbb{C}}$ is given in (17.36). For subsequent developments, it is useful to remove the ordering condition $\sigma \in \mathcal{D}_\sigma$ on the singular values, obtaining the (unordered) density function

$$f_{\Sigma}(\Sigma) = \frac{K_{\mathbb{C}}}{n!} \prod_{i=1}^{n} \sigma_i^{2(m-n)+1} \prod_{1 \le i < k \le n} (\sigma_i^2 - \sigma_k^2)^2 \tag{18.4}$$

defined on the domain $\{\sigma \in \mathbb{R}^n : \sigma_i \in (0,1), i = 1, \ldots, n\}$. We remark that the factorial term $(n!)$ in this equation is obtained by observing that the ordered case is one of the $n!$ possible permutations of the n unordered singular values. For convenience, we introduce the change of variables

$$\varsigma_i = \sigma_i^2, \quad i = 1, \ldots, n.$$

The Jacobian of the transformation from the random variable ς to σ is

$$J(\varsigma \to \sigma) = \frac{1}{2^n} \prod_{i=1}^{n} \varsigma_i^{-1/2}.$$

Then, applying Theorem 14.2 on the transformation between random variables, we obtain the density function of ς

$$f_{\varsigma}(\varsigma) = \frac{K_{\mathbb{C}}}{n! 2^n} \prod_{i=1}^{n} \varsigma_i^{m-n} \prod_{1 \le i < k \le n} (\varsigma_i - \varsigma_k)^2 \tag{18.5}$$

with domain $\{\varsigma \in \mathbb{R}^n : \varsigma_i \in (0,1), i = 1, \ldots, n\}$. This density function can be written in terms of the determinant of a Vandermonde matrix, as detailed in the following remark.

Remark 18.1 (Vandermonde determinant). Given $\varsigma = [\varsigma_1 \cdots \varsigma_n]^T$, define the vector

$$\mathcal{V}(\varsigma_i) \doteq \begin{bmatrix} 1 & \varsigma_i & \varsigma_i^2 & \cdots & \varsigma_i^{n-1} \end{bmatrix}^T, \quad i = 1, \ldots, n. \tag{18.6}$$

Then, the Vandermonde matrix $\mathcal{V}(\varsigma_1, \ldots, \varsigma_n)$ associated with vector ς is defined as

$$\mathcal{V}(\varsigma_1, \ldots, \varsigma_n) = \begin{bmatrix} \mathcal{V}(\varsigma_1) & \cdots & \mathcal{V}(\varsigma_n) \end{bmatrix}. \tag{18.7}$$

Similarly, for $i = 1, \ldots, n$ we define the *truncated* Vandermonde matrix as

$$\mathcal{V}(\varsigma_1, \varsigma_2, \ldots, \varsigma_i) = \begin{bmatrix} \mathcal{V}(\varsigma_1) & \cdots & \mathcal{V}(\varsigma_i) \end{bmatrix}. \tag{18.8}$$

For notational convenience, we write $\mathcal{V}_i$ to indicate $\mathcal{V}(\varsigma_1, \ldots, \varsigma_i)$. It is well known that the determinant of a Vandermonde matrix is given by

$$\det \mathcal{V}(\varsigma_1, \ldots, \varsigma_n) = \prod_{1 \le i < k \le n} (\varsigma_i - \varsigma_k).$$

Using this fact, it follows immediately that the density function (18.5) can be written as

$$f_\varsigma(\varsigma) = \frac{K_{\mathbb{C}}}{n!2^n}|\mathcal{V}(\varsigma_1,\ldots,\varsigma_n)|^2 \prod_{i=1}^{n} \varsigma_i^{m-n}. \tag{18.9}$$

◁

We now focus on the generation of random samples distributed according to (18.5). To this end, we apply the conditional density method introduced in Section 14.3.2. That is, we write the density (18.5) as

$$f_{\varsigma_1,\ldots,\varsigma_n}(\varsigma_1,\ldots,\varsigma_n) = f_{\varsigma_1}(\varsigma_1)f_{\varsigma_2|\varsigma_1}(\varsigma_2|\varsigma_1)\cdots f_{\varsigma_n|\varsigma_1,\ldots,\varsigma_{n-1}}(\varsigma_n|\varsigma_1,\ldots,\varsigma_{n-1})$$

where the conditional densities $f_{\varsigma_i|\varsigma_1,\ldots,\varsigma_{i-1}}(\varsigma_i|\varsigma_1,\ldots,\varsigma_{i-1})$ are defined as the ratio of marginal densities

$$f_{\varsigma_i|\varsigma_1,\ldots,\varsigma_{i-1}}(\varsigma_i|\varsigma_1,\ldots,\varsigma_{i-1}) = \frac{f_{\varsigma_1,\ldots,\varsigma_i}(\varsigma_1,\ldots,\varsigma_i)}{f_{\varsigma_1,\ldots,\varsigma_{i-1}}(\varsigma_1,\ldots,\varsigma_{i-1})}. \tag{18.10}$$

In turn, the marginal densities $f_{\varsigma_1,\ldots,\varsigma_i}(\varsigma_1,\ldots,\varsigma_i)$ are given by

$$f_{\varsigma_1,\ldots,\varsigma_i}(\varsigma_1,\ldots,\varsigma_i) = \int_0^1 \cdots \int_0^1 f_\varsigma(\varsigma_1,\ldots,\varsigma_n)d\varsigma_{i+1}\cdots d\varsigma_n. \tag{18.11}$$

Therefore, a random vector ς with density (18.5) can be obtained by generating sequentially the random variables ς_i, $i=1,\ldots,n$, where ς_i is distributed according to the univariate conditional density $f_{\varsigma_i|\varsigma_1,\ldots,\varsigma_{i-1}}(\varsigma_i|\varsigma_1,\ldots,\varsigma_{i-1})$. The following theorem provides a closed-form expression for the marginal density (18.11), without requiring symbolic computation of the integral.

Theorem 18.1. *The marginal density (18.11) is equal to*

$$f_{\varsigma_1,\ldots,\varsigma_i}(\varsigma_1,\ldots,\varsigma_i) =$$

$$K_{\mathbb{C}}\frac{(n-i)!}{n!2^n|H|^2}|\mathcal{V}(\varsigma_1,\ldots,\varsigma_i)^T H^T H\mathcal{V}(\varsigma_1,\ldots,\varsigma_i)|\prod_{k=1}^{i}\varsigma_k^{m-n} \tag{18.12}$$

with $\varsigma_k \in (0,1)$, $k=1,\ldots,i$, and where $\mathcal{V}(\varsigma_1,\ldots,\varsigma_i)$ is defined in (18.8), and $H \doteq R^{-T}$, being R the upper-triangular factor of the Cholesky decomposition $M = R^T R$ of the symmetric matrix with entries

$$[M]_{r,\ell} \doteq \frac{1}{r+\ell+m-n-1}, \quad r,\ell = 1,\ldots,n.$$

Proof. Following the discussion in Remark 18.1, we rewrite (18.5) in the form (18.9)

$$f_\varsigma(\varsigma) = \frac{K_{\mathbb{C}}}{n!2^n}|\mathcal{V}_n|^2 \prod_{k=1}^{n} \varsigma_k^{m-n}$$

where $\mathcal{V}_n = \mathcal{V}(\varsigma_1, \ldots, \varsigma_n)$ is the Vandermonde matrix. Then, for any given nonsingular matrix $H \in \mathbb{R}^{n,n}$, we have that

$$f_\varsigma(\varsigma) = \frac{K_\mathbb{C}}{n! 2^n |H|^2} |\mathcal{V}_n^T H^T H \mathcal{V}_n| \prod_{k=1}^n \varsigma_k^{m-n}. \tag{18.13}$$

Notice further that

$$H\mathcal{V}(\varsigma) = \begin{bmatrix} L_0(\varsigma) & L_1(\varsigma) & L_2(\varsigma) & \cdots & L_{n-1}(\varsigma) \end{bmatrix}^T$$

where $L_k(\varsigma)$, $k = 0, \ldots, n-1$, are polynomials of degree $n-1$ in the variable ς

$$L_k(\varsigma) = h_{k,0} + h_{k,1}\varsigma + h_{k,2}\varsigma^2 + \cdots + h_{k,n-1}\varsigma^{n-1} \tag{18.14}$$

and $h_{r-1,\ell-1} = [H]_{r,\ell}$ denotes the (r, ℓ) entry of H.

We observe in particular that the matrix H can be chosen such that the polynomials $L_k(\varsigma)$, $k = 0, 1, \ldots, n-1$, form an orthogonal polynomial basis on the interval $\varsigma \in [0, 1]$, with respect to the weight function ς^{m-n}. That is, for $k, \ell = 0, 1, \ldots, n-1$, we impose that

$$\int_0^1 L_k(\varsigma) L_\ell(\varsigma) \varsigma^{m-n} d\varsigma = \begin{cases} 1 & \text{if } k = \ell; \\ 0 & \text{otherwise.} \end{cases} \tag{18.15}$$

This condition can be written in matrix form as

$$HMH^T = I_n \tag{18.16}$$

where

$$M = \left(\int_0^1 \mathcal{V}(\varsigma) \mathcal{V}^T(\varsigma) \varsigma^{m-n} d\varsigma \right).$$

The integral term M is easily evaluated as

$$[M]_{r,\ell} = \frac{1}{r + \ell + m - n - 1}$$

for $r, \ell = 1, \ldots, n$. Let $M = R^T R$ be the Cholesky decomposition of M, where R is upper triangular. Then, the orthogonality condition (18.16) is satisfied for the choice $H = R^{-T}$, where the resulting matrix H is lower triangular. Define now, for $i = 1, \ldots, n$, the symmetric matrix

$$Z_i = Z_i(\varsigma_1, \ldots, \varsigma_i) \doteq \mathcal{V}_i^T H^T H \mathcal{V}_i. \tag{18.17}$$

It is straightforward to show that the matrix Z_i satisfies the conditions of the Dyson–Mehta theorem for the integral of certain determinants; see Appendix A.4. In particular, we have that the (r, ℓ) entry of Z_i is function of $\varsigma_r, \varsigma_\ell$, i.e. $[Z_i]_{r,\ell} = \psi(\varsigma_r, \varsigma_\ell)$, with

$$\psi(\varsigma_r, \varsigma_\ell) = \sum_{k=0}^{n-1} L_k(\varsigma_r) L_k(\varsigma_\ell).$$

The conditions of the Dyson–Mehta theorem are met for the function $\psi(\varsigma_r, \varsigma_\ell)$ with $d\mu(\varsigma) = \varsigma^{m-n} d\varsigma$. In particular, we have

$$\int_0^1 \psi(\varsigma_i, \varsigma_i) \varsigma_i^{m-n} d\varsigma_i = n.$$

Therefore, from Theorem A.3 in Appendix A.4 we obtain

$$\int_0^1 \det(Z_i(\varsigma_1, \ldots, \varsigma_i)) \varsigma_i^{m-n} d\varsigma_i = (n - i + 1) \det(Z_{i-1}(\varsigma_1, \ldots, \varsigma_{i-1})).$$

Applying this equation recursively, going backwards from n to $i + 1$, and noticing that $|Z_i| = \det(Z_i)$ since $\det(Z_i)$ is always positive, we have

$$\int_0^1 \cdots \int_0^1 |Z_n(\varsigma_1, \ldots, \varsigma_n)| (\varsigma_{i+1} \cdots \varsigma_n)^{m-n} d\varsigma_{i+1} \cdots d\varsigma_n = (n-i)! |Z_i(\varsigma_1, \ldots, \varsigma_i)|.$$

Then, by means of (18.13) we obtain the marginal density

$$f_{\varsigma_1, \ldots, \varsigma_i}(\varsigma_1, \ldots, \varsigma_i) = K_C \frac{(n-i)!}{n! 2^n |H|^2} |Z_i(\varsigma_1, \ldots, \varsigma_i)| \prod_{k=1}^i \varsigma_k^{m-n}. \tag{18.18}$$

The proof is then completed by substituting (18.17) in this expression.   □

Remark 18.2 (Relationship with orthogonal polynomials). In the proof of Theorem 18.1, the polynomials (18.14) of degree $n - 1$ in the variable ς

$$L_k(\varsigma) = h_{k,0} + h_{k,1}\varsigma + h_{k,2}\varsigma^2 + \cdots + h_{k,n-1}\varsigma^{n-1}$$

should satisfy the orthogonality condition (18.15). Indeed, such polynomial basis may be recognized as the family of Jacobi polynomials $G_{n-1}(p, q, \varsigma)$ with $p = q = (m - n + 1)$. Therefore, the entries of the matrix H may be derived in closed form using the formulas in [1] for the coefficients of $G_{n-1}(p, q, \varsigma)$, thus obtaining

$$[H]_{r,\ell} = (-1)^{r-\ell} \binom{r}{\ell} \frac{\ell \sqrt{m - n + 2r - 1}}{r!} \frac{\Gamma(m - n + r + \ell - 1)}{\Gamma(m - n + \ell)} \tag{18.19}$$

for $r = 1, \ldots, n$, $\ell = 1, \ldots, r$, and $[H]_{r,\ell} = 0$ for $r > \ell$. From this expression, we immediately obtain

$$|H| = \prod_{\ell=1}^n \frac{\sqrt{m - n + 2\ell - 1}}{(\ell - 1)!} \frac{\Gamma(m - n + 2\ell - 1)}{\Gamma(m - n + \ell)}.$$

◁

Theorem 18.1 provides a closed-form expression for the multiple integral (18.11). To apply the conditional density method, we need to compute recursively the conditional density (18.10) of the random variable ς_i when the values $\varsigma_1, \ldots, \varsigma_{i-1}$ are given. This is shown in the following corollary.

Corollary 18.1. *Let $i = 2, \ldots, n$. Then, the conditional density of ς_i given $\varsigma_1 = \varsigma_1$, $\varsigma_2 = \varsigma_2$, $\cdots$, $\varsigma_{i-1} = \varsigma_{i-1}$, is a polynomial of order $2(n-1)$ expressed as*

$$f_{\varsigma_i | \varsigma_1, \ldots, \varsigma_{i-1}}(\varsigma_i | \varsigma_1, \ldots, \varsigma_{i-1}) = (n - i + 1)^{-1} \varsigma_i^{m-n} \sum_{k=0}^{2(n-1)} b_{i,k} \varsigma_i^k. \tag{18.20}$$

The coefficients $b_{i,k} = b_{i,k}(\varsigma_i | \varsigma_1, \ldots, \varsigma_{i-1})$, $k = 0, 1, \ldots, 2(n-1)$, are given by

$$b_{i,k} \doteq \sum_{\{r + \ell = k + 2\}} [W_{i-1}]_{r,\ell} \tag{18.21}$$

where

$$W_{i-1} \doteq H^T \left(I - H \mathcal{V}_{i-1} Z_{i-1}^{-1} \mathcal{V}_{i-1}^T H^T \right) H; \tag{18.22}$$

$$Z_{i-1} \doteq \mathcal{V}_{i-1}^T H^T H \mathcal{V}_{i-1} \tag{18.23}$$

and H is given in (18.19). Moreover, the marginal density of ς_1 is given by the polynomial

$$f_{\varsigma_1}(\varsigma_1) = \frac{K_{\mathbb{C}}}{n 2^n |H|^2} \varsigma_1^{m-n} \sum_{k=0}^{2(n-1)} b_{1,k} \varsigma_1^k \tag{18.24}$$

where $b_{1,k} \doteq \sum_{\{r + \ell = k + 2\}} [W_0]_{r,\ell}$ and $W_0 = H^T H$.

Proof. As in (18.17) in the proof of Theorem 18.1, we define

$$Z_i = Z_i(\varsigma_1, \ldots, \varsigma_i) \doteq \mathcal{V}_i^T H^T H \mathcal{V}_i.$$

Next, recalling that $\mathcal{V}_i = [\mathcal{V}_{i-1} \; \mathcal{V}(\varsigma_i)]$, we obtain

$$|Z_i| = \left| \begin{matrix} Z_{i-1} & \mathcal{V}_{i-1}^T H^T H \mathcal{V}(\varsigma_i) \\ \mathcal{V}^T(\varsigma_i) H^T H \mathcal{V}_{i-1} & \mathcal{V}^T(\varsigma_i) H^T H \mathcal{V}(\varsigma_i) \end{matrix} \right|.$$

Using the Schur rule for this determinant, for $i = 2, \ldots, n$, we get

$$|Z_i(\varsigma_1, \ldots, \varsigma_i)| = |Z_{i-1}(\varsigma_1, \ldots, \varsigma_{i-1})| \mathcal{V}^T(\varsigma_i) W_{i-1} \mathcal{V}(\varsigma_i) \tag{18.25}$$

where

$$W_{i-1} = H^T \left(I - H \mathcal{V}_{i-1} Z_{i-1}^{-1} \mathcal{V}_{i-1}^T H^T \right) H.$$

The term $\mathcal{V}^T(\varsigma_i)W_{i-1}\mathcal{V}(\varsigma_i)$ can be written as a polynomial in the variable ς_i, with coefficients depending on $\varsigma_1, \ldots, \varsigma_{i-1}$. It is straightforward to verify that these coefficients are given by the sum of the elements of the anti-diagonals of W_{i-1}. That is, we have

$$\mathcal{V}^T(\varsigma_i)W_{i-1}\mathcal{V}(\varsigma_i) = \sum_{k=0}^{2(n-1)} b_{i,k}\varsigma_i^k \qquad (18.26)$$

where, for $k = 0, 1, \ldots, 2(n-1)$,

$$b_{i,k} = \sum_{\{r+\ell=k+2\}} [W_{i-1}]_{r,\ell}.$$

Moreover, $|Z_1(\varsigma_1)| = \mathcal{V}_1^T W_0 \mathcal{V}_1$, with $W_0 = H^T H$.

Combining the expressions (18.18), (18.25) and (18.26) we obtain that the marginal density (18.11) is given by

$$f_{\varsigma_1,\ldots,\varsigma_i}(\varsigma_1,\ldots,\varsigma_i) = K_{\mathbb{C}}\frac{(n-i)!}{n!2^n|H|^2}\varsigma_i^{m-n}\sum_{k=0}^{2(n-1)} b_{i,k}\varsigma_i^k.$$

Taking $i = 1$ in the latter expression, we prove (18.24). Equation (18.20) is then immediately obtained using the definition of conditional density (18.10).

$\square$

Remark 18.3 (Application of the conditional density method). Corollary 18.1 gives an expression of the conditional densities in the form of the polynomial (18.20), where the variables $\varsigma_1, \ldots, \varsigma_{i-1}$ are separated from the variable ς_i. In fact, at the ith step of the conditional density method the variables up to $i - 1$ are given, and only the dependence on ς_i is required. In this case, equation (18.20) represents a polynomial in the ς_i variable. The coefficients $b_{i,k} = b_{i,k}(\varsigma_1, \ldots, \varsigma_{i-1})$ can be easily computed according to (18.21), once the values of $\varsigma_1, \varsigma_2, \ldots, \varsigma_{i-1}$ are known. $\lhd$

Remark 18.4 (Computational improvements). We observe that, for $i \geq 2$, the factor Z_{i-1}^{-1} appearing in (18.22) can be computed recursively, so that no matrix inversion or determinant computation is required. In particular, we have

$$Z_i = \mathcal{V}_i^T H^T H \mathcal{V}_i = \begin{bmatrix} Z_{i-1} & \mathcal{V}_{i-1}^T H^T H \mathcal{V}(\varsigma_i) \\ \mathcal{V}^T(\varsigma_i)H^T H \mathcal{V}_{i-1} & \mathcal{V}^T(\varsigma_i)H^T H \mathcal{V}(\varsigma_i) \end{bmatrix}.$$

Using the block matrix inversion formula, we obtain

$$Z_i^{-1} = \begin{bmatrix} Z_{i-1}^{-1} + \Omega_i\Omega_i^T/\delta_i & -\Omega_i/\delta_i \\ -\Omega_i^T/\delta_i & 1/\delta_i \end{bmatrix} \qquad (18.27)$$

where $\Omega_i = Z_{i-1}^{-1}\mathcal{V}_{i-1}^T H^T H \mathcal{V}(\varsigma_i)$ and $\delta_i = \mathcal{V}^T(\varsigma_i)W_{i-1}\mathcal{V}(\varsigma_i) > 0$. $\lhd$

An explicit algorithm for the generation of samples of the singular values distributed according to the pdf (18.4) is reported next.

Algorithm 18.3 (Singular values generation).
Given n, m, $m \geq n$, this algorithm returns a random vector $\boldsymbol{\sigma} = [\sigma_1 \cdots \sigma_n]^T$ distributed according to the pdf (18.4).

1. Initialization.
   ▷ Set $i = 1$, $W_0 = H^T H$, where H is given in (18.19);
   ▷ Let $b_{1,k} \doteq \displaystyle\sum_{\{r+\ell=k+2\}} [W_0]_{r,\ell}$;
   ▷ Generate ς_1 according to the polynomial marginal density

$$f_{\varsigma_1}(\varsigma_1) = \frac{K_{\mathbb{C}}}{n 2^n |H|^2} \varsigma_1^{m-n} \sum_{k=0}^{2(n-1)} b_{1,k} \varsigma_1^k;$$

   ▷ Let $Z_1 = \mathcal{V}^T(\varsigma_1) W_0 \mathcal{V}(\varsigma_1)$;
2. Update.
   ▷ Set $\delta_i = \mathcal{V}^T(\varsigma_i) W_{i-1} \mathcal{V}(\varsigma_i)$ and $\Omega_i = Z_{i-1}^{-1} \mathcal{V}_{i-1}^T H^T H \mathcal{V}(\varsigma_i)$;
   ▷ Compute

$$Z_i^{-1} = \begin{bmatrix} Z_{i-1}^{-1} + \Omega_i \Omega_i^T / \delta_i & -\Omega_i / \delta_i \\ -\Omega_i^T / \delta_i & 1/\delta_i \end{bmatrix};$$

$$W_i = H^T \left(I - H \mathcal{V}_i Z_i^{-1} V_i^T H^T \right) H;$$

3. Generation.
   ▷ Set $b_{i,k} = \displaystyle\sum_{\{r+\ell=k+2\}} [W_{i-1}]_{r,\ell}, \quad k = 1, \ldots, 2(n-1)$;
   ▷ Generate $\varsigma_i \in (0,1)$ according to the polynomial density

$$f_{\varsigma_i|\varsigma_1,\ldots,\varsigma_{i-1}}(\varsigma_i|\varsigma_1,\ldots,\varsigma_{i-1}) = (n-i+1)^{-1} \varsigma_i^{m-n} \sum_{k=0}^{2(n-1)} b_{i,k} \varsigma_i^k;$$

4. Loop.
   ▷ If $i < n$, set $i = i + 1$ and goto 2;
5. Return $\boldsymbol{\sigma} = [\sqrt{\varsigma_1} \cdots \sqrt{\varsigma_n}]^T$.

In this algorithm, each ς_i is generated according to a univariate polynomial density. Standard and efficient algorithms for the generation of samples distributed according to a given polynomial density are available in the literature. Among these techniques we recall a classical one based on the inversion method presented in Algorithm 14.1.

18.4.2 Uniform sample generation of unitary matrices

In this section we concentrate on the generation of samples of U and V according to the Haar invariant distribution (17.28) and to the conditional Haar invariant distribution (17.30) respectively.

We first consider the problem of generating a random matrix $\mathbf{U}$ uniformly distributed in the unitary group $\mathcal{G}_{\mathcal{U}}^n$. From the properties of the Haar invariant distribution, we have that the distribution of $\mathbf{U}$ should be the same as the distribution of $W\mathbf{U}$, for any given unitary matrix W. Consider a random matrix $\mathbf{X} \in \mathbb{C}^{n,n} = \text{Re}(\mathbf{X}) + j\text{Im}(\mathbf{X})$, such that the entries of $\text{Re}(\mathbf{X})$ and $\text{Im}(\mathbf{X})$ are independent and normally distributed with zero mean and variance equal to one. The invariance property of the normal distribution under unitary transformations implies that, for any unitary matrix W, the distribution of $W\mathbf{X}$ is the same of the distribution of $\mathbf{X}$. Now, let $\mathbf{X} = \mathbf{QR}$ be the QR factorization of $\mathbf{X}$, where the diagonal entries of $\mathbf{R}$ are forced to be real and positive in order to make the representation unique. Then, since $W\mathbf{X} \sim \mathbf{X}$, it follows that $W\mathbf{Q} \sim \mathbf{Q}$. That is, $\mathbf{Q}$ is distributed according to the Haar invariant distribution.

This discussion suggests the following simple algorithm for the generation of samples according to the Haar invariant distribution.

Algorithm 18.4 (Generation of Haar unitary matrices).
Given n, the algorithm returns a sample of the random unitary matrix $\mathbf{U} \in \mathbb{C}^{n,n}$ *distributed according to the Haar invariant distribution (17.28).*

1. Generation of Gaussian $\mathbf{Y}$.
    ▷ Generate $\mathbf{Y}^{\mathbb{R}}, \mathbf{Y}^{\mathbb{I}} \in \mathbb{R}^{n,n}$, where each entry of $\mathbf{Y}^{\mathbb{R}}$ and $\mathbf{Y}^{\mathbb{I}}$ is distributed according to $\mathcal{N}_{0,1}$;
    ▷ Construct $\mathbf{Y} = \mathbf{Y}^{\mathbb{R}} + j\mathbf{Y}^{\mathbb{I}}$;
2. QR factorization.
    ▷ Factorize $\mathbf{Y}$ as $[\mathbf{Q}, \mathbf{R}] = QR(\mathbf{Y})$;
    ▷ Set $\mathbf{U} = \mathbf{Q} \, \text{diag}\left([e^{-j\phi_1} \cdots e^{-j\phi_n}]\right)$, where ϕ_i is the phase of the (i,i) entry of $\mathbf{R}$;
3. Return $\mathbf{U}$.

This algorithm is one of the simplest methods for generation of uniform unitary matrices. Other known methods are based, for example, on products of elementary Euler transformations; see e.g. [297]. We remark that the Haar invariant distribution may also be introduced for rectangular random matrices $\mathbf{V} \in \mathbb{C}^{m,n}$ having orthonormal columns. In this case, it can be observed that uniform samples of V can be obtained from uniform samples of a square $m \times m$ unitary matrix, using Algorithm 18.4, and simply neglecting the last $m - n$ columns. This fact follows from properties of the unitary group, see for instance [140].

Finally, we consider the generation of random matrices $\mathbf{V}$ with conditional Haar invariant distribution, i.e. uniform in the manifold $\mathcal{C}^{m,n}$ defined in (17.27). We notice that the normalization condition on the columns of $\mathbf{V}$ may be written as $\mathbf{V} = \tilde{\mathbf{V}}\boldsymbol{\Theta}$, where $\tilde{\mathbf{V}} \in \mathbb{C}^{m,n}$ is a (non-normalized) unitary matrix and $\boldsymbol{\Theta}$ is a diagonal unitary matrix $\boldsymbol{\Theta} = \mathrm{diag}\left([e^{-j\theta_1} \cdots e^{-j\theta_n}]\right)$, where θ_i is the phase of $[\tilde{\mathbf{V}}]_{1,i}$. It can be shown that if $\tilde{\mathbf{V}}$ is distributed according to the Haar invariant distribution, then $\mathbf{V}$ is distributed according to the conditional Haar invariant distribution. Samples of matrices drawn from the latter distribution may, therefore, be obtained by normalizing the samples drawn from the Haar invariant distribution.

We conclude this section by reporting the algorithm for direct generation of complex random matrices uniformly distributed in the spectral norm ball.

Algorithm 18.5 (Uniform generation in $\mathcal{B}_\sigma(r, \mathbb{C}^{n,m})$).
Given n, m, $m \geq n$, and r this algorithm returns a sample of the random matrix $\mathbf{X} \in \mathbb{C}^{n,m}$ with uniform distribution in the (complex) spectral norm ball of radius r.

1. Generation of $\boldsymbol{\Sigma}$.
   ▷ Generate $\boldsymbol{\sigma} = [\sigma_1 \cdots \sigma_n]^T$ using Algorithm 18.3;
   ▷ Construct $\boldsymbol{\Sigma} = \mathrm{diag}(\boldsymbol{\sigma})$;
2. Generation of $\mathbf{U}$ and $\mathbf{V}$.
   ▷ Generate $\mathbf{U} \in \mathbb{C}^{n,n}$ and $\mathbf{V} \in \mathbb{C}^{m,n}$ using Algorithm 18.4;
3. Return $\mathbf{X} = r\mathbf{U}\boldsymbol{\Sigma}\mathbf{V}^*$.

18.5 Uniform sample generation of real matrices

Similar to the developments of the complex case, we now study the generation of uniform matrix samples in the ball $\mathcal{B}_\sigma(\mathbb{R}^{n,m})$. This approach is based on the generation of the samples of the SVD factors $\mathbf{U}, \boldsymbol{\Sigma}, \mathbf{V}$ according to their respective densities derived in Corollary 17.4, and then on the construction of $\mathbf{X} = \mathbf{U}\boldsymbol{\Sigma}\mathbf{V}^T$. In the next section, we analyze the generation of $\boldsymbol{\Sigma}$, and subsequently we discuss a technique for generating $\mathbf{U}$ and $\mathbf{V}$ in Section 18.5.2.

18.5.1 Sample generation of singular values

If we assume that the random matrix $\mathbf{X} \in \mathbb{R}^{n,m}$, $m \geq n$, is uniformly distributed over the set $\mathcal{B}_\sigma(\mathbb{R}^{n,m})$, then it follows from Corollary 17.4 that the pdf of $\boldsymbol{\Sigma}$ is given by

$$f_{\boldsymbol{\Sigma}}(\Sigma) = K_{\mathbb{R}} \prod_{i=1}^{n} \sigma_i^{m-n} \prod_{1 \leq i < k \leq n} (\sigma_i^2 - \sigma_k^2), \qquad \sigma \in \mathcal{D}_\sigma \qquad (18.28)$$

where the (ordered) domain $\mathcal{D}_\sigma$ is defined in (17.5) and the constant $K_\mathbb{R}$ is given in (17.24). As in Section 18.4, we introduce the change of variables

$$\varsigma_i = \sigma_i^2, \quad i = 1, \ldots, n$$

so that the density may be written in terms of a Vandermonde determinant, see Remark 18.1. Therefore, we obtain

$$f_\varsigma(\varsigma_1, \ldots, \varsigma_n) = \frac{K_\mathbb{R}}{2^n} |\mathcal{V}(\varsigma_1, \ldots, \varsigma_n)| \prod_{i=1}^n \varsigma_i^\nu, \quad \nu = (m-n-1)/2 \qquad (18.29)$$

defined over the ordered domain

$$\mathcal{D}_\varsigma \doteq \{\varsigma \in \mathbb{R}^n : 1 > \varsigma_1 > \cdots > \varsigma_n > 0\}. \qquad (18.30)$$

Remark 18.5 (Ordering condition on the singular values). We recall that in the complex case the ordering condition $\sigma \in \mathcal{D}_\sigma$ has been removed to facilitate the computation of the marginal density. In the real case, the same approach is not helpful. This is due to the different form of the pdf, and in particular to the fact that in (18.29) the Vandermonde factor $|\mathcal{V}(\varsigma_1, \ldots, \varsigma_n)|$ appears without the square power, which is present in (18.9).   ◁

To generate a vector ς distributed according to the density (18.29), we apply the conditional density method described in Section 14.3.2. To this end, we need to compute the marginal density

$$f_{\varsigma_1, \ldots, \varsigma_i}(\varsigma_1, \ldots, \varsigma_i) = \frac{K_\mathbb{R}}{2^n} \mathcal{I}(\varsigma_1, \ldots, \varsigma_i) \prod_{k=1}^i \varsigma_k^\nu \qquad (18.31)$$

where

$$\mathcal{I}(\varsigma_1, \ldots, \varsigma_i) = \int \cdots \int |\mathcal{V}(\varsigma_1, \ldots, \varsigma_n)| \prod_{k=i+1}^n \varsigma_k^\nu d\varsigma_k \qquad (18.32)$$

with domain of integration $\{\varsigma_i > \cdots > \varsigma_n > 0\}$, and $\nu = (m-n-1)/2$.

The following theorem gives an explicit closed-form solution for the multiple integral (18.32).

Theorem 18.2. *The multiple integral (18.32) is given by*

$$\mathcal{I}(\varsigma_1, \ldots, \varsigma_i) = \varsigma_i^{\alpha_i} \det{}^{1/2} \left[\begin{array}{c|cc} Z(\varsigma_i) & \mathcal{V}(\varsigma_1, \ldots, \varsigma_{i-1}) & \\ & 0 & \\ \hline -\mathcal{V}^T(\varsigma_1, \ldots, \varsigma_{i-1}) \ 0 & 0 & \end{array} \right] \qquad (18.33)$$

for $i = 2, \ldots, n$, and

$$\mathcal{I}(\varsigma_1) = \varsigma_1^{\alpha_1} \det{}^{1/2} Z(\varsigma_1)$$

where $\alpha_i \doteq (\nu+1)(n-i)$, and

$$Z(\varsigma_i) \doteq \begin{cases} \begin{bmatrix} H(\varsigma_i) & \mathcal{V}(\varsigma_i) \\ -\mathcal{V}^T(\varsigma_i) & 0 \end{bmatrix} & \text{if } n-i \text{ even;} \\[2em] \begin{bmatrix} H(\varsigma_i) & \mathcal{V}(\varsigma_i) & h(\varsigma_i) \\ -\mathcal{V}^T(\varsigma_i) & 0 & 0 \\ -h^T(\varsigma_i) & 0 & 0 \end{bmatrix} & \text{if } n-i \text{ odd;} \end{cases} \qquad (18.34)$$

$$[H(\varsigma_i)]_{r,\ell} \doteq \frac{r-\ell}{(r+\nu)(\ell+\nu)(r+\ell+2\nu)} \varsigma_i^{r+\ell-2}, \quad r,\ell = 1,\dots,n; \quad (18.35)$$

$$h_\ell(\varsigma_i) \doteq \frac{\varsigma_i^{\ell-1}}{\ell+\nu}, \quad \ell = 1,\dots,n. \qquad (18.36)$$

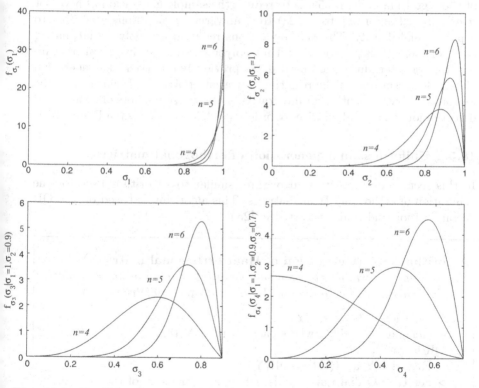

Fig. 18.1. Conditional probability densities of the first four singular values for uniform real square matrices of order $n = 4, 5, 6$.

Remark 18.6 (Theory of skew-symmetric matrices). The proof of this result, given in [59], is quite involved and it is based on the theory of skew-symmetric matrices and, in particular, on the so-called de Bruijn integral, see [79]. Notice that the matrix appearing in (18.33) is a skew-symmetric polynomial matrix of even order and, therefore, its determinant is always a perfect square in the entries of the matrix, see e.g. [281]. We remark that the square root of the determinant of a skew-symmetric matrix of even order is related to the so-called Pfaffian, see e.g. [186] for definitions and properties.   ◁

Remark 18.7 (Algorithm for generation of the singular values). Theorem 18.2 provides a closed-form expression for the marginal density (18.31), so that the conditional density method can be applied. This result has been exploited in [59] to develop an efficient recursive algorithm for the generation of the singular values. This algorithm is similar to Algorithm 18.3, but more complicated. In fact, four different cases, corresponding to the combinations of n and i in (18.31) being even or odd, need to be considered. The main feature of the algorithm given in [59] is to reduce the sample generation of a vector σ distributed according to (18.28) to n univariate generations according to a polynomial density. The only operations required are polynomial matrix additions and multiplications, and no polynomial matrix inversion or computation of determinants are needed. As previously observed, the generation of samples according to the resulting univariate polynomial density may be performed very efficiently. For illustrative purposes, some plots of conditional densities for square real matrices of order $n = 4, 5, 6$ are shown in Figure 18.1.

18.5.2 Uniform sample generation of orthogonal matrices

In this section we present an algorithm, similar to Algorithm 18.4, for the generation of orthogonal Haar matrices. This algorithm is based on the QR decomposition and is also reported in [254].

Algorithm 18.6 (Generation of Haar orthogonal matrices).
Given n, this algorithm returns a sample of the random orthogonal matrix $\mathbf{U} \in \mathbb{R}^{n,n}$ distributed according to the Haar invariant distribution (17.14).

1. Generation of Gaussian $\mathbf{Y}$.
    ▷ Generate $\mathbf{Y} \in \mathbb{R}^{n,n}$, where each entry of $\mathbf{Y}$ is $\mathcal{N}_{0,1}$;
2. QR factorization.
    ▷ Factorize $\mathbf{Y}$ as $[\mathbf{Q}, \mathbf{R}] = \mathrm{QR}(\mathbf{Y})$;
    ▷ Set $\mathbf{U} = \mathbf{Q} \, \mathrm{diag}\,([\mathbf{s}_1 \, \cdots \, \mathbf{s}_n])$, where $\mathbf{s}_i$ is the sign of the (i,i) entry of $\mathbf{R}$;
3. Return $\mathbf{U}$.

Regarding the generation of samples distributed according to the conditional Haar invariant distribution, i.e. uniform in the manifold $\mathcal{R}^{m,n}$ defined in (17.4), comments similar to those made for the complex case in Section 18.4.2 also apply here.

We conclude this chapter by reporting the algorithm for direct generation of real random matrices uniformly distributed in the spectral norm ball.

Algorithm 18.7 (Uniform generation in $\mathcal{B}_\sigma(r, \mathbb{R}^{n,m})$).
Given n, m, $m \geq n$, and r this algorithm returns a random matrix $\mathbf{X} \in \mathbb{R}^{n,m}$ with uniform distribution in the (real) spectral norm ball of radius r.

1. Generation of $\mathbf{\Sigma}$.
   ▷ Generate $\boldsymbol{\sigma} = [\boldsymbol{\sigma}_1 \cdots \boldsymbol{\sigma}_n]^T$ using the algorithm in [59];
   ▷ Construct $\mathbf{\Sigma} = \mathrm{diag}\,(\boldsymbol{\sigma})$;
2. Generation of $\mathbf{U}$ and $\mathbf{V}$.
   ▷ Generate $\mathbf{U} \in \mathbb{R}^{n,n}$ and $\mathbf{V} \in \mathbb{R}^{m,n}$ using Algorithm 18.6;
3. Return $\mathbf{X} = r\mathbf{U}\mathbf{\Sigma}\mathbf{V}^T$.

---

19. Applications of Randomized Algorithms

In this chapter we present further applications of randomized algorithms. In particular, we analyze congestion control of high-speed communication networks, robustness of flexible structures and iterative algorithms for quadratic stability of sampled-data quantized systems. While the first two applications deal with analysis problems, the objective of the last one is to design Lyapunov functions for probabilistic robust stabilization.

19.1 Stability and robustness of high-speed networks

High-speed communication networks have received increasing attention in the control literature, as evidenced by the appearance of several special issues devoted to this topic in leading journals in the field, such as [9, 54, 122]. Various approaches and solutions have been developed in this context, including modeling of TCP/IP traffic, congestion control for available bit rate (ABR) service in asynchronous transmission mode (ATM) networks, packet marking schemes for the Internet, application of low-order controllers for active queue management (AQM) and related problems. One of the critical issues at the heart of efficient operations of high-speed networks is *congestion control*. This involves the problem of regulating the source rates in a decentralized and distributed fashion, so that the available bandwidths on different links are used most efficiently while minimizing or totally eliminating loss of packets due to queues at buffers exceeding their capacities. This issue needs to be accomplished under variations in network conditions such as packet delays due to propagation as well as to queueing and bottleneck nodes.

In this section, based on [7], using randomized algorithms we perform a stability analysis of a model introduced in [6], which makes use of noncooperative game theory [30].

19.1.1 Network model

Fluid models, which replace discrete packets with continuous flows, are widely used in addressing a variety of network control problems, such as congestion control, routing and pricing. The topology of the network studied here is characterized by a set of nodes $\mathcal{N} = \{1, \ldots, N\}$ and a set of links $\mathcal{L} = \{1, \ldots, L\}$, with each link $\ell \in \mathcal{L}$ having a fixed capacity $C_\ell > 0$ and an associated buffer size $b_\ell \geq 0$. The set of users is denoted by $\mathcal{M} = \{1, \ldots, M\}$. Each user is associated with a unique connection between a source and a destination node. The connection is a path that connects various nodes, see Figure 19.1 for an illustration.

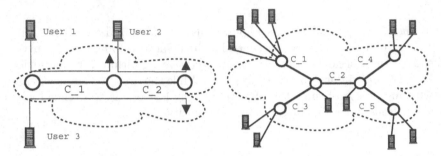

Fig. 19.1. Examples of two network topologies.

For the ith user, we consider a path $\overline{\mathcal{L}}_i$ which is a subset of $\mathcal{L}$. The non-negative flow x_i sent by the ith user over the path $\overline{\mathcal{L}}_i$ satisfies the bounds

$$0 \leq x_i \leq x_{i,\max}$$

where the upper bound $x_{i,\max}$ on the ith user's flow rate may be a user-specific physical limitation. This upper bound cannot exceed the minimum capacity of the links on the route

$$x_{i,\max} \leq \min_{\ell \in \overline{\mathcal{L}}_i} C_\ell.$$

As in [152], the model studied here makes use of a binary routing matrix R that describes the relation between the set of routes associated with the users and links. That is, for each user $i \in \mathcal{M}$ and link $\ell \in \mathcal{L}$ the entries of the routing matrix R are given by

$$[R]_{\ell,i} = \begin{cases} 1 \text{ if source } i \text{ uses link } \ell; \\ 0 \text{ otherwise.} \end{cases}$$

Using this matrix representation, we have the inequality

$$Rx \leq C$$

where $x = [x_1 \cdots x_M]^T$ is the users flow rate vector and $C = [C_1 \cdots C_L]^T$ is the link capacity vector. If the aggregate sending rate of users whose flows pass through link ℓ exceeds the capacity C_ℓ of the link, then the arriving packets are queued in the buffer b_ℓ of the link. The total flow on link ℓ is denoted as $\overline{x}_\ell(t)$ and is given by

$$\overline{x}_\ell(t) = \sum_{\{i:\ell\in\overline{\mathcal{L}}_i\}} x_i(t).$$

Ignoring boundary effects, the buffer level b_ℓ at link ℓ evolves in agreement with the differential equation

$$\dot{b}_\ell(t) = \overline{x}_\ell - C_\ell.$$

19.1.2 Cost function

For Internet-style networks, an important indication of congestion is the variation in queueing delay, defined as the difference between the actual delay experienced by a packet and the propagation delay of the connection. If the incoming flow rate to a router exceeds the capacity of the outgoing link, then packets are queued, generally on a first-come first-served basis, in the corresponding buffer of the router, thus leading to an increase in the round-trip time (RTT) of packets. Hence, the RTT on a congested path is longer than the base RTT, which is defined as the sum of propagation and processing delays on the path of a packet. The queueing delay τ_ℓ at a link can be modeled as

$$\dot{\tau}_\ell(x,t) = \frac{1}{C_\ell}\dot{b}_\ell(t) = \frac{1}{C_\ell}\left(\overline{x}_\ell(t) - C_\ell\right).$$

Thus, the queueing delay that a user experiences is the sum of queueing delays on its path

$$\overline{\tau}_i(x,t) = \sum_{\ell\in\overline{\mathcal{L}}_i} \tau_\ell(x,t).$$

The goal is to make use of variations in RTT to devise a congestion control and pricing scheme. Then, the cost function for the ith user at time t is the difference between a linear pricing function proportional (through a parameter q_i) to the queueing delay, and a strictly increasing logarithmic utility function multiplied by a user preference parameter u_i

$$J_i(x,t) = \sum_{\ell\in\overline{\mathcal{L}}_i} \left(q_i\tau_\ell(x,t)x_i\right) - u_i\log(x_i+1).$$

Since the users pick their flow rates in a way that would minimize their cost functions, we adopt a dynamic update model whereby each user changes the flow rate proportional to the gradient of the cost function with respect to the flow rate. Thus, the algorithm for the ith user is defined as

$$\dot{x}_i(t) = -\dot{J}_i(x,t) = \frac{u_i}{x_i+1} - q_i\bar{\tau}_i(t)$$

where we have ignored the effect of the ith user's flow on the delay $\bar{\tau}_i$.

Next, we observe that the users update their flow rates only at discrete time instances corresponding to multiples of RTT. Hence, we discretize this equation, obtaining

$$x_i(k+1) = x_i(k) + \kappa_i \left[\frac{u_i}{x_i(k)+1} - q_i \sum_{\ell \in \mathcal{L}_i} \tau_\ell(k) \right] \tag{19.1}$$

where $x_i(0) = 0$ and κ_i is a user-specific stepsize constant, which can be set to one without loss of generality. The queue model is discretized in a similar manner, obtaining

$$\tau_\ell(k+1) = \tau_\ell(k) + \frac{1}{C_\ell} \sum_{\{i:\ell \in \mathcal{L}_i\}} x_i(k) - 1 \tag{19.2}$$

where $\tau_\ell(0) = 0$.

19.1.3 Robustness analysis for symmetric single bottleneck

In the case of a single bottleneck node we essentially have a single link of interest, for which we denote the associated delay with $\tau = \tau_\ell$ and the capacity with $C = C_\ell$. Then, the unique equilibrium state of the system described by (19.1) and (19.2) is given by

$$x_i^* = \frac{u_i}{q_i\tau^*} - 1;$$

$$\tau^* = \frac{1}{C+M} \sum_{i=1}^{M} \frac{u_i}{q_i}. \tag{19.3}$$

Letting $\tilde{x}_i(k) \doteq x_i(k) - x_i^*$ and $\tilde{\tau} \doteq \tau(k) - \tau^*$, the system (19.1) and (19.2) with a single bottleneck link and $\kappa_i = 1$ can be rewritten around the equilibrium state as

$$\tilde{x}_i(k+1) = \tilde{x}_i(k) + \frac{u_i}{\tilde{x}_i(k) + x_i^* + 1} - q_i(\tilde{\tau}(k) + \tau^*);$$

$$\tilde{\tau}(k+1) = \tilde{\tau}(k) + \frac{1}{C} \sum_{i=1}^{M} \tilde{x}_i(k). \tag{19.4}$$

Linearizing this equation around $\tilde{x}^* = 0$ and $\tilde{\tau}^* = 0$, we easily obtain

$$\tilde{x}_i(k+1) = \left[1 - \frac{u_i}{(x_i^*+1)^2} \right] \tilde{x}_i(k) - q_i\tilde{\tau}(k);$$

$$\tilde{\tau}(k+1) = \tilde{\tau}(k) + \frac{1}{C} \sum_{i=1}^{M} \tilde{x}_i(k). \tag{19.5}$$

Letting

$$q = [q_1 \cdots q_M]^T;$$

$$v = \left[\frac{u_1}{(x_1^* + 1)^2} \cdots \frac{u_M}{(x_M^* + 1)^2} \right]^T$$

we rewrite (19.5) in matrix form

$$\begin{bmatrix} \tilde{x}(k+1) \\ \tilde{\tau}(k+1) \end{bmatrix} = A(q, v, C) \begin{bmatrix} \tilde{x}(k) \\ \tilde{\tau}(k) \end{bmatrix} \tag{19.6}$$

where the matrix $A(q, v, C)$ is given by

$$A(q, v, C) = \begin{bmatrix} 1 - v_1 & 0 & 0 & \cdots & -q_1 \\ 0 & 1 - v_2 & 0 & \cdots & -q_2 \\ 0 & 0 & 1 - v_3 & \cdots & -q_3 \\ \vdots & \vdots & \vdots & \ddots & \vdots \\ \frac{1}{C} & \frac{1}{C} & \frac{1}{C} & \cdots & 1 \end{bmatrix}.$$

This system is (locally) stable if and only if $A(q, v, C)$ is a Schur matrix, i.e. all its eigenvalues $\lambda(q, v, C)$ lie in the open unit circle. Thus, the objective is to determine conditions on the parameters q, v and C such that $|\lambda(q, v, C)| < 1$. This task proves to be prohibitively complex in general, and hence we consider first the special situation when the parameter v is symmetric across all M users. That is, we take $v_i = v$ for all $i = 1, \ldots, M$. In this case, the characteristic equation of the matrix $A(q, v, C)$ is given by

$$\det(\lambda I - A(q, v, C)) = (\lambda - 1 + v)^{M-1} \left[\lambda^2 - (2 - v)\lambda + 1 - v + \sum_{i=1}^{M} \frac{q_i}{C} \right]$$

and the matrix $A(q, v, C)$ has $M - 1$ repeated real eigenvalues at $1 - v$ and two (possibly complex) eigenvalues at

$$1 - \frac{v}{2} \pm \sqrt{\frac{v^2}{4} - \sum_{i=1}^{M} \frac{q_i}{C}}.$$

First, in order to gain further insight into stability properties, the eigenvalues of the matrix $A(q, v, C)$ for $q_i = 1,000, i = 1, \ldots, 20, v = 1$ and $C = 20,000$ are computed and shown in Figure 19.2. We note that this matrix is ill-conditioned with a condition number in the order of 10^5.

Finally, we conclude that the single bottleneck link system given by (19.4) is (locally) stable around its equilibrium state (19.3) if and only if the parameters $q_1, \ldots, q_M, v$ and C lie in the region defined by the inequalities

$$\frac{1}{C} \sum_{i=1}^{M} q_i < v < 2.$$

The general nonsymmetric case is studied next using randomized algorithms.

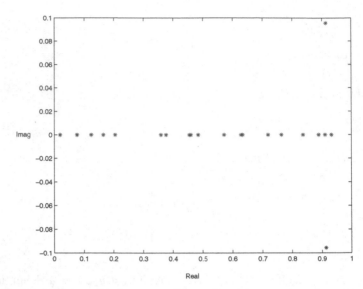

Fig. 19.2. Eigenvalues of the matrix $A(q, v, C)$ in the complex plane for the symmetric bottleneck case with 20 users.

19.1.4 Randomized algorithms for non symmetric case

We saw in the previous section that (local) stability and robustness can be studied analytically when the user utility preference parameters are the same for all users. If this is not the case, however, the eigenvalues of $A(q, v, C)$ cannot be expressed in closed form, and robustness of the system (19.6) is studied using randomized algorithms.

In particular, we investigate stability and robustness under parameter variations of the linearized single bottleneck link system (19.6) through randomization. That is, we investigate the effect of pricing and user parameters on the (local) stability of the system when $\mathbf{q}$ and $\mathbf{v}$ are random vectors with given pdfs $f_{\mathbf{q}}(q)$ and $f_{\mathbf{v}}(vu)$ and support $\mathcal{B}_q$ and $\mathcal{B}_v$, respectively. More precisely, for various values of C, the objective is to compute the probability

$$\mathrm{PR}\left\{\text{network stability}\right\} = \int_{\mathcal{B}_G} f_{\mathbf{q}}(q) f_{\mathbf{v}}(v) \mathrm{d}q \; \mathrm{d}v$$

where the good set is given by

$$\mathcal{B}_G = \{q \in \mathcal{B}_q, v \in \mathcal{B}_v : A(q, v, C) \text{ Schur}\}.$$

In particular, we performed simulations for the case of $M = 4$ users with parameter ranges

$$\mathcal{B}_q = \left\{q \in \mathbb{R}^4 : q_i \in [0, 1 \times 10^3], \; i = 1, \dots, 4\right\};$$
$$\mathcal{B}_v = \left\{v \in \mathbb{R}^4 : v_i \in [0, 1], \; i = 1, \dots, 4\right\} \tag{19.7}$$

and 22 fixed values of capacity in the interval $C \in [0, 3 \times 10^4]$. The values of C are chosen by performing a set of experiments regarding the stability of the network for different values of link capacity, obtaining

$$C = \{0.1, 0.5, 1, 2, 3, 4, 5, 6, 7, 8, 9, 10, 11, 12, 13, 14, 16, 18, 20, 22, 25, 30\} \times 10^3. \tag{19.8}$$

19.1.5 Monte Carlo simulation

We now present the results of the simulations, which are obtained by means of Algorithm 8.1. This randomized algorithm is based upon the Monte Carlo method discussed in Chapter 7. In particular, we consider random vectors with uniform probability density functions $f_{\mathbf{q}}(q)$ and $f_{\mathbf{v}}(v)$ and support sets given in (19.7). Since these sets are rectangles, linear congruential generators described in Section 14.1.1, or more sophisticated methods studied in the same chapter, can be immediately used for generation of pseudo-random samples of $\mathbf{q}$ and $\mathbf{v}$. First, we choose a level of confidence $\delta = 0.001$ and accuracy $\epsilon = 0.003$, and we determine the sample size N necessary to guarantee the required probabilistic levels. To this end, we use the Chernoff bound given in (9.14), obtaining $N \geq 4.23 \times 10^5$. Then, we choose $N = 450,000$ and construct the multisample

$$\mathbf{q}^{(1...N)} = \left\{ \mathbf{q}^{(1)}, \ldots, \mathbf{q}^{(N)} \right\}.$$

Similarly, we generate

$$\mathbf{v}^{(1...N)} = \left\{ \mathbf{v}^{(1)}, \ldots, \mathbf{v}^{(N)} \right\}.$$

Subsequently, for fixed values of C given in (19.8), we compute

$$A(\mathbf{q}^{(i)}, \mathbf{v}^{(i)}, C)$$

for $i = 1, \ldots, N$. Then, the empirical probability that the system (19.6) is stable is given by

$$\widehat{p}_N = \frac{N_G}{N}$$

where N_G is the number of "good" samples for which the system is stable. More precisely, we construct the indicator function of the good set

$$\mathbb{I}_{\mathcal{B}_G}(q, v, C) = \begin{cases} 1 \text{ if } q, v \in \mathcal{B}_G; \\ 0 \text{ otherwise.} \end{cases} \tag{19.9}$$

Then, the empirical probability is given by

$$\widehat{\mathbf{p}}_N = \frac{1}{N} \sum_{i=1}^{N} \mathbb{I}_{\mathcal{B}_G}(\mathbf{q}^{(i)}, \mathbf{v}^{(i)}, C).$$

Hence, we conclude that the inequality

$$|\text{PR}\{\text{network stability}\} - \widehat{\mathbf{p}}_N| \leq 0.003$$

holds with probability at least 0.999.

19.1.6 Quasi-Monte Carlo simulation

To validate the results, different sampling schemes for the parameters q and v are studied. That is, for comparison, we also compute an estimate of the volume of $\mathcal{B}_G$ using the quasi-Monte Carlo method (see Chapter 7 for a presentation of this method), which is a deterministic mechanism for generating samples which are "evenly distributed" within the sets of interest. In this case, no probability density function is specified for the parameter vectors q and v and the samples are constructed within the unit box. The samples obtained are then subsequently rescaled in the set $\mathcal{B}_q$ defined in (19.7). In particular, the Halton sequence, see Definition 7.5, is used for generation of the point sets

$$q^{(1...N)} = \{q^{(1)}, \ldots, q^{(N)}\};$$
$$v^{(1...N)} = \{v^{(1)}, \ldots, v^{(N)}\}.$$

These point sets have the property to minimize an upper bound on the star discrepancies $D_N^*(q^{(1...N)})$ and $D_N^*(v^{(1...N)})$, see Theorem 7.8. In particular, considering a binary base (i.e. $b_i = 2$ for $i = 1, 2, 3, 4$), a sample size $N = 300,000$ guarantees that $D_N^*(q^{(1...N)}) < 0.0421$, see the bound (7.20). Obviously, the same bound is obtained for the point set $v^{(1...N)}$. Notice that, in principle, these bounds on the discrepancy cannot be used to estimate the integration error (i.e. the error made when estimating $\text{Vol}(\mathcal{B}_G)$) using the Koksma–Hlawka inequality, see Theorem 7.5, since the function to be integrated is the indicator function of the set $\mathcal{B}_G$ and does not have bounded variation $V^{(n)}(g)$. Nevertheless, for comparison purposes, we still employ the QMC technique here, without relying on the theoretical bound of Theorem 7.5. Hence, for fixed values of C given in (19.8), we evaluate

$$A(q^{(i)}, v^{(i)}, C)$$

for $i = 1, \ldots, N$, and we compute

$$\frac{N_G}{N} = \frac{1}{N} \sum_{i=1}^{N} \mathbb{I}_{\mathcal{B}_G}(q^{(i)}, v^{(i)}, C)$$

where the sample size $N = 300,000$ is used.

The same approach for computing N_G/N is then followed using the Sobol' and Niederreiter sequences instead of the Halton sequence. Additional experiments regarding the stability of the network are performed using the quasi-Monte Carlo method for optimization, see Section 7.3. In this case, we construct an "optimal" grid minimizing the dispersions $d_N(q^{(1...N)})$ and $d_N(v^{(1...N)})$ according to the Sukharev sampling criterion, see Theorem 7.12. In particular, the point set $q^{(1...N)}$ is constructed as follows

$$q_i^{(k)} = 25(2k - 1)$$

for $i = 1, 2, 3, 4$ and $k = 1, \ldots, 20$. The sample size is therefore $N = 20^4 = 160,000$. It can be immediately verified, see Theorem 7.12, that this sample size guarantees $d_N(q^{(1...N)}) \geq 0.025$. Similarly, for $v^{(1...N)}$, we take

$$v_i^{(k)} = 0.025(2k - 1)$$

for $i = 1, 2, 3, 4$ and $k = 1, \ldots, 20$. Then, we compute

$$\frac{N_G}{N} = \frac{1}{N} \sum_{i=1}^{N} \mathbb{I}_{\mathcal{B}_G}(q^{(i)}, v^{(i)}, C)$$

where $N = 160,000$.

19.1.7 Numerical results

The results of the numerical experiments involving Monte Carlo and quasi-Monte Carlo methods are given in Figures 19.3 and 19.4, which show the network stability degradation versus capacity. In particular, for the Monte Carlo method the empirical probability $\hat{p}_N$ is plotted for the 22 values of capacity given in (19.8). For the quasi-Monte Carlo method, deterministic estimates N_G/N of $\text{Vol}(\mathcal{B}_G)$ for Halton, Sobol', Niederreiter sequences as well as for the optimal grid are also shown for the same values of capacity.

As a general comment, as expected, we observe that the stability of the system improves as capacity C increases. We also notice that the difference between the various sampling schemes is relatively small. However, the gridding method produces results which are slightly more "optimistic." Additional simulations show that these results are quite accurate even with a smaller sample size. One explanation for this phenomenon may be that the linearized system has relatively simple stability boundaries in the parameter space, see [7] for details. In the same paper, further results for a larger number of users are given, and stability properties of general network topologies with multiple bottleneck links are analyzed.

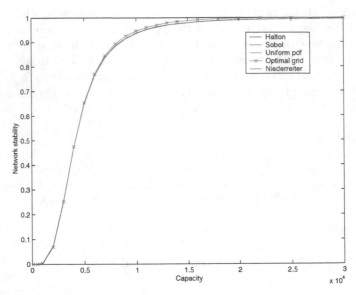

Fig. 19.3. Network stability versus capacity for $M = 4$ users using Monte Carlo and quasi-Monte Carlo methods.

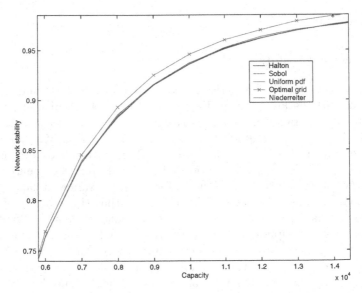

Fig. 19.4. A closer look at Figure 19.3 for larger values of network stability.

19.2 Probabilistic robustness of a flexible structure

We consider an example concerning a five-mass spring–damper model with four force actuators and four position sensors, as shown in Figure 19.5.

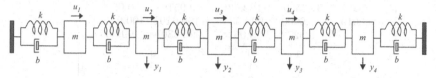

Fig. 19.5. Flexible structure with four noncolocated sensors and actuators.

This plant has a standard second-order representation of the form

$$M\ddot{\xi} + L\dot{\xi} + K\xi = E_a u;$$
$$y = E_s \xi$$

where $\xi \in \mathbb{R}^5$ is the mass displacement vector, $u \in \mathbb{R}^4$ is the input force vector, $y \in \mathbb{R}^4$ is the output displacement vector, and

$$E_a = \begin{bmatrix} 1\,0\,0\,0\,0 \\ 0\,1\,0\,0\,0 \\ 0\,0\,1\,0\,0 \\ 0\,0\,0\,1\,0 \end{bmatrix}^T, \quad E_s = \begin{bmatrix} 0\,1\,0\,0\,0 \\ 0\,0\,1\,0\,0 \\ 0\,0\,0\,1\,0 \\ 0\,0\,0\,0\,1 \end{bmatrix}$$

are the input and output influence matrices. The mass, damping and stiffness matrices, M, L, K, are given by

$$M = \operatorname{diag}\left([m\,m\,m\,m\,m]\right);$$

$$L = \begin{bmatrix} 2b & -b & 0 & 0 & 0 \\ -b & 2b & -b & 0 & 0 \\ 0 & -b & 2b & -b & 0 \\ 0 & 0 & -b & 2b & -b \\ 0 & 0 & 0 & -b & 2b \end{bmatrix};$$

$$K = \begin{bmatrix} 2k & -k & 0 & 0 & 0 \\ -k & 2k & -k & 0 & 0 \\ 0 & -k & 2k & -k & 0 \\ 0 & 0 & -k & 2k & -k \\ 0 & 0 & 0 & -k & 2k \end{bmatrix}.$$

The nominal values of the parameters are (in normalized units) $m = 1$, $k = 100$, and $b = 1$. A regulator $R(s)$ has been synthesized based on the nominal plant model, in order to improve the dynamic response of the plant. The state space representation of this regulator is given by

$$A_R = \begin{bmatrix} 0 & -0.1654 & -0.0761 & -0.0301 & -0.0095 \\ 0 & -0.3897 & -0.1951 & -0.0858 & -0.0301 \\ 0 & -0.1951 & -0.3983 & -0.1948 & -0.0764 \\ 0 & -0.0858 & -0.1948 & -0.3882 & -0.1644 \\ 0 & -0.0301 & -0.0764 & -0.1644 & -0.3114 \\ -200.0136 & 100.0057 & -0.0012 & -0.0139 & -0.0084 \\ 99.9777 & -199.6176 & 99.9514 & -0.0224 & -0.0139 \\ -0.0157 & 99.8529 & -199.6369 & 99.9510 & -0.0003 \\ -0.0019 & -0.0900 & 99.8533 & -199.6191 & 100.0042 \\ 0 & -0.0272 & -0.0578 & 99.9230 & -199.5654 \end{bmatrix}$$

$$\begin{bmatrix} 1 & 0 & 0 & 0 & 0 \\ 0 & 1 & 0 & 0 & 0 \\ 0 & 0 & 1 & 0 & 0 \\ 0 & 0 & 0 & 1 & 0 \\ 0 & 0 & 0 & 0 & 1 \\ -2.3073 & 0.8348 & -0.0769 & -0.0304 & -0.0096 \\ 0.8348 & -2.3845 & 0.8042 & -0.0864 & -0.0303 \\ -0.0769 & 0.8042 & -2.3947 & 0.8039 & -0.0767 \\ -0.0304 & -0.0864 & 0.8039 & -2.3860 & 0.8339 \\ 0 & 0 & 0 & 1 & -2.0000 \end{bmatrix} ;$$

$$B_R = \begin{bmatrix} 0.1654 & 0.0761 & 0.0301 & 0.0095 \\ 0.3897 & 0.1951 & 0.0858 & 0.0301 \\ 0.1951 & 0.3983 & 0.1948 & 0.0764 \\ 0.0858 & 0.1948 & 0.3882 & 0.1644 \\ 0.0301 & 0.0764 & 0.1644 & 0.3114 \\ -0.0073 & 0.0085 & 0.0167 & 0.0098 \\ -0.4009 & 0.0561 & 0.0378 & 0.0229 \\ 0.1167 & -0.3798 & 0.0715 & 0.0343 \\ 0.0719 & 0.1109 & -0.3885 & 0.0555 \\ 0.0272 & 0.0578 & 0.0770 & -0.4346 \end{bmatrix} ;$$

$$C_R = \begin{bmatrix} -0.0136 & -0.0016 & 0.0073 & 0.0029 & 0.0014 \\ -0.0223 & -0.0185 & 0.0074 & 0.0154 & 0.0090 \\ -0.0157 & -0.0305 & -0.0166 & 0.0225 & 0.0340 \\ -0.0019 & -0.0181 & -0.0357 & -0.0076 & 0.0596 \end{bmatrix}$$

$$\begin{bmatrix} -0.3073 & -0.1652 & -0.0769 & -0.0304 & -0.0096 \\ -0.1652 & -0.3845 & -0.1958 & -0.0864 & -0.0303 \\ -0.0769 & -0.1958 & -0.3947 & -0.1961 & -0.0767 \\ -0.0304 & -0.0864 & -0.1961 & -0.3860 & -0.1661 \end{bmatrix}$$

and $D_R = 0_{4,4}$. Notice that, although the uncontrolled system is structurally stable, the introduction of feedback may cause instabilities for some parameter values. Hence, the objective is to analyze the robustness properties of the feedback connection of the plant and the above regulator, with respect to variations of the plant parameters and unmodeled dynamics. Specifically, we consider parametric uncertainty on the damping and stiffness parameters

$$b = b_0 + 0.5q_1, \quad k = k_0 + 0.5q_2$$

with $|q_1| \leq 1$ and $|q_2| \leq 1$, and dynamic uncertainty with a frequency shape described by the weight function

$$W(s) = \frac{0.1035s + 1}{0.02071s + 1}.$$

With a standard procedure, we express the uncertain plant as a feedback connection of an augmented nominal plant (say P_0) and the uncertainty. The augmented plant P_0 has ten additional inputs and outputs, corresponding to the uncertain parameters b and k, each of which is repeated five times in the state space representation of the plant. The flexible structure with controller and uncertainty is then modeled in the classical $M-\Delta$ form, as shown in Figure 19.6, where the "M" part of the interconnection is enclosed in the dashed box.

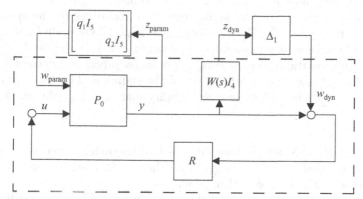

Fig. 19.6. Controlled flexible structure with uncertainty.

The matrix Δ representing the uncertainty in the system is structured and consists of two repeated real parameters q_1, q_2 and one full dynamic block $\Delta_1 \in \mathbb{C}^{4,4}$. That is, Δ is assumed to belong to the structured set

$$\mathbb{D} = \{\Delta : \Delta = \mathrm{bdiag}(q_1 I_5, q_2 I_5, \Delta_1)\}.$$

For this $M-\Delta$ system, lower and upper bounds $1/\mu_+$ and $1/\mu_-$ of the robustness margin $1/\mu$ have been computed with the Matlab μ Analysis and Synthesis Toolbox [21], obtaining

$$1/\mu_+ = 1.172, \quad 1/\mu_- = 1.185.$$

This deterministic analysis hence shows that the system interconnection is certainly stable for all structured perturbations Δ having norm smaller than $1/\mu_+$, see Section 3.7.

Next, we proceed to a probabilistic analysis of the robust stability properties of the system for perturbations whose radius goes beyond the deterministic margin $1/\mu_+$. That is, we study how the probability of stability degrades with increasing radius of the uncertainty. Formally, for fixed $\rho > 0$, we consider the structured set

$$\mathcal{B}_{\mathbb{D}}(\rho) = \{\Delta \in \mathbb{D} : \bar{\sigma}(\Delta) \le \rho\}$$

and assume that the uncertainty Δ is a random matrix with uniform probability distribution in this set. Hence, letting (A, B, C, D) be a state space representation of system M, we define

$$\text{PR}\{\text{stability}\} = \text{PR}\{\Delta \in \mathcal{B}_{\mathbb{D}}(\rho) : A + B\Delta(I - D\Delta)^{-1}C \text{ is stable}\}$$

and, for given $p^* \in [0, 1]$, we define the *probabilistic stability margin*

$$\rho(p^*) \doteq \sup\{\rho : \text{PR}\{\text{stability}\} \ge p^*\}.$$

Given a probability level p^*, the probabilistic stability margin $\rho(p^*)$ gives the maximum size of the perturbation Δ, measured according to the spectral norm, so that the probability $\text{PR}\{\text{stability}\}$ is at least p^*. Once $\text{PR}\{\text{stability}\}$ is estimated by means of a randomized algorithm (see e.g. Section 8.2), the next step is to construct the *probability degradation function*, i.e. the plot of the probability of stability as a function of the radius ρ. This plot may be compared with the classical worst-case stability margin $1/\mu_+$, obtaining

$$\rho(p^*) \ge 1/\mu_+$$

for any $p^* \in [0, 1]$. This fact, in turn, implies that the margin computed with probabilistic methods is always larger than the classical worst-case margin, at the expense of a risk expressed in probability.

Taking $\epsilon = \delta = 0.02$, by means of the Chernoff bound (see Section 9.3)

$$N \ge \frac{\log \frac{2}{\delta}}{2\epsilon^2}$$

we obtained $N \ge 23{,}026$. Then, we estimated the probability degradation function for 40 equispaced values of ρ in the range $[0.15, 2.1]$. For each grid point $\rho_k \in [0.15, 2.1]$ the probability of stability is estimated as

$$\hat{\mathbf{p}}_N(\rho_k) = \frac{1}{N} \sum_{i=1}^{N} \mathbb{I}(\Delta^{(i)})$$

where

$$\mathbb{I}(\Delta^{(i)}) = \begin{cases} 1 \text{ if } A + B\Delta^{(i)}(I - D\Delta^{(i)})^{-1}C \text{ is stable}; \\ 0 \text{ otherwise.} \end{cases}$$

where $\Delta^{(i)}$ is extracted uniformly at random in the set $\mathcal{B}_{\mathbb{D}}(\rho_k)$. The accuracy of this estimation is such that

$$\text{PR}\{|\hat{\mathbf{p}}_N(\rho_k) - \text{PR}\{\text{stability}\}| \le 0.02\} \ge 0.98.$$

The plot of the obtained realizations of $\hat{\mathbf{p}}_N(\rho)$ as a function of ρ is shown in Figure 19.7 together with the deterministic robustness margin $1/\mu_+$.

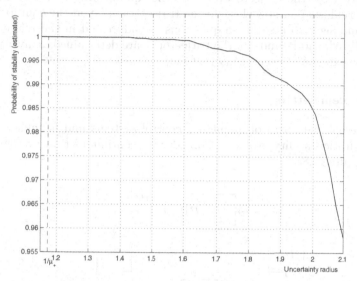

Fig. 19.7. Degradation of the probability of stability for the controlled structure.

From this plot we observe, for instance, that if a 1% loss of probabilistic performance may be tolerated, then the stability margin may be increased by approximately 64% with respect to its deterministic counterpart. In fact, the risk-adjusted stability margin for a probability level 0.99 is $\rho(0.99) \approx 1.93$. In addition, we notice that the estimated probability is equal to one up to $\rho \approx 1.4$. We conclude that, in this example, even if the upper and lower bounds of μ approximately coincide, so that $1/\mu$ is a nonconservative deterministic measure of robustness, this measure turns out to be quite conservative in a probabilistic sense.

19.3 Stability of quantized sampled-data systems

In this section we study the application of RAs for quadratic stability of sampled-data systems with memoryless quantizers, see [141] for a more detailed analysis. Quantization involved in control systems has recently become an active research topic, see e.g. [52, 106, 142]. The need for quantization inevitably arises when digital networks are part of the feedback loop and it is of interest to reduce the data rate necessary for the transmission of control signals. Then, a fundamental issue is to determine the minimum information to achieve the control objectives. Clearly, if a quantized discrete-time signal takes only a finite number of fixed values, then the trajectories may go close to an equilibrium but not converge, so that asymptotic stability is not achieved. Then, various problems may be posed. For example, it is of interest to clarify how close the trajectories get to the equilibrium point and, if the sampling period is large, how close do the trajectories stay at the equilibrium between sampling instants. These questions are addressed in the references previously listed, and bounds on the trajectories are determined analytically, albeit at the expense of crude conservatism.

19.3.1 Problem setting

We now present the setup of quantized control systems and formulate the related quadratic stability problem. Consider the continuous-time system depicted in Figure 19.8.

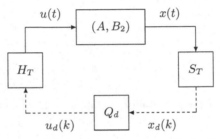

Fig. 19.8. Quantized sampled-data system where S_T is the sampler, H_T is the zeroth order hold and Q_d is the quantizer.

The pair (A, B_2), which is assumed stabilizable, represents a linear time-invariant plant with the state equation

$$\dot{x} = Ax + B_2 u \tag{19.10}$$

with initial state $x(0) = x_0 \in \mathbb{R}^{n_s}$ given but arbitrary. We study the case when A is not stable, because otherwise the problem becomes trivial. The output of the sampler S_T is a discrete-time signal given by

$$x_d(k) = x(kT)$$

where k is a positive integer and $T > 0$ is the sampling period. The output of zeroth order hold H_T is a continuous-time signal defined by

$$u(t) = u_d(k)$$

where $t \in [kT, (k+1)T)$.

We now introduce the definitions of cell and (memoryless) quantizer.

Definition 19.1 (Cell of the quantizer). *Given a countable index set $\mathcal{I}$, a partition $\{Q_i\}_{i \in \mathcal{I}} \subset \mathbb{R}^{n_s}$ consists of bounded sets (called the cell of the quantizer) such that:*

1. $Q_i \cap Q_k = \emptyset$ for $i \neq k$ and $\cup_{i \in \mathcal{I}} Q_i = \mathbb{R}^{n_s}$;
2. $0 \in Q_0$;
3. $0 \notin \partial Q_i, i \in \mathcal{I}$.

Definition 19.2 (Memoryless quantizer). *Given a set of cells $\{Q_i\}_{i \in \mathcal{I}}$ and a set of inputs $\{u_i\}_{i \in \mathcal{I}}$, a quantizer Q_d is a mapping from $\mathbb{R}^{n_s}$ to $\{u_i\}_{i \in \mathcal{I}}$ defined by*

$$Q_d(x) = u_i \text{ if } x \in Q_i, i \in \mathcal{I}. \tag{19.11}$$

In words, a quantizer Q_d maps a state $x \in \mathbb{R}^{n_s}$ in a cell Q_i to the corresponding input u_i. Various types of quantizer are studied, including the uniform and the logarithmic quantizer, see Figure 19.9 for an illustration of the latter.

Next, following the quadratic stabilizability approach described in Section 4.2.2, we define a quadratic Lyapunov function $V(x) = x^T P x$, for positive definite $P \succ 0$. The time derivative of this function along trajectories of the system (19.10) is given by

$$\dot{V}(x, u) = \frac{d}{dt} V(x(t)) = (Ax + B_2 u)^T P x + x^T P(Ax + B_2 u).$$

Furthermore, for positive definite $R \succ 0$, the set

$$\left\{ x \in \mathbb{R}^{n_s} : \dot{V}(x, u) \leq -x^T R x \right\}$$

is the set of states at which the Lyapunov function decreases when the control u is applied. In fact, taking the state feedback $u = Kx$, we compute

$$\dot{V}(x, u) = x^T (A + B_2 K)^T P x + x^T P(A + B_2 K)x.$$

Hence, we can choose

$$R = (A + B_2 K)^T P + P(A + B_2 K).$$

Based on this *control Lyapunov function* approach, we study quadratically attractive sets for quantized sampled-data systems. This study leads to the construction of a specific randomized algorithm. We now state the definition of a quadratically attractive set.

Definition 19.3 (Quadratically attractive set). *Given $\bar{r} > 0$, positive definite matrices $P, R \succ 0$ and a fixed quantizer Q_d, for the closed-loop system in Figure 19.8, the ball $\mathcal{B}_{\|\cdot\|_2}(r) = \{x \in \mathbb{R}^{n_s} : \|x\|_2 \leq r\}$ of radius $r > 0$, is quadratically attractive from $\mathcal{B}_{\|\cdot\|_2}(\bar{r})$ with respect to P, R if every trajectory of the system (19.10) with $x(0) \in \mathcal{B}_{\|\cdot\|_2}(\bar{r})$ satisfies either*

$$\dot{V}(x(t), u(t)) \leq -x(t)^T R x(t)$$

or

$$x(t) \in \mathcal{E}\left(0, r^2 \lambda_{\min} P^{-1}\right) = \{x \in \mathbb{R}^{n_s} : x^T P x \leq r^2 \lambda_{\min}\}. \tag{19.12}$$

for all $t \geq 0$, where $\lambda_{\min}$ denotes the smallest eigenvalue of P.

We note that quadratic attractiveness coincides with (asymptotic) stability if $\underline{r} = 0$ and $\bar{r} = \infty$. We also observe that this definition of attractiveness for sampled-data system is in the continuous-time domain. Hence, the Lyapunov function $V(x(t))$ must decrease at a certain rate even between sampling instants and the ellipsoid $\mathcal{E}\left(0, \underline{r}^2 \lambda_{\min} P^{-1}\right) = \{x \in \mathbb{R}^{n_s} : V(x) \leq \underline{r}^2 \lambda_{\min}\}$ is an invariant set. In particular, $\mathcal{E}\left(0, \underline{r}^2 \lambda_{\min} P^{-1}\right)$ is the largest level set of $V(x) = x^T P x$ contained in $\mathcal{B}_{\|\cdot\|_2}(\underline{r})$. We now formally define the quadratic stability problem of quantized sampled-data systems.

Problem 19.1 (Quadratic stability of quantized systems). Given $\bar{r} > 0$, a fixed quantizer Q_d, a sampling period $T > 0$ and $R \succ 0$, find $P \succ 0$ and $\underline{r} > 0$ such that, for the closed-loop system in Figure 19.8, the ball $\mathcal{B}_{\|\cdot\|_2}(\underline{r})$ is quadratically attractive from the ball $\mathcal{B}_{\|\cdot\|_2}(\bar{r})$ with respect to P, R.

Remark 19.1 (Quadratic attractiveness of quantized systems). The setup of this problem is similar to that in [142], where stabilization of linear sampled-data systems with memoryless quantizers is studied. More precisely, the quantizer design problem considered in this reference can be roughly summarized as follows: given $\bar{r} > 0$, state feedback $u = Kx$ such that $A + B_2 K$ is stable, and matrices $P, R \succ 0$, design a quantizer Q_d with sampling period T and $\underline{r}$ for quadratic attractiveness with respect to P, R. As discussed previously, we can easily verify that, taking the state feedback $u = Kx$, the closed-loop continuous-time system $\dot{x} = (A + B_2 K)x$ is quadratically stable with respect to P, R, where $R = (A + B_2 K)^T P + P(A + B_2 K)$.

The class of quantizers considered in [142] is somewhat restricted compared with that in Definition 19.2, but it allows the finding of an analytic solution. However, the drawback of the approach proposed in [142] lies in the conservatism in the design, especially in the derivation of T and $\underline{r}$. The analysis method based on randomized algorithms provides a way to obtain less conservative estimates of the performance of the designed system, at the expense of obtaining a probabilistic solution instead of a guaranteed one. ◁

We denote by $\phi(x_0, u, t)$ the state of the system (19.10) at time t corresponding to the initial conditions $x_0 = x(0) \in \mathbb{R}^{n_s}$ and the constant control input u. That is, we have

$$\phi(x_0, u, t) = e^{At} x_0 + \left[\int_0^t e^{A\tau} B_2 \, d\tau\right] u.$$

Similar to (19.12), we also consider the ellipsoid

$$\mathcal{E}\left(0, \bar{r}^2 \lambda_{\max} P^{-1}\right) = \{x \in \mathbb{R}^{n_s} : V(x) \leq \bar{r}^2 \lambda_{\max}\}$$

where $\lambda_{\max}$ is the largest eigenvalue of P. We note that $\mathcal{E}\left(0, \bar{r}^2 \lambda_{\max} P^{-1}\right)$ is the smallest ellipsoid containing $\mathcal{B}_{\|\cdot\|_2}(\bar{r})$. In addition, we consider two ellipsoids $\underline{\mathcal{E}} \subseteq \bar{\mathcal{E}} \subset \mathbb{R}^{n_s}$ which provide estimates of the invariant sets

$\mathcal{E}\left(0, \underline{r}^2\lambda_{\min}P^{-1}\right)$ and $\mathcal{E}\left(0, \overline{r}^2\lambda_{\max}P^{-1}\right)$. A sketch of these sets and a partition of a logarithmic quantizer is given in Figure 19.9. We notice that the partition cells are strips orthogonal to the subspace spanned by K^T and become wider as the distance from the origin grows.

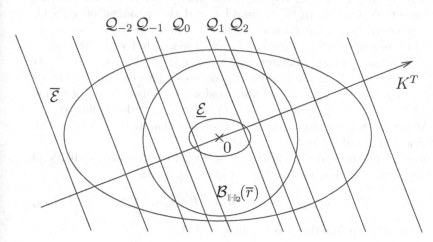

Fig. 19.9. Invariant sets and partition of a logarithmic quantizer.

We now state without proof a simple sufficient condition for quadratic attractiveness, see [141] for details.

Lemma 19.1 (Quadratic attractiveness). *Suppose that $P \succ 0$ and $\underline{r} > 0$ satisfy the following conditions:*

1. $\mathcal{E} \subset \mathcal{E}\left(0, \underline{r}^2\lambda_{\min}P^{-1}\right) \subset \mathcal{E}\left(0, \overline{r}^2\lambda_{\max}P^{-1}\right) \subset \overline{\mathcal{E}};$
2. *For every $x_0 \in \overline{\mathcal{E}}$ and $t \in [0,T]$, we have*

$$\phi(x_0, Q_d(x_0), t) \in \left\{x \in \mathbb{R}^{n_s} : \dot{V}(x, Q_d(x_0)) \leq -x^T R x\right\} \cup \underline{\mathcal{E}}. \quad (19.13)$$

Then, the ball $\mathcal{B}_{\|\cdot\|_2}(\underline{r})$ is quadratically attractive from $\mathcal{B}_{\|\cdot\|_2}(\overline{r})$ with respect to P, R for the closed-loop system shown in Figure 19.8.

There are several consequences of this lemma. First, since the ellipsoid $\mathcal{E}\left(0, \overline{r}^2\lambda_{\min}P^{-1}\right)$ is an invariant set contained in $\overline{\mathcal{E}}$, all trajectories starting in the ball $\mathcal{B}_{\|\cdot\|_2}(\overline{r})$ remain in this set. A second consequence is that the region where the Lyapunov function $V(x(t))$ increases is contained in $\underline{\mathcal{E}}$. Therefore, all trajectories of the system (19.10) enter $\mathcal{E}\left(0, \underline{r}^2\lambda_{\min}P^{-1}\right)$, because this is also an invariant set.

19.3.2 Randomized algorithm and violation function

In this section, we describe the iterative algorithm used for constructing a quadratic Lyapunov function and for finding $P \succ 0$ such that the second condition in Lemma 19.1 holds. This algorithm falls in the general category of randomized algorithms for robust performance synthesis described in Section 8.3.2 and is a variation on the sequential algorithms described in Chapters 11 and 12. However, one of the main differences with the algorithms described previously is that we require probability density functions having supporting sets depending on the ellipsoid $\bar{\mathcal{E}}$ and the quantization cells $\mathcal{Q}_i$ and we perform random generation in both state and time. In particular, we consider a probability density function $f_{\mathbf{x},\mathbf{i}}(x,i)$ associated with the state x and the cell index i of the quantizer and a pdf $f_{\mathbf{t}}(t) > 0$ for $t \in [0,T]$. At the kth iteration, the algorithm randomly generates a pair of state and index $(\mathbf{x}^{(k)}, \mathbf{i}^{(k)})$ according to $f_{\mathbf{x},\mathbf{i}}(x,i)$ and also randomly generates a set of ℓ time instants $\{\mathbf{t}^{(i)}\}_{i=0}^{\ell-1} \subset [0,T]$ according to $f_{\mathbf{t}}(t)$. An update rule similar to that used in Algorithm 8.4, and based on a violation function, may be introduced; see Definition 11.1 and related discussions.

We now define the specific violation function for quadratic stability of sampled-data systems. For $P, R \succ 0, x, i$ and $t \in [0,T]$, let

$$v(P,x,i,t) = \dot{V}(\phi(x,u_i,t),u_i) + \phi(x,u_i,t)^T R \phi(x,u_i,t). \tag{19.14}$$

Clearly, this violation function has the property that

$$v(P,x,i,t) \leq 0 \tag{19.15}$$

if and only if

$$\phi(x,u_i,t) \in \left\{ x \in \mathbb{R}^{n_s} : \dot{V}(x,u_i) \leq -x^T R x \right\}.$$

Therefore, checking whether the state of the system at time t enters the set of states $\left\{ x \in \mathbb{R}^{n_s} : \dot{V}(x,u_i) \leq -x^T R x \right\}$ at which the Lyapunov function decreases can be executed by a verification of the sign of the violation function. Then, at each step of the algorithm, for randomly generated $(\mathbf{x}, \mathbf{i})$ and $\mathbf{t}$, a matrix P is sought for achieving condition (19.15). Since v is linear in P, its gradient can be easily computed in closed form. Taking an arbitrary initial matrix P, an update in P may be computed using a gradient-based or ellipsoid algorithm, see Chapter 11 for details.

Remark 19.2 (Randomized algorithm for quantized systems). In [141], the various steps of the algorithm are described precisely. In particular, it is shown that convergence to a feasible solution $P \succ 0$ in a finite number of iterations is achieved with probability one, provided that a solution exists and minor technical assumptions are satisfied. This feasible solution is a matrix $P \succ 0$ which meets the two conditions given by Lemma 19.1.

The algorithm provides a systematic way to analyze the quantized system with less conservatism than other approaches, but with the drawback that probabilistic results, instead of guaranteed solutions, are found. On the other hand, the algorithm is based on the simple sufficient condition of Lemma 19.1

and has certain redundancy. For example, the lemma requires that trajectories for all $x_0 \in \overline{\mathcal{E}}$ and all $t \in [0,T]$ satisfy (19.13). Therefore, the number of trajectories is generally very high and the randomized algorithm may become practically intractable. In order to improve its efficiency, in [141] it is shown that the number of random trajectories which are generated can be greatly reduced. In particular, it is shown that it is not necessary to define density functions and perform randomization of the set $\overline{\mathcal{E}}$, but it suffices to consider the boundaries of $\mathcal{E}, \overline{\mathcal{E}}$ and of the quantization cells $\mathcal{Q}_i$.

More generally, in [141] the specific structure of the quantized sampled-data systems is exploited in order to reduce the computational complexity of the randomized algorithm. For nonlinear systems, deterministic computational methods are sought for obtaining less conservative invariant sets, see e.g. [41]. Hence, this method can be viewed as an alternative to finding probabilistically invariant sets for quantized systems. ◁

19.3.3 Numerical experiments

The randomized algorithm has been utilized for the magnetic ball levitation system shown in Figure 19.10. Details regarding this apparatus are given in [142], where the quadratic attractiveness for the sampled-data system with a logarithmic quantizer designed by an analytic method is analyzed. This method, however, is known to be subject to conservatism. In this section we present numerical results regarding probabilistic bounds obtained by means of a randomized algorithm, and clarify the extent of conservatism of the analytic deterministic condition.

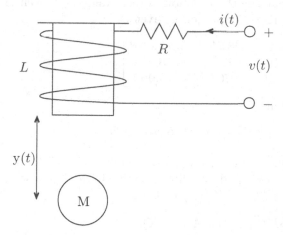

Fig. 19.10. Magnetic ball levitation system.

In Figure 19.10, a steel ball of mass M is levitated by the electromagnet. The position y of the ball is kept at an equilibrium through controlling the voltage v. The current in the coil is i, and the resistance and inductance of the magnet are R and L respectively. The system is linearized around an equilibrium $[y_0 \ \dot{y}_0 \ i_0]^T$ for the nominal voltage $v_0 = 10$ V. The resulting state is given by

$$x = [y - y_0 \ \dot{y} - \dot{y}_0 \ i - i_0]^T$$

and the system matrices A and B_2 are

$$A = \begin{bmatrix} 0 & 1 & 0 \\ 2\frac{Rg}{v_0}\sqrt{\frac{Mg}{\kappa}} & 0 & -2\frac{Rg}{v_0} \\ 0 & 0 & -\frac{R}{L} \end{bmatrix}, \quad B_2 = \begin{bmatrix} 0 \\ 0 \\ \frac{1}{L} \end{bmatrix}$$

where $M = 0.068$ kg, $R = 10\,\Omega$, $L = 0.41$ H, $\kappa = 3.3\times10^{-5}$ Nm^2/A^2 and $g = 9.8$ m/s^2.

The sampled-data controller designed in [142] for this system is now described. First, the optimal state feedback K is determined solving a Riccati equation corresponding to the linear quadratic regulator

$$A^T P_0 + P_0 A - P_0 B_2 Q_{uu}^{-1} B_2^T P_0 + Q_{xx} = 0$$

where $Q_{uu} = 0.878$ and

$$Q_{xx} = \begin{bmatrix} 1.15\times10^7 & 2.17\times10^5 & -5.52\times10^4 \\ 2.17\times10^5 & 4.11\times10^3 & -1.04\times10^3 \\ -5.52\times10^4 & -1.04\times10^3 & 265 \end{bmatrix}.$$

Further details regarding the selection of these values are given in [142]. The solution P_0 of the Riccati equation is given by

$$P_0 = \begin{bmatrix} 7.80\times10^5 & 1.48\times10^4 & -3.75\times10^3 \\ 1.48\times10^4 & 279 & -70.9 \\ -3.75\times10^3 & -70.9 & 18.0 \end{bmatrix}$$

whose eigenvalues are

$$\lambda(P_0) = [7.81\times10^5 \ 2.09\times10^{-4} \ 9.40\times10^{-5}]^T.$$

Then, a logarithmic quantizer Q_d and a sampling period T are designed so that the ball $\mathcal{B}_{\|\cdot\|_2}(\underline{r})$ is quadratically attractive from $\mathcal{B}_{\|\cdot\|_2}(\overline{r})$ with respect to $P_0, \gamma R_0$, where $\underline{r} = 32$, $\overline{r} = 10$, $\gamma = 0.10$ and

$$R_0 = (A + B_2 K)^T P_0 + P_0(A + B_2 K).$$

We notice that the presence of sampling and quantization requires sacrificing the decay rate by introducing γ.

We now describe the quantizer considered here. That is, the index set is $\mathcal{I} = \{0, \pm1, \pm2, \ldots\}$ and the partition cells $\mathcal{Q}_i, i \in \mathcal{I}$, are given by

$$Q_i = \begin{cases} \{x : Kx \in (-\alpha, \alpha)\} & \text{if } i = 0; \\ \{x : Kx \in [\text{sgn}(i)\alpha\delta^{|i|-1}, \text{sgn}(i)\alpha\delta^{|i|})\} & \text{otherwise} \end{cases}$$

where $\alpha = 0.451$ and $\delta = 1.78$. The control input values are

$$u_i = \text{sgn}(i)\beta\delta^{|i|-1}$$

where $\beta = 0.652$ and $i \in \mathcal{I}$. The designed sampling period is $T_s = 3.07 \times 10^{-3}$. We observe that the data considered here are different than the original in [142] for a change in the coordinate system.

We now discuss the conservatism in the deterministic design considered. First, although $\underline{r} = 32$ is a relatively large value, we observed in simulation that the trajectories of the resulting system generally entered a ball of radius 0.02. We notice that the different orders of magnitude are partially due to the largest eigenvalue of P_0, which is equal to 7.81×10^5 and makes its level sets very "narrow." Moreover, even when the size of T was doubled, the trajectories still entered a ball of a similar radius. To obtain less conservative results, we had to run the algorithm several times consecutively, starting each run with the matrix P resulting from the previous run. For each run, we used 10,000 samples in state and, for each sampled state, four samples in time. When there was no update for two runs in a row, we modified $\underline{r}$ and/or T.

We now describe the runs with $T = 1.5T_s$. In the first run, we set $P^{(1)} = P_0$ and we also took $\underline{\mathcal{E}}$ and $\overline{\mathcal{E}}$ to be level sets of $P^{(1)}$ so that $\mathcal{B}_{\|\cdot\|_2}(\overline{r}) \subset \overline{\mathcal{E}}$ and $\underline{\mathcal{E}} \subset \mathcal{B}_{\|\cdot\|_2}(\underline{r})$ with $\underline{r} = 6.0$. Two updates in the cells of the quantizer were observed and $P^{(2)}$ was obtained. In the next run, we started with $P^{(2)}$ and the corresponding sets $\overline{\mathcal{E}}$ and $\underline{\mathcal{E}}$. In this case there was no update in this run nor in the next. Therefore, we set $\underline{r}$ to 0.50 and continued in this manner. The results of the runs are summarized in Table 19.1. After 19 runs with 28 updates, we obtained $\underline{r} = 0.053$. The resulting matrix is given by

$$P^{(19)} = \begin{bmatrix} 7.80 \times 10^5 & 1.48 \times 10^4 & -3.75 \times 10^3 \\ 1.48 \times 10^4 & 290 & -70.3 \\ -3.75 \times 10^3 & -70.3 & 21.2 \end{bmatrix}$$

with eigenvalues

$$\lambda(P^{(19)}) = [7.81 \times 10^5 \quad 11.3 \quad 3.20]^T.$$

We remark that the largest eigenvalues of $P^{(1)}$ and $P^{(19)}$ almost coincide, but the others are much larger in $P^{(19)}$. This implies that the level sets of $P^{(19)}$ are "rounder," and hence a smaller attractive ball could be obtained.

Subsequently, a time response was calculated for the initial state

$$x(0) = [0.70 \times 10^{-3} \quad 0 \quad 0]^T.$$

The plot of the Euclidean norm of $x(t)$ is depicted in Figure 19.11 as a function of t. The horizontal line is shown for $\underline{r} = 0.053$. The trajectory of

Table 19.1. The results of the 19 runs for different values of $\underline{r}$ and $T = 1.5T_s$.

Run	$\underline{r}$	Number of updates	Cell indices at updates
1	6	2	$0, -10$
2, 3	6	0	none
4	0.5	15	$0, 1, -14$
5	0.5	9	$-6, -11, 13$
6	0.5	1	12
7	0.5	1	13
8, 9	0.5	0	none
10-19	0.25-0.053	0	none

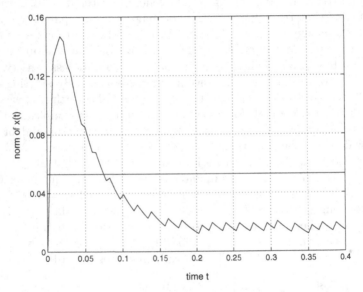

Fig. 19.11. Plot of Euclidean norm of $x(t)$ as a function of t.

the system goes below this line; for this trajectory we plotted $V(x(t))$ and the related violation function $v(P, x, i, t)$ in Figure 19.12 where the solid lines are for $P = P^{(19)}$ and the dashed lines for $P = P^{(1)}$.

In the top plot the horizontal line corresponds to the size of the level set $\mathcal{E}\left(0, \underline{r}^2 \lambda_{\min} P^{-1}\right)$ corresponding to $P^{(19)}$ entered by all trajectories. A similar line for the level set corresponding to $P^{(1)}$ is plotted as well, but this is not visible because it is too close to zero. For quadratic attractiveness, the violation function in the bottom plot should be smaller than zero whenever $V(x(t))$ is above the horizontal line in the top plot for every t. The solid lines satisfy this, but not the dashed lines. In a similar way, we obtained $\underline{r} = 0.040$ for $T = T_s$ and $\underline{r} = 0.062$ for $T = 1.7T_s$.

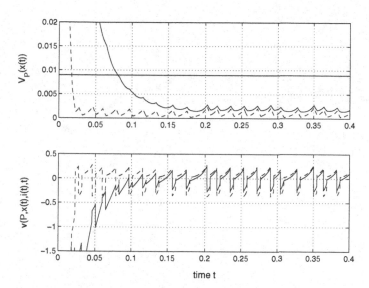

Fig. 19.12. Top: plot of $V(x(t))$ for $P^{(19)}$ (solid) and $V(x(t))$ for $P^{(1)}$ (dashed). Bottom: plot of $v(P, x, i, t)$ for $P^{(19)}$ (solid) and $v(P, x, i, t)$ for $P^{(1)}$ (dashed).

Appendix A

A.1 Transformations between random matrices

We next give a generalization of Theorem 14.2 to the case of functions of random matrices, see for instance [209].

Theorem A.1 (Functions of random matrices). *Let* $\mathbf{X}$ *and* $\mathbf{Y}$ *be two random matrices with the same number of free elements* $(x_1, \ldots, x_p)$ *and* $(y_1, \ldots, y_p)$ *respectively. Let the pdf of* $\mathbf{X}$ *be* $f_{\mathbf{X}}(X)$, *and let* $\mathbf{X}, \mathbf{Y}$ *be related by a one-to-one transformation* $\mathbf{Y} = g(\mathbf{X})$. *Let* $h(\cdot) \doteq g^{-1}(\cdot)$, *then the pdf* $f_{\mathbf{Y}}(Y)$ *is*

$$f_{\mathbf{Y}}(Y) = f_{\mathbf{X}}(h(Y)) J(X \to Y)$$

where the Jacobian $J(X \to Y)$ *is defined as*

$$J(X \to Y) \doteq \begin{vmatrix} \frac{\partial x_1}{\partial y_1} & \frac{\partial x_2}{\partial y_1} & \cdots & \frac{\partial x_p}{\partial y_1} \\ \frac{\partial x_1}{\partial y_2} & \frac{\partial x_2}{\partial y_2} & \cdots & \frac{\partial x_p}{\partial y_2} \\ \vdots & \vdots & & \vdots \\ \frac{\partial x_1}{\partial y_p} & \frac{\partial x_2}{\partial y_p} & \cdots & \frac{\partial x_p}{\partial y_p} \end{vmatrix}.$$

For the purpose of calculation, it is sometimes desirable to express the Jacobian in terms of the free elements of X and Y, for example $J(X \to Y)$ might be written as $J(x_1, \ldots, x_p \to y_1, \ldots, y_p)$. More generally, if the matrices $X_1, \ldots, X_k$ and $Y_1, \ldots, Y_m$ satisfy the equations $Y_i = g_i(x_1, \ldots, x_k)$, $i = 1, \ldots, m$, and $(X_1, \ldots, X_k)$, $(Y_1, \ldots, Y_m)$ have the free elements $(x_1, \ldots, x_p)$ and $(y_1, \ldots, y_p)$ respectively, then the Jacobian of the transformation from $(X_1, \ldots, X_k)$ to $(Y_1, \ldots, Y_m)$ will be denoted as $J(X_1, \ldots, X_k \to Y_1, \ldots, Y_m)$.

Remark A.3 (Many-to-few mappings). To handle the case when a transformation maps the random variables $\mathbf{x} \in \mathbb{R}^n$ to $\mathbf{y} \in \mathbb{R}^m$, with $m < n$ (i.e. the transformation is not one-to-one, but maps many to fewer variables) we may proceed as follows. Let the original transformation be

$$y_i = g_i(x_1, \ldots, x_n), \quad i = 1, \ldots, m; \quad m < n.$$

If additional slack functions $y_i = g_i(x_1, \ldots, x_n)$, $i = m+1, \ldots, n$, can be determined such that the transformation between $x_1, \ldots, x_n$ and the augmented set of variables $\widetilde{y} \doteq [y_1 \cdots y_m \, y_{m+1} \cdots y_n]^T$ satisfy the hypotheses of Theorem 14.2, then the pdf of $\mathbf{y}$ can be obtained by computing the marginal density

$$f_{\mathbf{y}}(y_1, \ldots, y_m) = \int \cdots \int f_{\mathbf{x}}(g^{-1}(y_1, \ldots, y_n)) J(x \to \widetilde{y}) dy_{m+1} \cdots dy_n.$$

◁

A.2 Jacobians of transformations

We report here several rules for the computation of Jacobians of matrix transformations. More comprehensive results related to Jacobians are given for instance in [80, 124, 209].

Rule A.1 (Chain rule for Jacobians).

$$J(Y \to X) = J(Y \to Z) J(Z \to X).$$

Rule A.2 (Jacobian of the derivatives). *Given a matrix transformation (linear or not) $Y = F(X)$, then the transformation of the differentials, $dY = dF(X)$ is linear, and*

$$J(Y \to X) = J(dY \to dX).$$

Rule A.3 (Jacobian of $Y = AX$). *The Jacobian of the linear matrix transformation*

$$Y = AX$$

where $X \in \mathbb{R}^{n,m}$, $A \in \mathbb{R}^{n,n}$, is given by

$$J(Y \to X) = |A|^m.$$

Similarly, the Jacobian of the matrix transformation $Y = XB$, with $B \in \mathbb{R}^{m,m}$, is given by

$$J(Y \to X) = |B|^n.$$

Rule A.4 (Jacobian of real $Y = AXB$). *The Jacobian of the matrix transformation*

$$Y = AXB$$

where $X \in \mathbb{R}^{n,m}$, $A \in \mathbb{R}^{n,n}$, $B \in \mathbb{R}^{m,m}$, is given by

$$J(Y \to X) = |A|^m |B|^n.$$

If A and B are orthogonal, then $J(Y \to X) = 1$.

Rule A.5 (Jacobian of complex $Y = AXB$). *The Jacobian of the matrix transformation*

$$Y = AXB$$

where $X \in \mathbb{C}^{n,m}$, $A \in \mathbb{C}^{n,n}$, $B \in \mathbb{C}^{m,m}$, is given by

$$J(Y \to X) = |\widetilde{A}|^{2m}|\widetilde{B}|^{2n}$$

where

$$\widetilde{A} = \begin{bmatrix} \mathrm{Re}(A) & -\mathrm{Im}(A) \\ \mathrm{Im}(A) & \mathrm{Re}(A) \end{bmatrix}, \quad \widetilde{B} = \begin{bmatrix} \mathrm{Re}(B) & -\mathrm{Im}(B) \\ \mathrm{Im}(B) & \mathrm{Re}(B) \end{bmatrix}.$$

Proof. To prove this, notice that, by Rule A.1, $J(Y \to X) = J(Y \to Z)J(Z \to X)$, where $Z = AX$. Write then the linear equation $Z = AX$ in terms of the real and imaginary parts

$$\begin{bmatrix} \mathrm{Re}(Z) \\ \mathrm{Im}(Z) \end{bmatrix} = \widetilde{A} \begin{bmatrix} \mathrm{Re}(X) \\ \mathrm{Im}(X) \end{bmatrix}$$

then, by Rule A.3, $J(Z \to X) = |\widetilde{A}|^{2m}$. Similarly, $J(Y \to Z) = |\widetilde{B}|^{2n}$.   □

Notice that if A is unitary, then it can be easily seen that $\widetilde{A}$ is orthogonal. Therefore, for A, B unitary, $J(Y \to X) = 1$.

A.3 Selberg integral

We present the solution of the so-called Selberg integral, derived in [243].

Theorem A.2 (Selberg integral). *For any positive integer n, let*

$$\varphi(x) = \varphi(x_1, \ldots, x_n) = \prod_{1 \leq i < k \leq n} (x_i - x_k)$$

if $n > 1$ and $\varphi(x) = 1$ for $n = 1$, and

$$\Phi(x) = |\varphi(x)|^{2\gamma} \prod_{i=1}^{n} x_i^{\alpha-1}(1 - x_i)^{\beta-1}.$$

Then

$$\int_0^1 \cdots \int_0^1 \Phi(x)\mathrm{d}x_1 \cdots \mathrm{d}x_n = \prod_{i=0}^{n-1} \frac{\Gamma(1 + \gamma + i\gamma)\Gamma(\alpha + i\gamma)\Gamma(\beta + i\gamma)}{\Gamma(1 + \gamma)\Gamma(\alpha + \beta + (n + i - 1)\gamma)}.$$

for any $\alpha, \beta, \gamma \in \mathbb{C}$ such that

$$\mathrm{Re}(\alpha) > 0, \quad \mathrm{Re}(\beta) > 0, \quad \mathrm{Re}(\gamma) > -\min\left\{\frac{1}{n}, \frac{\mathrm{Re}(\alpha)}{n-1}, \frac{\mathrm{Re}(\beta)}{n-1}\right\}.$$

A.4 Dyson–Mehta integral

The next theorem reports a result on the computation of the integral of certain determinants. The proof of this theorem can be found in [186].

Theorem A.3 (Dyson–Mehta). *Let $Z_n \in \mathbb{R}^{n,n}$ be a $n \times n$ symmetric matrix such that:*

1. *$[Z_n]_{i,j} = \psi(x_i, x_j)$, i.e. $[Z_n]_{i,j}$ depends only on x_i and x_j;*

2. *$\int \psi(x, x)\mathrm{d}\mu(x) = c$;*

3. *$\int \psi(x, y)\psi(y, z)\mathrm{d}\mu(y) = \psi(x, z)$*

where $\mathrm{d}\mu(x)$ is a suitable measure and c is a constant. Then

$$\int \det(Z_n)\mathrm{d}\mu(x_n) = (c - n + 1)\det(Z_{n-1}) \tag{A.16}$$

where Z_{n-1} is the $(n-1) \times (n-1)$ matrix obtained from Z_n by removing the row and the column containing x_n.

List of Symbols

We denote vector or scalar variables with lower case letters and matrix variables with upper case. Boldface indicates random variables and matrices.

$\triangleleft$	end of remark
$\square$	end of proof
$\star$	end of example

Vector Spaces and Cones

$\mathbb{R}^n$	space of real n-dimensional vectors
$\mathbb{C}^n$	space of complex n-dimensional vectors
$\mathbb{F}^n$	space of n-dimensional vectors with entries from $\mathbb{R}$ or $\mathbb{C}$
$\mathbb{R}^{n,m}$	space of real n-by-m matrices
$\mathbb{C}^{n,m}$	space of complex n-by-m matrices
$\mathbb{F}^{n,m}$	space of n-by-m matrices with entries from $\mathbb{R}$ or $\mathbb{C}$
$\mathbb{R}^n_+$	nonnegative orthant
$\mathbb{S}^n$	space of n-by-n real symmetric matrices
$\mathbb{S}^n_+$	cone of n-by-n positive semidefinite real symmetric matrices
$\mathbb{SK}^n$	space of n-by-n real skew-symmetric matrices
$\mathbb{H}^n$	space of n-by-n complex Hermitian matrices
$\mathbb{HK}^n$	space of n-by-n complex skew-Hermitian matrices
$\mathcal{G}^n_{\mathcal{O}}$	group of orthogonal matrices in $\mathbb{R}^{n,n}$; (17.3)
$\mathcal{G}^n_{\mathcal{U}}$	group of unitary matrices in $\mathbb{C}^{n,n}$; (17.26)
$\mathcal{H}^{n,m}_2$	$\mathcal{H}_2$ space of n-by-m transfer functions; Definition 3.3
$\mathcal{RH}^{n,m}_2$	$\mathcal{RH}_2$ space of n-by-m transfer functions; Definition 3.3
$\mathcal{H}^{n,m}_\infty$	$\mathcal{H}_\infty$ space of n-by-m transfer functions; Definition 3.2
$\mathcal{RH}^{n,m}_\infty$	$\mathcal{RH}_\infty$ space of n-by-m transfer functions; Definition 3.1
$[0,1]^n$	n-dimensional unit cube

Basic Operations

$\lceil x \rceil$	minimum integer greater or equal to $x \in \mathbb{R}$
$\lfloor x \rfloor$	largest integer smaller or equal to $x \in \mathbb{R}$
I_n	n-by-n identity matrix
$0_{n,m}$	n-by-m zero matrix
X^T	transpose of X
X^*	Hermitian of X
X^{-1}	inverse of (nonsingular) X
rank X	rank of matrix X
$X^\perp$	orthogonal complement of X, i.e. a matrix of maximum rank such that $X^T X^\perp = 0$, $X^{\perp T} X^\perp = I$
det X	determinant of X

$\lvert X \rvert$	absolute value of the determinant of X
$[X]_{i,k}$	(i,k) entry of X
$\mathrm{Re}(x), \mathrm{Im}(x)$	real and imaginary parts of $x \in \mathbb{C}$
$\langle x, y \rangle$	inner product of vectors x and y
$\mathrm{Tr}\, X$	trace of X
$\mathrm{vec}(X)$	column vectorization of matrix X; (3.8)
$\mathrm{diag}\,(x)$	diagonal matrix formed with the entries of vector x
$\mathrm{bdiag}(X_1, \ldots, X_n)$	block diagonal matrix formed with $X_1, \ldots, X_n$;
$\rho_\lambda(X)$	spectral radius of X
$[X]_S$	projection of X onto the (closed) set S; (11.1)
$[X]_+$	projection of $X \in \mathbb{S}^n$ onto the cone $\mathbb{S}^n_+$; (11.2)
$X \succ 0$	positive definite symmetric matrix
$X \succeq 0$	positive semidefinite symmetric matrix
$X \prec 0$	negative definite symmetric matrix
$X \preceq 0$	negative semidefinite symmetric matrix
$\mathbb{I}_S(\cdot)$	indicator function of the set S
$\mathrm{Vol}(S)$	volume of the set S; (3.14)
$\mathrm{Surf}(S)$	surface of the set S
$\mathrm{Card}\,(S)$	cardinality of the finite set S
$\Gamma(\cdot)$	Gamma function

Vector Norms and Balls

$\lVert x \rVert_p$	ℓ_p norm of the vector x; (3.1)
$\mathcal{B}_{\lVert \cdot \rVert_p}(\rho, \mathbb{F}^n)$	ball of radius ρ in the ℓ_p norm in $\mathbb{F}^n$; (3.2)
$\partial \mathcal{B}_{\lVert \cdot \rVert_p}(\rho, \mathbb{F}^n)$	boundary of $\mathcal{B}_{\lVert \cdot \rVert_p}(\rho, \mathbb{F}^n)$; (3.3)
$\lVert x \rVert_2^W$	weighted ℓ_2 norm of the vector x; (3.4)
$\mathcal{B}_{\lVert \cdot \rVert_2^W}(\rho, \mathbb{R}^n)$	ball of radius ρ in the ℓ_2^W norm in $\mathbb{F}^n$; (3.5)
$\mathcal{E}\,(m, W)$	ellipsoid of center m and shape matrix $W \succ 0$; (3.6)

Matrix Norms and Balls

$\lVert X \rVert_p$	ℓ_p Hilbert–Schmidt norm of the matrix X; (3.7)
$\lVert\!\lVert X \rVert\!\rVert_p$	ℓ_p-induced norm of the matrix X; (3.9)
$\mathcal{B}_{\lVert\!\lVert \cdot \rVert\!\rVert_p}(\rho, \mathbb{F}^{n,m})$	ball of radius ρ in the ℓ_p-induced norm in $\mathbb{F}^{n,m}$; (3.12)
$\mathcal{B}_\sigma(\rho, \mathbb{F}^{n,m})$	ball of radius ρ in the spectral norm in $\mathbb{F}^{n,m}$; (3.13)

Probability

$(\Omega, \mathcal{S}, \mathrm{PR}\,\{S\})$	probability space
$f_{\mathbf{X}}(X)$	probability density function of $\mathbf{X}$
$F_{\mathbf{X}}(X)$	cumulative distribution function of $\mathbf{X}$
$\mathrm{E}_{\mathbf{X}}(J(X))$	expected value of $J(X)$ taken with respect to $\mathbf{X}$
$\mathrm{Var}(\mathbf{x})$	variance of the random variable $\mathbf{x}$
$\mathrm{Cov}(\mathbf{x})$	covariance matrix of the random vector $\mathbf{x}$
$f_{\mathbf{x}_1, \ldots, \mathbf{x}_i}$	marginal density function; (2.1)

$f_{\mathsf{x}_i | x_1 \dots x_{i-1}}$ conditional density function; (2.2)

$J(x \to y)$ Jacobian of the function $x = h(y)$; (14.7)

Density Functions

$\mathcal{N}_{\bar{x}, \sigma^2}$ normal density with mean $\bar{x}$ and variance σ^2; (2.3)

$\mathcal{N}_{\bar{x}, W}$ multivariate normal density with mean $\bar{x}$ and covariance W; (2.4)

$\mathcal{U}_{[a,b]}$ uniform density in the interval $[a, b]$; (2.5)

$\mathcal{U}_S$ uniform density over the set S; (2.6)

$G_{a,b}$ Gamma density with parameters a, b; (2.10)

$\overline{G}_{a,c}$ generalized Gamma density with parameters a, c; (2.11)

Robust Control

$\widetilde{\mathbb{D}}$ structured operator uncertainty set; (3.25)

$\mathbb{D}$ structured matrix uncertainty set; (3.27)

$\mathcal{B}_{\widetilde{\mathbb{D}}}(\rho)$ ball of radius ρ in $\widetilde{\mathbb{D}}$; (3.26)

$\mathcal{B}_{\mathbb{D}}(\rho)$ ball of radius ρ in $\mathbb{D}$; (3.28)

$\mathcal{B}_q(\rho)$ ball of radius ρ of parametric uncertainty; (3.45)

$\mathcal{F}_u(\cdot)$ upper linear fractional transformation; (3.29)

$\mathcal{F}_l(\cdot)$ lower linear fractional transformation; (4.3)

$\mu_{\mathbb{D}}(M)$ structured singular value of the matrix M; (3.38)

$r_{\mathbb{R}}, r_{\mathbb{C}}$ real and complex stability radii; (3.36)

$r_{\mathbb{D}}$ stability radius under structured perturbations; (3.40)

$J(\Delta)$ performance function for analysis

$J(\Delta, \theta)$ performance function for design

$\mathcal{B}_G, \mathcal{B}_B$ good and bad sets; (6.2)

Randomization and Learning

$p(\gamma)$ probability of performance $J(\Delta) \le \gamma$; (6.6)

$\Delta^{(1 \dots N)}$ multisample $\Delta^{(1)}, \dots, \Delta^{(N)}$ of Δ; (7.1)

$\widehat{\mathbf{p}}_N(\gamma)$ empirical probability of performance; (7.2)

$\widehat{\mathbf{E}}_N(J(\Delta))$ empirical mean of $J(\Delta)$; (7.5)

$\mathrm{degrad}(\rho)$ performance degradation function; (6.13)

$x^{(1 \dots N)}$ deterministic point set; (7.13)

$D_N(\mathcal{S}, x^{(1 \dots N)})$ discrepancy of $x^{(1 \dots N)}$ with respect to $\mathcal{S}$; (7.14)

$d_N(x^{(1 \dots N)})$ dispersion of $x^{(1 \dots N)}$; (7.22)

$\mathbb{S}_{\mathcal{J}}(N)$ shatter coefficient of the family $\mathcal{J}$; (10.3)

$\mathrm{VC}(\mathcal{J})$ VC dimension of the family $\mathcal{J}$; Definition 10.2

$\mathrm{P\text{-}DIM}(\mathcal{J})$ P dimension of the family $\mathcal{J}$; Definition 10.3

References

1. M. Abramowitz and I.A. Stegun (editors). *Handbook of Mathematical Functions*. Dover, New York, 1970.

2. J.E. Ackermann, H.Z. Hu, and D. Kaesbauer. Robustness analysis: a case study. *IEEE Transactions on Automatic Control*, AC-35:352–356, 1990.

3. S. Agmon. The relaxation method for linear inequalities. *Canadian Journal of Mathematics*, 6:382–392, 1954.

4. A.V. Aho, J.E. Hopcroft, and J.D. Ullman. *The Design and Analysis of Computer Algorithms*. Addison-Wesley, Reading, 1974.

5. J.H. Ahrens and U. Dieter. Computer methods for sampling from Gamma, beta, Poisson and binomial distributions. *Computing*, 12:223–246, 1974.

6. T. Alpcan and T. Başar. A utility-based congestion control scheme for Internet-style networks with delay. In *Proceedings IEEE Infocom*, 2003.

7. T. Alpcan, T. Başar, and R. Tempo. Randomized algorithms for stability and robustness analysis of high speed communication networks. In *Proceedings IEEE Conference on Control Applications*, pages 397–403, 2003.

8. N.S. Altman. Bitwise behavior of random number generators. *SIAM Journal on Scientific and Statistical Computing*, 9:941–949, 1988.

9. V. Anantharam and J. Walrand. Special issue on control methods for communication networks – editorial. *Automatica*, 35:1891, 1990.

10. B.D.O. Anderson, N.K. Bose, and E.I. Jury. Output feedback stabilization and related problems – solution via decision methods. *IEEE Transactions on Automatic Control*, AC-20:53–66, 1975.

11. B.D.O. Anderson and J.B. Moore. *Optimal Control: Linear Quadratic Methods*. Prentice-Hall, Englewood Cliffs, 1990.

12. T.W. Anderson. *An Introduction to Multivariate Statistical Analysis*. Wiley, New York, 1958.

13. T.W. Anderson and D.A. Darling. Asymptotic theory of certain "goodness-of-fit" criteria based on stochastic processes. *Annals of Mathematical Statistics*, 23:193–212, 1952.

14. P. Apkarian and R.J. Adams. Advanced gain-scheduling techniques for uncertain systems. *IEEE Transactions on Control Systems Technology*, CST-6:21–32, 1998.

15. P. Apkarian and P. Gahinet. A convex characterization of gain-scheduled $\mathcal{H}_\infty$ controllers. *IEEE Transactions on Automatic Control*, AC-40:853–864, 1995.

16. D. Applegate and R. Kannan. Sampling and integration of near log-concave functions. In *Proceedings of the ACM Symposium on Theory of Computing*, pages 156–163, 1991.

17. T. Asai and S. Hara. A unified approach to LMI-based reduced order self-scheduling control synthesis. *Systems & Control Letters*, 36:75–86, 1999.

18. E.-W. Bai, K.M. Nagpal, and R. Tempo. Bounded-error parameter estimation: noise models and recursive algorithms. *Automatica*, 32:985–999, 1996.

19. E.-W. Bai, R. Tempo, and M. Fu. Worst-case properties of the uniform distribution and randomized algorithms for robustness analysis. *Mathematics of Control, Signals, and Systems*, 11:183–196, 1998.

20. V. Balakrishnan, S. Boyd, and S. Balemi. Branch and bound algorithm for computing the minimum stability degree of parameter-dependent linear systems. *International Journal of Robust and Nonlinear Control*, 1:295–317, 1992.

21. G.J. Balas, J.C. Doyle, K. Glover, A. Packard, and R. Smith. *μ-Analysis and Synthesis Toolbox*. MUSYN Inc. and The MathWorks Inc., Natick, 1993.

22. K. Ball. An elementary introduction to modern convex geometry. In *Flavors of Geometry (S. Levy editor)*, pages 1–58. Cambridge University Press, Cambridge, 1997.

23. I. Bárány and Z. Füredi. Computing the volume is difficult. *Discrete and Computational Geometry*, 2:319–326, 1987.

24. B.R. Barmish. *New Tools for Robustness of Linear Systems*. Macmillan, New York, 1994.

25. B.R. Barmish and H.I. Kang. A survey of extreme point results for robustness of control systems. *Automatica*, 29:13–35, 1993.

26. B.R. Barmish, P.P. Khargonekar, Z. Shi, and R. Tempo. Robustness margin need not be a continuous function of the problem data. *Systems & Control Letters*, 15:91–98, 1990.

27. B.R. Barmish and C.M. Lagoa. The uniform distribution: a rigorous justification for its use in robustness analysis. *Mathematics of Control, Signals, and Systems*, 10:203–222, 1997.

28. B.R. Barmish and P.S. Shcherbakov. On avoiding vertexization of robustness problems: the approximate feasibility concept. *IEEE Transactions on Automatic Control*, AC-47:819–824, 2002.

29. A.C. Bartlett, C.V. Hollot, and L. Huang. Root locations of an entire polytope of polynomials: it suffices to check the edges. *Mathematics of Control, Signals, and Systems*, 1:61–71, 1988.

30. T. Başar and G. J. Olsder. *Dynamic Noncooperative Game Theory*. SIAM, Philadelphia, 1999.

31. G. Becker and A. Packard. Robust performance of linear parametrically varying systems using parametrically-dependent linear feedback. *Systems & Control Letters*, 23:205–215, 1994.

32. M. Bellare, S. Goldwasser, and D. Micciancio. *"Pseudo-Random" Number Generation within Cryptographic Algorithms: The DSS Case*. Springer-Verlag, New York, 1997.

33. R. Bellman. *Dynamic Programming*. Princeton University Press, Princeton, 1957.

34. A. Ben-Tal and A. Nemirovski. Robust convex optimization. *Mathematics of Operations Research*, 23:769–805, 1998.

35. G. Bennett. Probability inequalities for the sum of independent random variables. *Journal of the American Statistical Association*, 57:33–45, 1962.

36. J. Bernoulli. *Ars Conjectandi*. Paris, 1713.

37. S.N. Bernstein. *The Theory of Probabilities*. Gostehizdat Publishing House, Moscow, 1946 (in Russian).

38. D. Bertsimas and J. Sethuraman. Moment problems and semidefinite optimization. In *Handbook of Semidefinite Programming (H. Wolkowicz, R. Saigal and L. Vandenberghe editors)*, pages 469–509. Kluwer, Boston, 2000.

39. S.P. Bhattacharyya. *Robust Stabilization Against Structured Perturbations*. Springer-Verlag, New York, 1987.

40. S.P. Bhattacharyya, H. Chapellat, and L.H. Keel. *Robust Control: The Parametric Approach*. Prentice-Hall, Upper Saddle River, 1995.

41. F. Blanchini. Set invariance in control. *Automatica*, 35:1747–1767, 1999.

42. V.D. Blondel and J.N. Tsitsiklis. A survey of computational complexity results in systems and control. *Automatica*, 36:1249–1274, 2000.

43. L. Blum, M. Blum, and M. Shub. A simple unpredictable pseudo-random number generator. *SIAM Journal on Computing*, 15:364–383, 1986.

44. L. Blum, F. Cucker, M. Shub, and S. Smale. *Complexity and Real Computation*. Springer-Verlag, New York, 1997.

45. B. Bollobás. Volume estimates and rapid mixing. In *Flavors of Geometry (S. Levy editor)*, pages 151–194. Cambridge University Press, Cambridge, 1997.

46. V. A. Bondarko and V. A. Yakubovich. The method of recursive aim inequalities in adaptive control theory. *International Journal of Adaptive Control and Signal Processing*, 6:141–160, 1992.

47. S. Boucheron, G. Lugosi, and P. Massart. Concentration inequalities using the entropy method. *Annals of Probability*, 31:1583–1614, 2003.

48. S. Boyd and C.H. Barrat. *Linear Controller Design – Limits of Performance*. Prentice-Hall, Englewood Cliffs, 1991.

49. S. Boyd, L. El Ghaoui, E. Feron, and V. Balakrishnan. *Linear Matrix Inequalities in System and Control Theory*. SIAM, Philadelphia, 1994.

50. R.P. Braatz, P.M. Young, J.C. Doyle, and M. Morari. Computational complexity of μ calculation. *IEEE Transactions on Automatic Control*, AC-39:1000–1002, 1994.

51. M.S. Branicky, S.M. LaValle, K. Olson, and L. Yang. Quasi-randomized path planning. In *Proceedings IEEE Conference on Robotics and Automation*, pages 1481–1487, 2001.

52. R.W. Brockett and D. Liberzon. Quantized feedback stabilization of linear systems. *IEEE Transactions on Automatic Control*, AC-45:1279–1289, 2000.

53. S.H. Brooks. A discussion of random methods for seeking maxima. *Operations Research*, 6:244–251, 1958.

54. L.G. Bushnell. Special issue on networks and control – editorial. *IEEE Control Systems Magazine*, 21:22–23, 2001.

55. G. Calafiore and M.C. Campi. Interval predictors for unknown dynamical systems: an assessment of reliability. In *Proceedings IEEE Conference on Decision and Control*, pages 4766–4771, 2002.

56. G. Calafiore and M.C. Campi. Robust convex programs: randomized solutions and applications in control. In *Proceedings IEEE Conference on Decision and Control*, pages 2423–2428, 2003.

57. G. Calafiore and M.C. Campi. Uncertain convex programs: randomized solutions and confidence levels. *Mathematical Programming*, DOI: 10.1007/s10107-003-0499-y, 2004.

58. G. Calafiore and F. Dabbene. Loop gain under random feedback. In *Proceedings IEEE Conference on Decision and Control*, pages 5016–5019, 2001.

59. G. Calafiore and F. Dabbene. A probabilistic framework for problems with real structured uncertainty in systems and control. *Automatica*, 38:1265–1276, 2002.

60. G. Calafiore, F. Dabbene, and R. Tempo. Radial and uniform distributions in vector and matrix spaces for probabilistic robustness. In *Topics in Control and its Applications (D.E. Miller and L. Qiu editors)*, pages 17–31. Springer-Verlag, New York, 1999.

61. G. Calafiore and B.T. Polyak. Stochastic algorithms for exact and approximate feasibility of robust LMIs. *IEEE Transactions on Automatic Control*, AC-46:1755–1759, 2001.

62. G.C. Calafiore, F. Dabbene, and R. Tempo. Randomized algorithms for probabilistic robustness with real and complex structured uncertainty. *IEEE Transactions on Automatic Control*, AC-45:2218–2235, 2000.

63. S. Cambanis, S. Huang, and G. Simons. On the theory of elliptically contoured distributions. *Journal of Multivariate Analysis*, 11:368–385, 1981.

64. P. Chebychev. Sur les valeurs limites des intégrales. *Journal de Mathematique Pure et Appliquee*, 19:157–160, 1874.

65. C.-T. Chen. *Linear Systems Theory and Design*. Oxford University Press, New York, 1999.

66. X. Chen and K. Zhou. Order statistics and probabilistic robust control. *Systems & Control Letters*, 35:175–182, 1998.

67. H. Chernoff. A measure of asymptotic efficiency for tests of a hypothesis based on the sum of observations. *Annals of Mathematical Statistics*, 23:493–507, 1952.

68. R.Y. Chiang and M.G. Safonov. *The Robust Control Toolbox*. The MathWorks Inc., Natick, 1996.

69. K.L. Chung. *A Course on Probability Theory*. Academic Press, London, 2001.

70. W.J. Conover. *Practical Nonparametric Statistics*. Wiley, New York, 1980.

71. S.A. Cook. The complexity of theorem proving procedures. In *Proceedings of the ACM Symposium on Theory of Computing*, pages 117–128, 1971.

72. R. Couture and P. L'Ecuyer. Special issue on uniform random number generation – editorial. *ACM Transactions on Modeling and Computer Simulation*, 8:1–2, 1998.

73. T.M. Cover. Geometrical and statistical properties of system of linear inequalities with applications in pattern recognition. *IEEE Transactions on Electronic Computers*, 14:326–334, 1965.

74. G.E. Coxson and C.L. De Marco. The computational complexity of approximating the minimal perturbation scaling to achieve instability in an interval matrix. *Mathematics of Control, Signals, and Systems*, 7:279–291, 1994.

75. M.A. Dahleh and I.J. Diaz-Bobillo. *Control of Linear Systems: A Linear Programming Approach*. Prentice-Hall, Englewood Cliffs, 1995.

76. G.B. Dantzig and G. Infanger. Multi-stage stochastic linear programs for portfolio optimization. *Annals of Operations Research*, 45:59–76, 1993.

77. M. Davis. Hilbert's tenth problem is unsolvable. *Mathematical Monthly*, 80:233–269, 1973.

78. P.J. Davis and P. Rabinowitz. *Methods of Numerical Integration*. Academic Press, New York, 1984.

79. N.G. de Bruijn. On some multiple integrals involving determinants. *Journal of the Indian Mathematical Society*, 19:133–151, 1955.

80. W.L. Deemer and I. Olkin. The Jacobians of certain matrix transformations useful in multivariate analysis. *Biometrika*, 38:345–367, 1951.

81. A. Dembo and O. Zeitouni. *Large Deviations Techniques and Applications*. Jones and Bartlett, Boston, 1993.

82. J.W. Demmel. The component-wise distance to the nearest singular matrix. *SIAM Journal on Matrix Analysis and Applications*, 13:10–19, 1992.

83. L. Devroye, L. Györfi, and G. Lugosi. *A Probabilistic Theory of Pattern Recognition*. Springer-Verlag, New York, 1996.

84. L.P. Devroye. *Non-Uniform Random Variate Generation*. Springer-Verlag, New York, 1986.

85. L.P. Devroye. Random variate generation for multivariate unimodal densities. *ACM Transactions on Modeling and Computer Simulation*, 7:447–477, 1997.

86. P. Diaconis and P. Hanlon. Eigen-analysis for some examples of the Metropolis algorithm. *Contemporary Mathematics*, 138:99–117, 1992.

87. P. Djavdan, H.J.A.F. Tulleken, M.H. Voetter, H.B. Verbruggen, and G.J. Olsder. Probabilistic robust controller design. In *Proceedings IEEE Conference on Decision and Control*, pages 2164–2172, 1989.

88. J.L. Doob. *Stochastic Processes*. Wiley, New York, 1990.

89. P. Dorato, D. Famularo, C.T. Abdallah, and W. Yang. Robust nonlinear feedback design via quantifier elimination theory. *International Journal of Robust and Nonlinear Control*, 9:817–822, 1999.

90. P. Dorato, Li Kun, E.B. Kosmatopoulos, P.A. Ioannou, and H. Ryaciotaki-Boussalis. Quantified multivariate polynomial inequalities. The mathematics of practical control design problems. *IEEE Control Systems Magazine*, 20:48–58, 2000.

91. P. Dorato, R. Tempo, and G. Muscato. Bibliograpy on robust control. *Automatica*, 29:201–213, 1993.

92. J. Doyle. Analysis of feedback systems with structured uncertainties. *IEE Proceedings*, 129(D):242–250, 1982.

93. J.C. Doyle, K. Glover, P.P. Khargonekar, and B.A. Francis. State-space solutions to standard $\mathcal{H}_2$ and $\mathcal{H}_\infty$ control problems. *IEEE Transactions on Automatic Control*, AC-34:831–847, 1989.

94. R.M. Dudley. Central limit theorems for empirical measures. *Annals of Probability*, 6:899–929, 1978.

95. R.M. Dudley. Balls in $\mathbb{R}^k$ do not cut all subsets of $k+2$ points. *Advances in Mathematics*, 31:306–308, 1979.

96. R.M. Dudley. *A Course on Empirical Processes*. Springer-Verlag, New York, 1984.

97. R.M. Dudley. *Uniform Central Limit Theorems*. Cambridge University Press, Cambridge, 1999.

98. G.E. Dullerud and F. Paganini. *A Course in Robust Control Theory: A Convex Approach*. Springer-Verlag, New York, 2000.

99. M.E. Dyer, A.M. Frieze, and R. Kannan. A random polynomial-time algorithm for approximating the volume of convex bodies. *Journal of the ACM*, 38:1–17, 1991.

100. A. Edelman. *Eigenvalues and Condition Numbers of Random Matrices*. Ph.D. Dissertation, Massachusetts Institute of Technology, Cambridge, 1989.

101. A. Edelman, E. Kostlan, and M. Shub. How many eigenvalues of a random matrix are real? *Journal of the American Mathematical Society*, 7:247–267, 1994.

102. B. Efron and C. Stein. The jackknife estimate of variance. *Annals of Statistics*, 9:586–596, 1981.

103. L. El Ghaoui and S.-I. Niculescu (editors). *Advances in Linear Matrix Inequality Methods in Control*. SIAM, New York, 2000.

104. L. El Ghaoui, F. Oustry, and M. AitRami. A cone complementary linearization algorithm for static output-feedback and related problems. *IEEE Transactions on Automatic Control*, AC-42:1171–1176, 1997.

105. L. El Ghaoui, F. Oustry, and H. Lebret. Robust solutions to uncertain semidefinite programs. *SIAM Journal on Optimization*, 9:33–52, 1998.

106. N. Elia and S.K. Mitter. Stabilization of linear systems with limited information. *IEEE Transactions on Automatic Control*, AC-46:1384–1400, 2001.

107. A.T. Fam. The volume of the coefficient space stability domain of monic polynomials. In *Proceedings International Symposium on Circuits and Systems*, pages 1780–1783, 1989.

108. K.T. Fang, S. Kotz, and K.W. Ng. *Symmetric Multivariate and Related Distributions*. Chapman & Hall, New York, 1990.

109. H. Faure. Discrépance de suites associées à un système de numération (en dimension s). *Acta Arithmetica*, 41:337–351, 1982.

110. V.N. Fomin. *Mathematical Theory of Learning Recognizing Systems*. LGU, Leningrad, 1976 (in Russian).

111. A. Frieze, J. Hastad, R. Kannan, J.C. Lagarias, and A. Shamir. Reconstructing truncated linear variables satisfying linear congruences. *SIAM Journal on Computing*, 17:262–280, 1988.

112. Y. Fujisaki, F. Dabbene, and R. Tempo. Probabilistic robust design of LPV control systems. *Automatica*, 39:1323–1337, 2003.

113. Y. Fujisaki and Y. Oishi. Guaranteed cost regulator design: Probabilistic solution and randomized algorithm. In *Proceedings of the International Symposium on Mathematical Theory of Networks and Systems*, 2004.

114. K. Fukunaga. *Introduction to Statistical Pattern Recognition*. Academic Press, Boston, 1990.

115. P. Gahinet. Explicit controller formulas for LMI-based $\mathcal{H}_\infty$ synthesis. *Automatica*, 32:1007–1014, 1996.

116. P. Gahinet and P. Apkarian. A linear matrix inequality approach to $\mathcal{H}_\infty$ control. *International Journal of Robust and Nonlinear Control*, 4:421–448, 1994.

117. F.R. Gantmacher. *The Theory of Matrices*. American Mathematical Society, Providence, 1959.

118. M.R. Garey and D.S. Johnson. *Computers and Intractability: A Guide to the Theory of NP-Completeness*. Freeman, New York, 1979.

119. J.E. Gentle. *Random Number Generation and Monte Carlo Methods*. Springer-Verlag, New York, 1998.

120. V.L. Girko. *Theory of Random Determinants*. Kluwer, Dordrecht, 1990.

121. K.-C. Goh, M.G. Safonov, and J.H. Ly. Robust synthesis via bilinear matrix inequalities. *International Journal of Robust and Nonlinear Control*, 6:1079–1095, 1996.

122. W. Gong and T. Başar. Special issue on systems and control methods for communication networks – editorial. *IEEE Transactions on Automatic Control*, AC-47:877–879, 2002.

123. M. Green and D.J.N. Limebeer. *Linear Robust Control*. Prentice-Hall, Englewood Cliffs, 1995.

124. A.K. Gupta and D.K. Nagar. *Matrix Variate Distributions*. CRC Press, Boca Raton, 1999.

125. A.K. Gupta and D. Song. Characterization of p-generalized normality. *Journal of Multivariate Analysis*, 60:61–71, 1997.

126. J.H. Halton. On the efficiency of certain quasi-random sequences of points in evaluating multi-dimensional integrals. *Numerische Mathematik*, 2:84–90, 1960; Berichtigung, ibid., 2:196, 1960.

127. W.K. Hastings. Monte Carlo sampling methods using Markov chains and their applications. *Biometrika*, 57:97–109, 1970.

128. D. Haussler. Decision theoretic generalizations of the PAC model for neural net and other learning applications. *Information and Computation*, 100:78–150, 1992.

129. P. Hellekalek. Good random number generators are (not so) easy to find. *Mathematics and Computers in Simulation*, 46:487–507, 1998.

130. P. Hellekalek and G. Larcher (editors). *Random and Quasi-Random Point Sets*. Springer-Verlag, New York, 1998.

131. R. Hernandez and S. Dormido. Kharitonov's theorem extension to interval polynomials which can drop in degree: a Nyquist approach. *IEEE Transactions on Automatic Control*, AC-41:1009–1012, 1996.

132. J.S. Hicks and R.F. Wheeling. An efficient method for generating uniformly distributed points on the surface of an n-dimensional sphere. *Communications of the ACM*, 2:17–19, 1959.

133. D. Hinrichsen and A.J. Pritchard. Stability radii of linear systems. *Systems & Control Letters*, 7:1–10, 1986.

134. E. Hlawka. Funktionen von beschränkter variation in der Theorie der Gleichverteilung. *Annali di Matematica Pura e Applicata*, 61:325–333, 1954.

135. W. Hoeffding. Probability inequalities for sums of bounded random variables. *Journal of the American Statistical Association*, 58:13–30, 1963.

136. H.P. Horisberger and P.R. Belanger. Regulators for linear time invariant plants with uncertain parameters. *IEEE Transactions on Automatic Control*, AC-21:705–708, 1976.

137. R.A. Horn and C.R. Johnson. *Topics in Matrix Analysis*. Cambridge University Press, Cambridge, 1991.

138. I. Horowitz. Survey of quantitative feedback theory (QFT). *International Journal of Control*, 53:255–291, 1991.

139. C.H. Houpis and S.J. Rasmussen. *Quantitative Feedback Theory*. Marcel Dekker, New York, 1999.

140. L.K. Hua. *Harmonic Analysis of Functions of Several Complex Variables in the Classical Domains*. American Mathematical Society, Providence, 1979.

141. H. Ishii, T. Başar, and R. Tempo. Randomized algorithms for quadratic stability of quantized sampled-data systems. *Automatica*, 40:839–846, 2004.

142. H. Ishii and B.A. Francis. *Limited Data Rate in Control Systems with Networks*. Springer-Verlag, New York, 2002.

143. T. Iwasaki and R.E. Skelton. All controllers for the general $\mathcal{H}_\infty$ control problem: LMI existence conditions and state-space formulas. *Automatica*, 30:1307–1317, 1994.

144. M. Jerrum and A. Sinclair. The Markov chain Monte Carlo method: an approach to approximate counting and integration. In *Approximation algorithms for NP-hard problems (D.S. Hochbaum editor)*, pages 482–520. PWS Publishing, Boston, 1996.

145. U. Jönsson and A. Rantzer. Optimization of integral quadratic constraints. In *Advances in Linear Matrix Inequality Methods in Control (L. El Ghaoui and S.-I. Niculescu editors)*, pages 109–127. SIAM, New York, 2000.

146. S. Kaczmarz. Angenäherte Aufslösung von Systemen linearer Gleichunger. *Bulletin International de l'Academie Polonaise des Sciences*, Lett. A:355–357, 1937 (English translation: Approximate solution of systems of linear equations. *International Journal of Control*, 57:1269-1271, 1993).

147. A.A. Kale and A.L. Tits. On Kharitonov's theorem without invariant degree assumption. *Automatica*, 36:1075–1076, 2000.

148. S. Kanev, B. De Schutter, and M. Verhaegen. An ellipsoid algorithm for probabilistic robust controller design. *Systems & Control Letters*, 49:365–375, 2003.

149. R. Kannan, L. Lovász, and M. Simonovits. Random walks and an $O^*(n^5)$ volume algorithm for convex bodies. *Random Structures and Algorithms*, 11:1–50, 1997.

150. M. Karpinski and A. Macintyre. Polynomial bounds for VC dimension of sigmoidal and general Pfaffian neural networks. *Journal of Computational System Science*, 54:169–176, 1997.

151. L.H. Keel and S.P. Bhattacharyya. A linear programming approach to controller design. In *Proceedings IEEE Conference on Decision and Control*, pages 2139–2148, 1997.

152. F.P. Kelly, A.K. Maulloo, and D.K.H. Tan. Rate control in communication networks: shadow prices, proportional fairness and stability. *Journal of the Operational Research Society*, 49:237–252, 1998.

153. L.G. Khachiyan. The problem of computing the volume of polytopes is NP-hard. *Uspekhi Mat. Nauk*, 44:179–180, 1989 (in Russian).

154. P. Khargonekar and A. Tikku. Randomized algorithms for robust control analysis and synthesis have polynomial complexity. In *Proceedings IEEE Conference on Decision and Control*, pages 3470–3475, 1996.

155. P.P. Khargonekar, I.R. Petersen, and K. Zhou. Robust stabilization of uncertain linear systems: Quadratic stabilizability and $\mathcal{H}_\infty$ control theory. *IEEE Transactions on Automatic Control*, AC-35:356–361, 1990.

156. V.L. Kharitonov. Asymptotic stability of an equilibrium position of a family of systems of linear differential equations. *Differentsial'nye Uravneniya*, 14:2086–2088, 1978 (in Russian).

157. H. Kimura. *Chain Scattering Approach to $\mathcal{H}_\infty$ Control*. Birkhäuser, Boston, 1997.

158. D.E. Knuth. *The Art of Computer Programming*, volume 2, Seminumerical Algorithms. Addison-Wesley, Reading, 1981.

159. J.F. Koksma. Een algemeene stelling uit de theorie der gelijkmatige verdeeling modulo 1. *Mathematica B (Zutphen)*, 11:7–11, 1942-1943.

160. V. Koltchinskii, C.T. Abdallah, M. Ariola, P. Dorato, and D. Panchenko. Improved sample complexity estimates for statistical learning control of uncertain systems. *IEEE Transactions on Automatic Control*, AC-46:2383–2388, 2000.

161. İ.E. Köse, F. Jabbari, and W.E. Schmitendorf. A direct characterization of $\mathcal{L}_2$-gain controllers for LPV systems. *IEEE Transactions on Automatic Control*, AC-43:1302–1307, 1998.

162. M.V. Kothare, V. Balakrishnan, and M. Morari. Robust constrained model predictive control using linear matrix inequalities. *Automatica*, 32:1361–1379, 1996.

163. H.J. Kushner and G.G. Yin. *Stochastic Approximation and Recursive Algorithms and Applications*. Springer-Verlag, New York, 2003.

164. H. Kwakernaak. Special issue on robust control – editorial. *Automatica*, 29:3, 1993.

165. H. Kwakernaak and R. Sivan. *Linear Optimal Control Systems*. Wiley, New York, 1972.

166. H. Kwakernaak and R. Sivan. *Modern Signals and Systems*. Prentice-Hall, Englewood Cliffs, 1991.

167. C.M. Lagoa and B.R. Barmish. Distributionally robust Monte Carlo simulation: A tutorial survey. In *Proceedings of the IFAC World Congress*, pages 1327–1338, 2002.

168. C.M. Lagoa, P.S. Shcherbakov, and B.R. Barmish. Probabilistic enhancement of classical robustness margins: the unirectangularity concept. *Systems & Control Letters*, 35:31–43, 1998.

169. A. Lanzon, B.D.O. Anderson, and X. Bombois. Selection of a single uniquely specifiable $\mathcal{H}_\infty$ controller in the chain-scattering framework. *Automatica*, 40:985–994, 2004.

170. P. L'Ecuyer. Uniform random number generation. *Annals of Operations Research*, 53:77–120, 1994.

171. P. L'Ecuyer, F. Blouin, and R. Couture. A search for good multiple recursive random number generators. *ACM Transactions on Modeling and Computer Simulation*, 3:87–98, 1993.

172. D.H. Lehmer. Mathematical methods in large-scale computing units. In *Proceedings of the Second Symposium on Large-Scale Digital Calculation Machinery*, pages 141–146, 1951.

173. D.J. Leith and W.E. Leithead. Survey of gain-scheduling analysis and design. *International Journal of Control*, 73:1001–1025, 2000.

174. L. Lovász. Random walks on graphs: a survey. In *Combinatorics, Paul Erdös is Eighty (D. Miklós, V. T. Sós and T. Szönyi editors)*, pages 353–398. János Bolyai Mathematical Society, Budapest, 1996.

175. L. Lovász. Hit-and-run mixes fast. *Mathematical Programming*, 86:443–461, 1999.

176. L. Lovász, M. Grötschel, and A. Schrijver. *Geometric Algorithms and Combinatorial Optimization*. Springer-Verlag, New York, 1993.

177. G. Lugosi. Pattern classification and learning theory. In *Principles of nonparametric learning (L. Györfi editor)*, pages 1–56. Springer-Verlag, New York, 2002.

178. A.J. Macintyre and E.D. Sontag. Finiteness results for sigmoidal "neural" networks. In *Proceedings of the ACM Symposium on Theory of Computing*, pages 325–334, 1993.

179. M. Mansour. Discrete-time and sampled-data stability tests. In *The Control Handbook (W.S. Levine editor)*, pages 146–151. CRC Press, Boca Raton, 1996.

180. A. Markov. *On Certain Applications of Algebraic Continued Fractions*. Ph.D. Dissertation, St Petersburg, 1884 (in Russian).

181. A. Marshall and I. Olkin. Multivariate Chebyshev inequalities. *Annals of Mathematical Statistics*, 31:1001–1014, 1960.

182. I. Masubuchi, A. Ohara, and N. Suda. LMI-based controller synthesis: A unified formulation and solution. *International Journal of Robust and Nonlinear Control*, 8:669–686, 1998.

183. Yu. Matiyasevich. Enumerable sets are Diophantine. *Doklady Akademii Nauk SSSR*, 191:279–282, 1970 (in Russian).

184. M. Matsumoto and T. Nishimura. Mersenne twister: a 623-dimensionally equidistributed uniform pseudo-random number generator. *ACM Transactions on Modeling and Computer Simulation*, 8:3–30, 1998.

185. A. Megretski and A. Ranzer. System analysis via integral quadratic constraints. *IEEE Transactions on Automatic Control*, AC-42:819–830, 1997.

186. M.L. Mehta. *Random Matrices*. Academic Press, Boston, 1991.

187. K.L. Mengersen and R.L. Tweedie. Rates of convergence of the Hastings and Metropolis algorithms. *Annals of Statistics*, 24:101–121, 1996.

188. N. Metropolis, A.W. Rosenbluth, M.N. Rosenbluth, A. Teller, and H. Teller. Equations of state calculations by fast computing machines. *Journal of Chemical Physics*, 21:1087–1091, 1953.

189. N. Metropolis and S.M. Ulam. The Monte Carlo method. *Journal of the American Statistical Association*, 44:335–341, 1949.

190. S.P. Meyn and R.L. Tweedie. *Markov Chains and Stochastic Stability*. Springer-Verlag, New York, 1996.

191. R.J. Minnichelli, J.J. Anagnost, and C.A. Desoer. An elementary proof of Kharitonov's stability theorem with extensions. *IEEE Transactions on Automatic Control*, AC-34:995–998, 1989.

192. R. Motwani and P. Raghavan. *Randomized Algorithms*. Cambridge University Press, Cambridge, 1995.

193. T.S. Motzkin and I.J. Schoenberg. The relaxation method for linear inequalities. *Canadian Journal of Mathematics*, 6:393–404, 1954.

194. M.E. Muller. A note on a method for generating random points uniformly distributed on n-dimensional spheres. *Communications of the ACM*, 2:19–20, 1959.

195. K. Mulmuley. *Computational Geometry: An Introduction through Randomization Algorithms*. Prentice-Hall, Englewood Cliffs, 1994.

196. A. Nemirovski. Several NP-hard problems arising in robust stability analysis. *Mathematics of Control, Signals, and Systems*, 6:99–195, 1993.

197. A. Nemirovski. On tractable approximations of randomly perturbed convex constraints. In *Proceedings IEEE Conference on Decision and Control*, pages 2419–2422, 2003.

198. A.S. Nemirovski and D.B. Yudin. *Problem Complexity and Method Efficiency in Optimization*. Wiley, New York, 1983.

199. Y. Nesterov and A.S. Nemirovski. *Interior Point Polynomial Algorithms in Convex Programming*. SIAM, Philadelphia, 1994.

200. Y. Nesterov and J.-Ph. Vial. *Confidence level solutions for stochastic programming*. Technical Report, Université Catholique de Louvain, 2000.

201. M.P. Newlin and P.M. Young. Mixed μ problems and branch and bound techniques. *International Journal of Robust and Nonlinear Control*, 7:145–164, 1997.

202. H. Niederreiter. Point sets and sequences with small discrepancy. *Monatshefte für Mathematik*, 104:273–337, 1987.

203. H. Niederreiter. *Random Number Generation and Quasi-Monte Carlo Methods*. SIAM, Philadelphia, 1992.

204. H. Niederreiter. New developments in uniform pseudorandom number and vector generation. In *Monte Carlo and Quasi-Monte Carlo Methods in Scientific Computing (H. Niederreiter and P.J.-S. Shiue editors)*, pages 87–120. Springer-Verlag, New York, 1995.

205. H. Niederreiter. Some current issues in quasi-Monte Carlo methods. *Journal of Complexity*, 23:428–433, 2003.

206. Y. Oishi. Probabilistic design of a robust state-feedback controller based on parameter-dependent Lyapunov functions. In *Proceedings IEEE Conference on Decision and Control*, pages 1920–1925, 2003.

207. Y. Oishi and H. Kimura. Model-set identification based on learning-theoretic inequalities. In *Proceedings of the IFAC World Congress*, 2002.

208. Y. Oishi and H. Kimura. Computational complexity of randomized algorithms for solving parameter-dependent linear matrix inequalities. *Automatica*, 39:2149–2156, 2003.

209. I. Olkin. Note on the Jacobians of certain matrix transformations useful in multivariate analysis. *Biometrika*, 40:43–46, 1953.

210. I.S. Pace and S. Barnett. Numerical comparison of root-location algorithms for constant linear systems. In *Recent Mathematical Developments in Control (D.J. Bell editor)*, pages 373–392. Academic Press, London, 1973.

211. A. Packard and J. Doyle. The complex structured singular value. *Automatica*, 29:71–109, 1993.

212. F. Paganini and E. Feron. Linear matrix inequality methods for robust $\mathcal{H}_2$ analysis: a survey with comparisons. In *Advances in Linear Matrix Inequality Methods in Control (L. El Ghaoui and S.-I. Niculescu editors)*, pages 129–151. SIAM, New York, 2000.

213. C.H. Papadimitriou. *Computational Complexity*. Addison-Wesley, Reading, 1994.

214. A. Papoulis and S.U. Pillai. *Probability, Random Variables and Stochastic Processes*. McGraw-Hill, New York, 2002.

215. I.R. Petersen and D.C. McFarlane. Optimal guaranteed cost control and filtering for uncertain linear systems. *IEEE Transactions on Automatic Control*, AC-39:1971–1977, 1994.

216. S. Poljak and J. Rohn. Checking robust nonsingularity is NP-hard. *Mathematics of Control, Signals, and Systems*, 6:1–9, 1993.

217. D. Pollard. *Convergence of Stochastic Processes*. Springer-Verlag, New York, 1984.

218. D. Pollard. *Empirical Processes: Theory and Applications*, volume 2. NSF-CBMS Regional Conference Series in Probability and Statistics, Institute of Mathematical Statistics, 1990.

219. B.T. Polyak. Gradient methods for solving equations and inequalities. *Zh. Vychisl. Mat. i Mat. Fiz.*, 4:995–1005, 1964 (in Russian).

220. B.T. Polyak and P.S. Shcherbakov. Random spherical uncertainty in estimation and robustness. *IEEE Transactions on Automatic Control*, AC-45:2145–2150, 2000.

221. B.T. Polyak and R. Tempo. Probabilistic robust design with linear quadratic regulators. *Systems & Control Letters*, 43:343–353, 2001.

222. I. Popescu. *Applications of Optimization in Probability, Finance and Revenue Management*. Ph.D. Dissertation, Massachusetts Institute of Technology, Cambridge, 1999.

223. A. Prékopa. *Stochastic Programming*. Kluwer, Dordrecht, 1995.

224. L. Qiu, B. Bernhardsson, A. Rantzer, E.J. Davison, P.M. Young, and J.C. Doyle. A formula for computation of the real stability radius. *Automatica*, 31:879–890, 1995.

225. L.R. Ray and R.F. Stengel. A Monte Carlo approach to the analysis of control system robustness. *Automatica*, 29:229–236, 1993.

226. R. Reemtsen and J.-J. Rückmann (editors). *Semi-Infinite Programming*. Kluwer, Dordrecht, 1998.

227. R.Y. Rubinstein. *Monte-Carlo Optimization, Simulation and Sensitivity of Queueing Networks.* Wiley, New York, 1986.

228. W.J. Rugh. *Linear System Theory.* Prentice-Hall, Upper Saddle River, 1996.

229. W.J. Rugh and J.S. Shamma. Research on gain-scheduling. *Automatica,* 36:1401–1425, 2000.

230. M.G. Safonov. Stability margins of diagonally perturbed multivariable feedback systems. *IEE Proceedings,* 129(D):251–256, 1982.

231. M.G. Safonov and M.K.H. Fan. Special issue on multivariable stability margin – editorial. *International Journal of Robust and Nonlinear Control,* 7:97–103, 1997.

232. M. Sampei, T. Mita, and M. Nakamichi. An algebraic approach to $\mathcal{H}_\infty$ output feedback control problems. *Systems & Control Letters,* 14:13–24, 1990.

233. N. Sauer. On the density of families of sets. *Journal of Combinatorial Theory,* 13(A):145–147, 1972.

234. C. Scherer. *The Riccati Inequality and State Space $\mathcal{H}_\infty$-Optimal Control.* Ph.D. Dissertation, University of Würzburg, 1990.

235. C. Scherer. Mixed $\mathcal{H}_2/\mathcal{H}_\infty$ control. In *Trends in Control: A European Perspective (A. Isidori editor),* pages 173–216. Springer-Verlag, New York, 1995.

236. C. Scherer, P. Gahinet, and M. Chilali. Multiobjective output feedback control via LMI optimization. *IEEE Transactions on Automatic Control,* AC-42:896–911, 1997.

237. C.W. Scherer. $\mathcal{H}_\infty$ control for plants with zeros on the imaginary axis. *SIAM Journal on Control and Optimization,* 30:123–142, 1992.

238. C.W. Scherer. $\mathcal{H}_\infty$ optimization without assumptions on finite or infinte zeros. *SIAM Journal on Control and Optimization,* 30:143–166, 1992.

239. C.W. Scherer. LPV control and full block multipliers. *Automatica,* 37:361–375, 2001.

240. A. Schrijver. *Theory of Linear and Integer Programming.* Wiley, New York, 1998.

241. F.C. Schweppe. *Uncertain Dynamical Systems.* Prentice-Hall, Englewood Cliffs, 1973.

242. A. Seidenberg. A new decision method for elementary algebra. *Annals of Mathematics,* 60:365–374, 1954.

243. A. Selberg. Bemerkninger om et multiplet integral. *Norsk Matematisk Tidsskrift,* 26:71–78, 1944.

244. R.E. Skelton, T. Iwasaki, and K.M. Grigoriadis. *A Unified Algebraic Approach to Linear Control Design.* Taylor & Francis, London, 1998.

245. S. Skogestad and I. Postlethwaite. *Multivariable Feedback Control: Analysis and Design.* Wiley, New York, 1996.

246. R.L. Smith. Efficient Monte-Carlo procedures for generating points uniformly distributed over bounded regions. *Operations Research,* 32:1296–1308, 1984.

247. G.W. Snedecor and W.G. Cochran. *Statistical Methods.* Iowa State Press, Ames, 1989.

248. I.M. Sobol'. The distribution of points in a cube and the approximate evaluation of integrals. *Zh. Vychisl. Mat. i Mat. Fiz.,* 7:784–802, 1967 (in Russian).

249. D. Song and A.K. Gupta. L_p-norm uniform distribution. *Proceedings of the American Mathematical Society*, 125:595–601, 1997.

250. E.D. Sontag. VC dimension of neural networks. In *Neural Networks and Machine Learning (C.M. Bishop editor)*. Springer-Verlag, New York, 1998.

251. J.C. Spall. Estimation via Markov chain Monte Carlo. *IEEE Control Systems Magazine*, 23:34–45, 2003.

252. R.F. Stengel. Some effects of parameter variations on the lateral-directional stability of aircraft. *AIAA Journal of Guidance and Control*, 3:124–131, 1980.

253. R.F. Stengel. *Stochastic Optimal Control: Theory and Application*. Wiley, New York, 1986.

254. G.W. Stewart. The efficient generation of random orthogonal matrices with an application to condition estimators. *SIAM Journal on Numerical Analysis*, 17:403–409, 1980.

255. T.J. Stieltjes. Recherches sur les fractions continues. *Annales de la Faculté de Sciences de Toulouse*, 8:1–122, 1894.

256. T.J. Stieltjes. Recherches sur les fractions continues. *Annales de la Faculté de Sciences de Toulouse*, 9:5–47, 1895.

257. A.G. Sukharev. Optimal strategies of the search for an extremum. *Zh. Vychisl. Mat. i Mat. Fiz.*, 11:910–924, 1971 (in Russian).

258. M. Sznaier, T. Amishima, P. A. Parrilo, and J. Tierno. A convex approach to robust $\mathcal{H}_2$ performance analysis. *Automatica*, 38:957–966, 2002.

259. M. Sznaier, H. Rotstein, B. Juanyu, and A. Sideris. An exact solution to continuous-time mixed $\mathcal{H}_2/\mathcal{H}_\infty$ control problems. *IEEE Transactions on Automatic Control*, AC-45:2095–2101, 2000.

260. M. Talagrand. New concentration inequalities in product spaces. *Inventiones Mathematicae*, 126:505–563, 1996.

261. A. Tarski. *A Decision Method for Elementary Algebra and Geometry*. University of California Press, Berkeley, 1951.

262. R.C. Tausworthe. Random numbers generated by linear recurrence modulo two. *Mathematics of Computation*, 19:201–209, 1965.

263. R. Tempo, E.-W. Bai, and F. Dabbene. Probabilistic robustness analysis: explicit bounds for the minimum number of samples. In *Proceedings IEEE Conference on Decision and Control*, pages 3424–3428, 1996.

264. R. Tempo, E.-W. Bai, and F. Dabbene. Probabilistic robustness analysis: explicit bounds for the minimum number of samples. *Systems & Control Letters*, 30:237–242, 1997.

265. R. Tempo and F. Blanchini. Robustness analysis with real parametric uncertainty. In *The Control Handbook (W.S. Levine editor)*, pages 495–505. CRC Press, Boca Raton, 1996.

266. M.J. Todd. Semidefinite optimization. *Acta Numerica*, 10:515–560, 2001.

267. O. Toker. On the complexity of the robust stability problem for linear parameter varying systems. *Automatica*, 33:2015–2017, 1997.

268. O. Toker. On the complexity of purely complex μ computation and related problems in multidimensional systems. *IEEE Transactions on Automatic Control*, AC-43:409–414, 1998.

269. Y.L. Tong. *Probability Inequalities in Multivariate Distributions*. Academic Press, New York, 1980.

270. J.F. Traub, G.W. Wasilkowski, and H. Woźniakowski. *Information-Based Complexity*. Academic Press, New York, 1988.

271. J.F. Traub and A.G. Werschulz. *Complexity and Information*. Cambridge University Press, Cambridge, 1998.

272. J.G. Truxal. Control systems - some unusual design problems. In *Adaptive Control Systems (E. Mishkin and L. Braun editors)*, pages 91–118. McGraw-Hill, New York, 1961.

273. J.S. Tyler and F.B. Tuteur. The use of a quadratic performance index to design multivariable invariant plants. *IEEE Transactions on Automatic Control*, AC-11:84–92, 1966.

274. V.A. Ugrinovskii. Randomized algorithms for robust stability and guaranteed cost control of stochastic jump parameter systems with uncertain switching policies. *Journal of Optimization Theory and Applications*, 2004, in press.

275. S. Uryasev (editor). *Probabilistic Constrained Optimization: Methodology and Applications*. Kluwer, New York, 2000.

276. J.V. Uspensky. *Introduction to Mathematical Probability*. McGraw-Hill, New York, 1937.

277. J.G. van der Corput. Verteilungsfunktionen i, ii. *Nederl. Akad. Wetensch. Proceedings*, 38(B):813–821, 1058–1066, 1935.

278. L. Vandenberghe and S. Boyd. Semidefinite programming. *SIAM Review*, 38:49–95, 1996.

279. V.N. Vapnik. *Statistical Learning Theory*. Wiley, New York, 1996.

280. V.N. Vapnik and A.Ya. Chervonenkis. On the uniform convergence of relative frequencies to their probabilities. *Theory of Probability and Its Applications*, 16:264–280, 1971.

281. R. Vein and P. Dale. *Determinants and their Applications in Mathematical Physics*. Springer-Verlag, New York, 1999.

282. M. Vidyasagar. Statistical learning theory and randomized algorithms for control. *IEEE Control Systems Magazine*, 18:69–85, 1998.

283. M. Vidyasagar. Randomized algorithms for robust controller synthesis using statistical learning theory. *Automatica*, 37:1515–1528, 2001.

284. M. Vidyasagar. *Learning and Generalization: With Applications to Neural Networks*. Springer-Verlag, New York, 2002.

285. M. Vidyasagar and V. Blondel. Probabilistic solutions to some NP-hard matrix problems. *Automatica*, 37:1397–1405, 2001.

286. J. von Neumann. Various techniques used in connection with random digits. *U.S. Nat. Bur. Stand. Appl. Math. Ser.*, pages 36–38, 1951.

287. R.S. Wenocur and R.M. Dudley. Some special Vapnik–Chervonenkis classes. *Discrete Mathematics*, 33:313–318, 1981.

288. J.C. Willems and R. Tempo. The Kharitonov theorem with degree drop. *IEEE Transactions on Automatic Control*, AC-44:2218–2220, 1999.

289. F. Wu and K.M. Grigoriadis. LPV systems with parameter-varying time delays: analysis and control. *Automatica*, 37:221–229, 2001.

290. K.Y. Yang, S.R. Hall, and E. Feron. Robust $\mathcal{H}_2$ control. In *Advances in Linear Matrix Inequality Methods in Control (L. El Ghaoui and S.-I. Niculescu editors)*, pages 155–174. SIAM, New York, 2000.

291. D.C. Youla and M. Saito. Interpolation with positive-real functions. *Journal of the Franklin Institute*, 284:77–108, 1967.

292. P.M. Young. The rank one mixed μ problem and "Kharitonov-type" analysis. *Automatica*, 30:1899–1911, 1994.

293. G. Zames. Feedback and optimal sensitivity: model reference transformations, multiplicative seminorms and approximate inverses. *IEEE Transactions on Automatic Control*, AC-26:301–320, 1981.

294. A.A. Zhigljavsky. *Theory of Global Random Search*. Kluwer, Dordrecht, 1991.

295. K. Zhou, J.C. Doyle, and K. Glover. *Robust and Optimal Control*. Prentice-Hall, Upper Saddle River, 1996.

296. G. Zhu, M.A. Rotea, and R. Skelton. A convergent algorithm for the output covariance constraint control problem. *SIAM Journal on Control and Optimization*, 35:341–361, 1997.

297. K. Zyczkowski and M. Kus. Random unitary matrices. *Journal of Physics*, 27:4235–4245, 1994.

Index

ARE, *see* Riccati equation
ARI, *see* Riccati inequality

bad set, *see* set
ball, *see* norm, ball
bilinear matrix inequality, 68
bit model, 61
BMI, *see* bilinear matrix inequality
bound
    Bernoulli, 122, 125, 126
    Chernoff, 109, 122–128, 304
    worst-case, 127–130
bounded real lemma, 25, 45, 46, 49

cdf, *see* distribution, function
central controller, *see* $\mathcal{H}_\infty$
chi-square test, 208, 209
communication networks, 291–299
computational complexity, 60–65
    of RAs, *see* randomized algorithms
conditional density, *see* density
conditional density method, 215,
        278–282, 286–287
confidence intervals, 125, 126
convergence
    almost everywhere, 14
    in probability, 14
convex set, 23
covariance matrix, 12
curse of dimensionality, 67, 94, 215

decidable problem, 60
defining function, 223, 247, 253
density
    ℓ_2^W radial, 231–235
    ℓ_p induced radial, 248–269
    ℓ_p radial, 223–231, 247–248
    binomial, 13
    chi-square, 13
    conditional, 12
    conditional method, *see* conditional
        density method

exponential, *see* density, Laplace
    function, 10
    Gamma, 14, 205, 213, 214
    generalized Gamma, 14, 205, 224,
        239–246
    joint, 11
    Laplace, 14, 206, 224
    marginal, 12
    normal, 13, 223, 231
    polynomial, 207, 208
    uniform, 13, 224, 248
    Weibull, 14, 205, 213
    Wishart, 269, 270
discrepancy, 98–103
    extreme, 99–101, 105
    star, 99–101
dispersion, 104–106
distribution
    function, 10
    joint, 11
distribution-free robustness, 87–89
D–K iteration, *see* μ synthesis
Dyson–Mehta integral, 279, 280, 320

edge theorem, 41
ellipsoid algorithm, 172–174
empirical
    maximum, 97, 127
    mean, 92, 131–137
    probability, *see* probability
EXP-complete, 63
expected value, 11
extreme discrepancy, *see* discrepancy

flexible structure, 301–305

gain scheduling, *see* linear parameter
        varying
Gamma function, 13
Gaussian, *see* density, normal
good set, *see* set
Gramian